Biocatalysts and Enzyme Technology

Biocatalysts and Enzyme Technology

Edited by Orion Wright

New York

Published by Syrawood Publishing House,
750 Third Avenue, 9th Floor,
New York, NY 10017, USA
www.syrawoodpublishinghouse.com

Biocatalysts and Enzyme Technology
Edited by Orion Wright

International Standard Book Number: 978-1-64740-390-4 (Hardback)

Cataloging-in-publication Data

Biocatalysts and enzyme technology / edited by Orion Wright.
p. cm.
Includes bibliographical references and index.
ISBN 978-1-64740-390-4
1. Biocatalysis. 2. Enzymes--Biotechnology. 3. Enzymes. 4. Biotechnology.
5. Bioengineering. I. Wright, Orion.
TP248.65.E59 B57 2023
660.634--dc23

TABLE OF CONTENTS

PREFACE

The main aim of this book is to educate learners and enhance their research focus by presenting diverse topics covering this vast field. This is an advanced book which compiles significant studies by distinguished experts in the area of analysis. This book addresses successive solutions to the challenges arising in the area of application, along with it; the book provides scope for future developments.

Enzymes are proteins that speed-up metabolism and chemical reactions that occur inside the human bodies. These proteins build or create some substances, while others are broken down. Enzyme technology refers to the branch of industrial biotechnology which encompasses biocatalysis, fundamental and applied enzymology, molecular modeling, structural biology, and diagnostics. The application of whole cells or isolated enzymes, as catalysts in organic reactions is termed as biocatalysis. The material-intensive process of enzyme purification is accomplished effectively by using whole-cell biocatalysts that contain active pathways or enzymes. The identification of the design of whole-cell biocatalysts and enzymes has been facilitated by recent developments in molecular biology, metabolic engineering, synthetic biology and computational techniques. This book unravels the recent studies on biocatalysts and enzyme technology. It will help the readers in keeping pace with the rapid changes in these fields of study.

It was a great honour to edit this book, though there were challenges, as it involved a lot of communication and networking between me and the editorial team. However, the end result was this all-inclusive book covering diverse themes in the field.

Finally, it is important to acknowledge the efforts of the contributors for their excellent chapters, through which a wide variety of issues have been addressed. I would also like to thank my colleagues for their valuable feedback during the making of this book.

Editor

1

Biocatalytic Oxidation of Alcohols

Hendrik Puetz [1], Eva Puchľová [2], Kvetoslava Vranková [2] and Frank Hollmann [1,*]

[1] Department of Biotechnology, Delft University of Technology, van der Maasweg 9, 2629 HZ Delft, The Netherlands; Hendrik.pue@live.de

[2] Axxence Slovakia s.r.o, Mickiewiczova 9, 81107 Bratislava, Slovakia; evka.puchlova@axxence.sk (E.P.); kvetka.vrankova@axxence.sk (K.V.)

* Correspondence: f.hollmann@tudelft.nl

Abstract: Enzymatic methods for the oxidation of alcohols are critically reviewed. Dehydrogenases and oxidases are the most prominent biocatalysts, enabling the selective oxidation of primary alcohols into aldehydes or acids. In the case of secondary alcohols, region and/or enantioselective oxidation is possible. In this contribution, we outline the current state-of-the-art and discuss current limitations and promising solutions.

Keywords: alcohols; alcohol oxidation; alcohol dehydrogenases; alcohol oxidases; kinetic resolution; deracemization

---

1. Introduction

1.1. Why Use Biocatalysis for Alcohol Oxidations

The (catalytic) oxidation of alcohols is a mature yet still very active field of chemical research. Within the scope of Green Chemistry, well-established textbook methods are substituted with more efficient catalytic methods, shifting from problematic oxidants such as $AgNO_3$, K_2CrO_4 or $KMnO_4$ to environmentally more acceptable oxidants such as O_2 or H_2O_2 [1,2].

Biocatalysis could play a major role in this transition. Arguments frequently used in favor of biocatalysis are the mild reaction conditions and the renewable origin and biodegradability of enzymes. More importantly, however, enzymes are very selective catalysts enabling precision chemistry avoiding tedious protection group chemistry. At the same time, large parts of the chemical community tend to ignore enzymes as potential tools for synthesis planning, which is due to perceived and real limitations of enzyme catalysis.

In this contribution, we will briefly outline the current state-of-the-art in biocatalytic alcohol oxidations, highlighting synthetic opportunities but also critically discussing current limitations.

1.2. Biocatalysis for Alcohol Oxidation: Perceived and Real Limitations

Arguments frequently held against biocatalysis in general are its limited availability, narrow product scope, poor stability of the catalysts, and high price [3]. While this situation may have been true two decades ago, there has been tremendous progress alleviating or even solving many of the issues held against enzymes. Some of these will be discussed in the following sections.

1.2.1. Availability of Oxidative Enzymes

Some 30 years ago, oxidative biocatalysis was largely restricted to natural diversity, i.e., enzymes available from natural resources. The famous alcohol dehydrogenase from horse liver (HLADH) [4,5] is just one prominent example of this. Then, HLADH was indeed obtained from horse liver resulting in ethical issues and variations of availability and quality. With the rise of recombinant protein expression technology, cost-efficient and scalable enzyme production has become possible [6–8]. Various commercial enzyme suppliers offer oxidoreductases in quantities ranging from small to bulk.

In addition, the diversity of natural enzymes has been increasing (and continues doing so) considerably with new oxidative enzymes identified from metagenome libraries [9–12] and new habitats and organisms [13–16].

1.2.2. Substrate Scope

Natural enzymes have been optimized by natural evolution to serve the host organisms' purpose, which does not necessarily coincide with the needs of an organic chemist aiming at the selective oxidation of a given target molecule. Next to screening natural diversity for more suitable enzymes, protein engineering has become a very powerful tool to tailor the properties of a given enzyme such as cofactor specificity, thermo and solvent stability, (enantio)selectivity, and more [17–33].

For example, vanillyl alcohol oxidase has been engineered intensively by the groups of van Berkel and Fraaije to e.g., engineer the substrate specificity or the stereoselectivity of the hydroxylation reaction [34–38].

Another nice example exemplifying the power of protein engineering comes from the Alcalde lab [39]. Here, the aryl alcohol oxidase is from *Pleurotus eryngii*. Directed evolution resulted in enzyme mutants with higher stability and activity but also increased expression levels [21,40]. The wild-type enzyme shows only little activity toward secondary benzylic alcohols, which can be overcome by semi-rational design [20,41]; the resulting oxidase mutants were highly selective in the kinetic resolution of a range of secondary benzyl alcohols.

1.2.3. Stability

Compared to common chemical catalysts, the thermal stability of biocatalysts indeed is generally much lower. However, considering the high catalytic efficiency of enzymes at temperatures below 100 °C, the question arises as to why this should be an issue at all. High temperatures are generally applied to accelerate chemical transformations. However, a (bio)catalytic reaction proceeds sufficiently fast already; at more ambient temperatures, highly thermostable catalysts (operating at temperature ranges between 100 and 500 °C as commonly applied in chemical transformations) are not necessary.

Nevertheless, if activity and stability at elevated temperatures is desired, oxidoreductases from (hyper)thermophilic host organisms [13] such as *Pyrococcus* [42], *Thermus* [43,44], or *Sulfolobus* [45] are available.

1.2.4. Biocatalyst Costs

Finally, the seemingly high costs are spuriously held against enzymes. If purchased from a specialty chemical supplier, enzymes are indeed very expensive due to the usually small production scale. However, it should be kept in mind that enzyme production costs are subject to economy of scale and typical enzyme costs if produced at large scale are as low as 250 € kg^{-1} enzyme [46]! A simple calculation reveals the catalyst performance needed to achieve a given cost contribution of the enzyme to the final product (Figure 1). For example, using an enzyme cost of 250 € kg^{-1}, only 100,000 turnovers are needed to attain an enzyme cost contribution of less than 1 € $kg^{-1}{}_{product}$. Obviously, this is an over-simplistic view, and other factors contribute to the cost structure of a given production process.

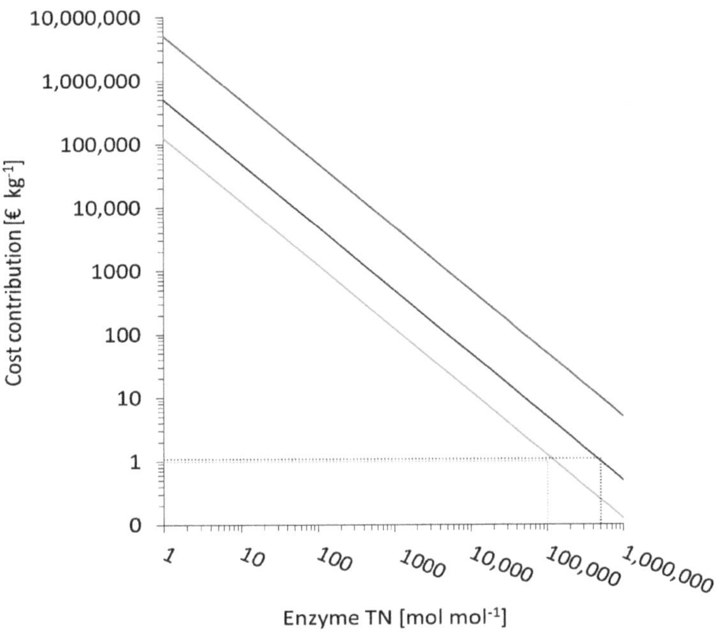

Figure 1. Cost contribution of a biocatalyst to the final product depending on the number of catalytic cycles performed (turnover number, TN = $mol_{product} \times mol^{-1}{}_{enzyme}$). Assumptions: molecular weight of the product: 200 g × mol^{-1}; molecular weight of the enzyme: 200 kDa, **green**: enzyme costs = 250 € kg^{-1}, **blue**: enzyme costs = 1000 € kg^{-1}, **red**: enzyme costs = 10,000 € kg^{-1}.

Amongst the real limitations of biocatalysis is the still very common use of aqueous reaction mixtures. As the majority of reactants of interest are rather hydrophobic, aqueous reaction media support only concentrations in the lower millimolar range. Such low reagent concentrations are very unattractive from an economical point of view (Figure 2) as they also imply high operational costs (and follow-up cost for downstream processing handling of large volumes). Furthermore, large amounts of contaminated waste water will be generated, which have to be treated prior release into the environment, causing further costs and consuming energy and resources.

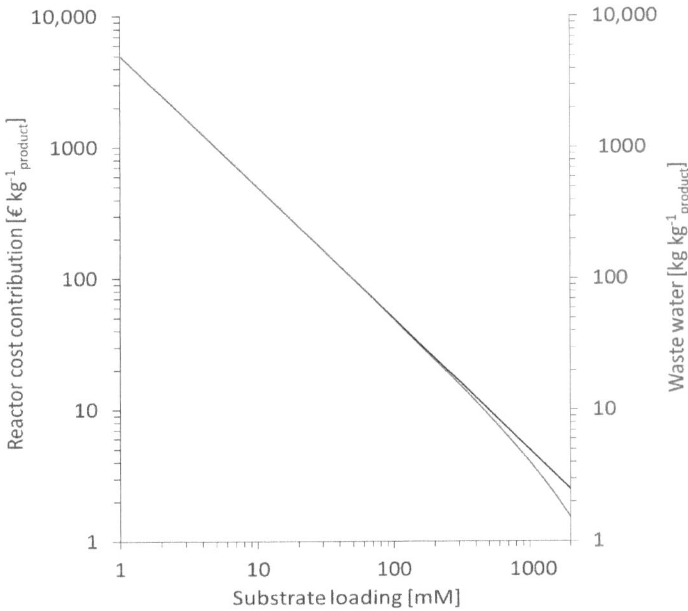

Figure 2. Estimation of the cost contribution of a reactor cost contribution (**black**) and waste water formed (**red**, based on a simple E-factor calculation). Assumptions: full conversion of the starting material into the product (Mw = 200 g mol^{-1}) and generic reactor costs of 1 € L^{-1} reaction volume.

Fortunately, concepts such as non-aqueous or biphasic reaction mixtures are increasingly used, rendering biocatalytic oxidations more attractive from a preparative point of view.

1.3. *The Catalysts Used*

Various enzyme systems are available for the oxidation of alcohols. Next to the widely used alcohol dehydrogenases (ADHs) and alcohol oxidases (AlcOxs), so-called laccase-mediator systems (LMS) are worth mentioning also.

Alcohol dehydrogenases (also frequently denoted as ketoreductases, KREDs) utilize the oxidized nicotinamide cofactors ($NAD(P)^+$) as a hydride acceptor for the oxidation of alcohols (Scheme 1). The catalytic mechanism starts with the (reversible) binding of the oxidized nicotinamide cofactor (not shown) followed by the (likewise reversible) coordination of the alcohol starting material to the enzyme active site (Scheme 1, step (1)) and its deprotonation [47]. Coordination of the alcohol and $NAD(P)^+$ to a metal ion (often Zn^{2+}) ensures precise positioning of the alcohol–C–H bond to the pyridinium ring of $NAD(P)^+$ facilitating the hydride transfer (Scheme 1, step (2)). Finally, the carbonyl product leaves the enzyme active site (Scheme 1, step (3)), leaving the bound, reduced NAD(P)H behind (which can diffuse out of the enzyme active site or stay for a reductive round, which is the reversal of the just described oxidation reaction).

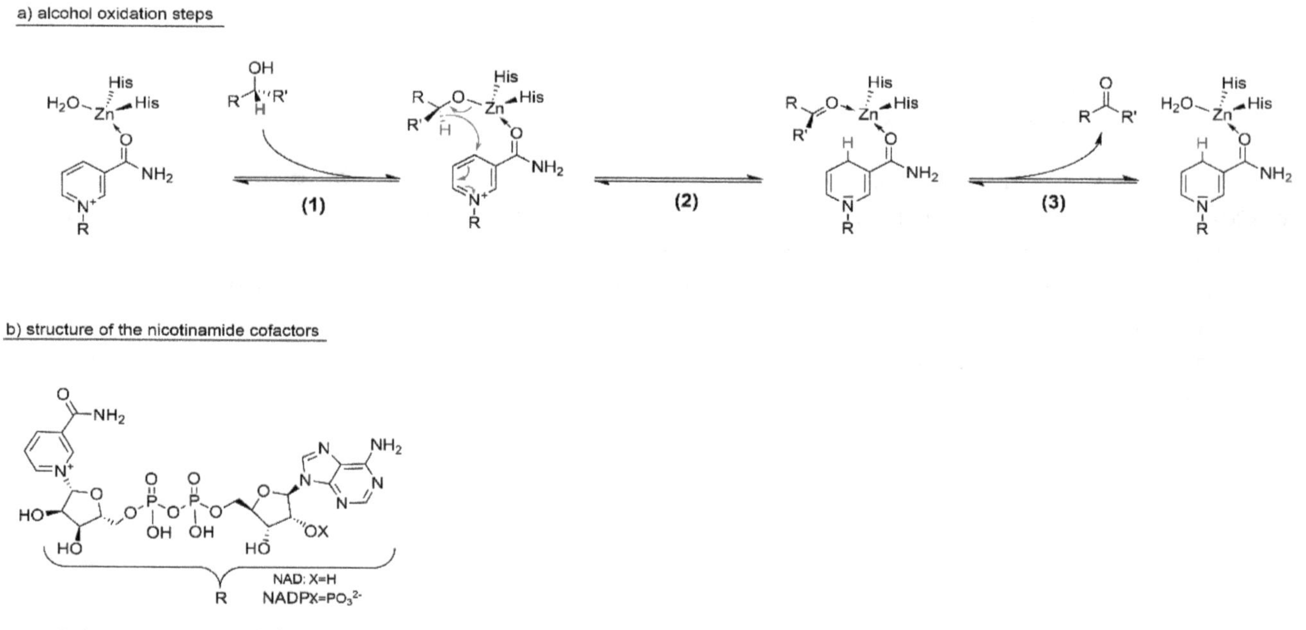

Scheme 1. Simplified mechanism of alcohol dehydrogenases (ADH)-catalyzed alcohol oxidation (**a**) all reaction steps are fully reversible. (**b**) Structure of the oxidized nicotinamide cofactors (NAD^+ and $NADP^+$).

The reversibility of the single reaction steps also explains the fact that ADHs can be used in both directions. In fact, the majority of ADH reports deal with the (enantioselective) reduction of prochiral ketones [27,48].

Cofactor Regeneration Strategies

The catalytic mechanism shown in Scheme 1 also implies that the oxidation of one equivalent of alcohol also results in the consumption of one equivalent of the oxidized nicotinamide ($NAD(P)^+$) yielding its reduced form (NAD(P)H). The still relatively high costs and frequently observed inhibitory effects (by both the oxidized and the reduced cofactor) on the biocatalysts prohibit their use in stoichiometric amounts. Therefore, over the years, a range of in situ $NAD(P)^+$ regeneration methods have been developed (Table 1). In essence, they allow using the costly nicotinamide cofactor in catalytic amounts only as its active, oxidized form is continuously regenerated at the expense of a cosubstrate being reduced. In principle, two regeneration approaches can be distinguished: (1) the so-called substrate-coupled approach and the (2) enzyme-coupled approach.

The substrate-coupled approach exploits the reversibility of the ADH-catalyzed oxidation reaction by using the production ADH for $NAD(P)^+$ regeneration (driven by the ADH-catalyzed reduction of a cosubstrate, as shown in Figure 3a). Overall, this approach represents a biocatalytic variant of the chemical Oppenauer oxidation [49–52].

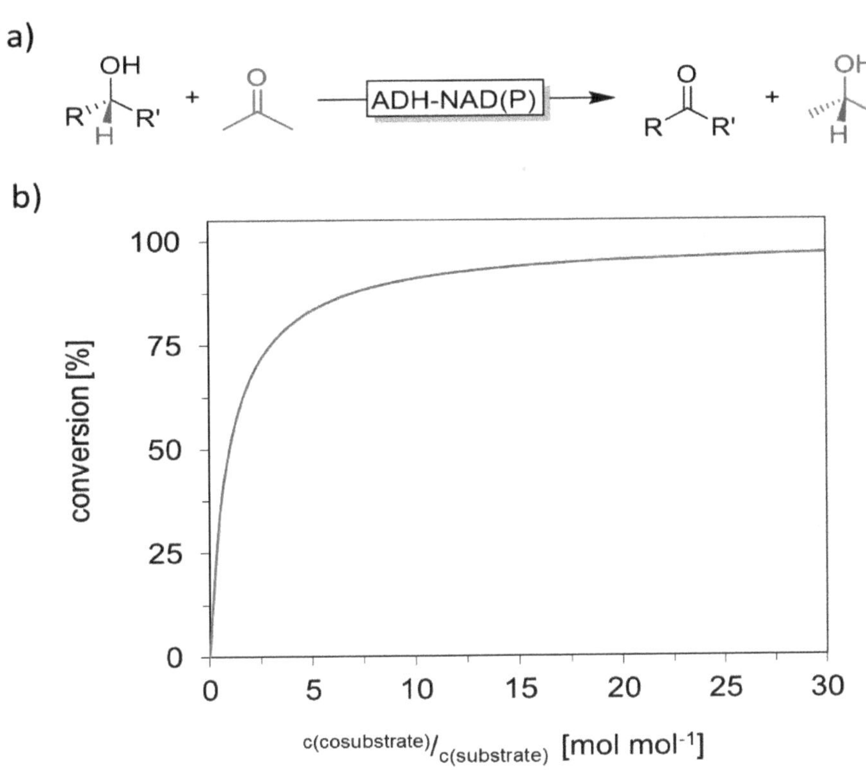

Figure 3. Substrate-coupled oxidation of alcohols. (**a**) Overall reaction using acetone as oxidant; (**b**) dependency of the equilibrium substrate conversion on the molar surplus of the cosubstrate (acetone) assuming an equilibrium constant K of 1.

Advantages of the substrate-coupled alcohol oxidation approach are that (1) the production enzyme also serves as a regeneration enzyme (no need for a second $NAD(P)^+$ regeneration catalyst), (2) the nicotinamide cofactor does not have to leave the enzyme active site for regeneration and thereby is less exposed to buffer-related degradation [53]. This also implies that ADH-catalyzed oxidation can principally be performed under non-aqueous conditions.

A disadvantage of the substrate-coupled approach is the low thermodynamic driving force of the overall reaction as the chemical composition of the products (alcohol and ketone) is essentially the same as that of the starting materials (alcohol and ketone). As a result of this, the equilibrium of the reaction is rather unfavorable, and additional measures are needed to shift the equilibrium. In some cases, removing one of the reaction products via extraction or distillation is possible. However, more common is to supply the (usually cheaper) cosubstrate in excess (Figure 3b). On the one hand, the surplus cosubstrate can be seen as a cosolvent, facilitating the solubilization of hydrophobic reagents. However, on the other hand, this surplus also represents an environmental burden and generates additional wastes [54].

A promising solution has been proposed by Kroutil and coworkers by using α-halo ketones as cosubstrates (Scheme 2) [55]. For example, chloroacetone enabled the authors to quasi-irreversibly oxidize a range of racemic alcohols using just 1.5 equivalents of the 'smart cosubstrate'. The authors hypothesized that the coproduct was stabilized by intramolecular hydrogen bonding, thereby shifting the equilibrium. Unfortunately, chloroacetone is a strong lachrymator (e.g., used in pepper spray), rendering it unattractive for many applications (such as the synthesis of consumer products).

Scheme 2. One-way oxidation of alcohols using the ADH from Sphingobium yanoikuyae (SyADH) as a non-selective biocatalyst and chloroacetone as a 'smart cosubstrate'.

The enzyme-coupled regeneration approach relies on the cooperation of two catalysts, the NAD(P)$^+$-dependent ADH (production enzyme) and an NAD(P)$^+$-regenerating catalyst. At first sight, this approach seems more complicated than the above-discussed substrate-coupled approach. However, it allows us to make use of molecular oxygen as a terminal electron acceptor, thereby making use of the high thermodynamic driving force of oxygen reduction (Table 1). Hydrogen peroxide (which is generally dismutated into O_2 and H_2O by the addition of a catalase) or water is formed as a by-product, which from a waste perspective is very attractive. To facilitate the aerobic oxidation of NAD(P)H, a range of enzymatic and non-enzymatic systems have been developed (Table 1).

For example, NADH oxidases directly re-oxidize NADH into NAD$^+$ while reducing molecular oxygen to H_2O_2 or H_2O [56–64].

The so-called laccase mediator systems (LMSs) comprise combinations of laccases and chemical redox dyes to aerobically regenerate NAD(P)$^+$ from NAD(P)H. Compared to the aforementioned NADH oxidases, LMSs excel by their indifference with respect to the cofactor (NADP$^+$ or NAD$^+$) regenerated, since the NAD(P)H oxidation step is performed by an unselective chemical mediator (Scheme 3) [65–71]. Quite often, the reduced mediator itself reacts with molecular oxygen, making co-catalysis by laccase superfluous [72–77].

Table 1. Selection of aerobic NAD(P)$^+$ regeneration systems to drive ADH-catalyzed oxidation reactions.

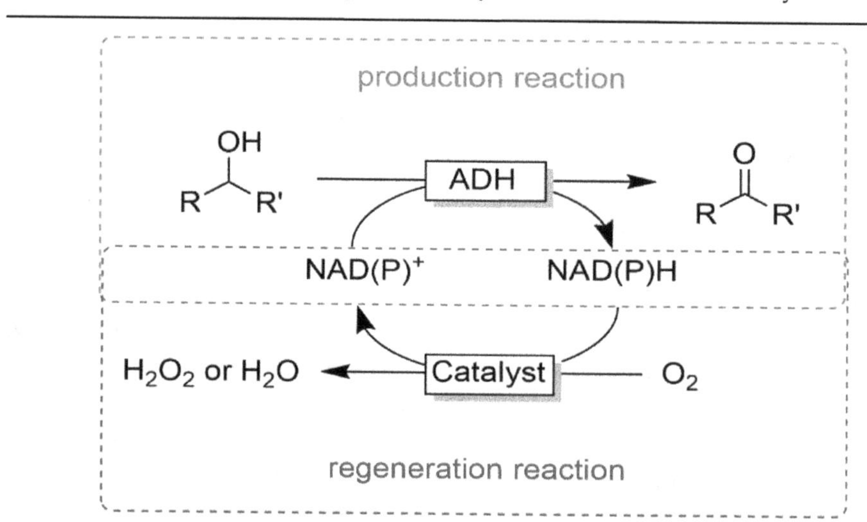

Cosubstrate	Coproduct	Catalyst	Ref.
O_2	H_2O_2	(Modified) flavins	[72–77]
	H_2O	LMS	[65–71]
		NADH oxidase	[56,57,61,63,64,78–81]
		Other oxidative enzymes	[82–86]

LMS: laccase mediator system.

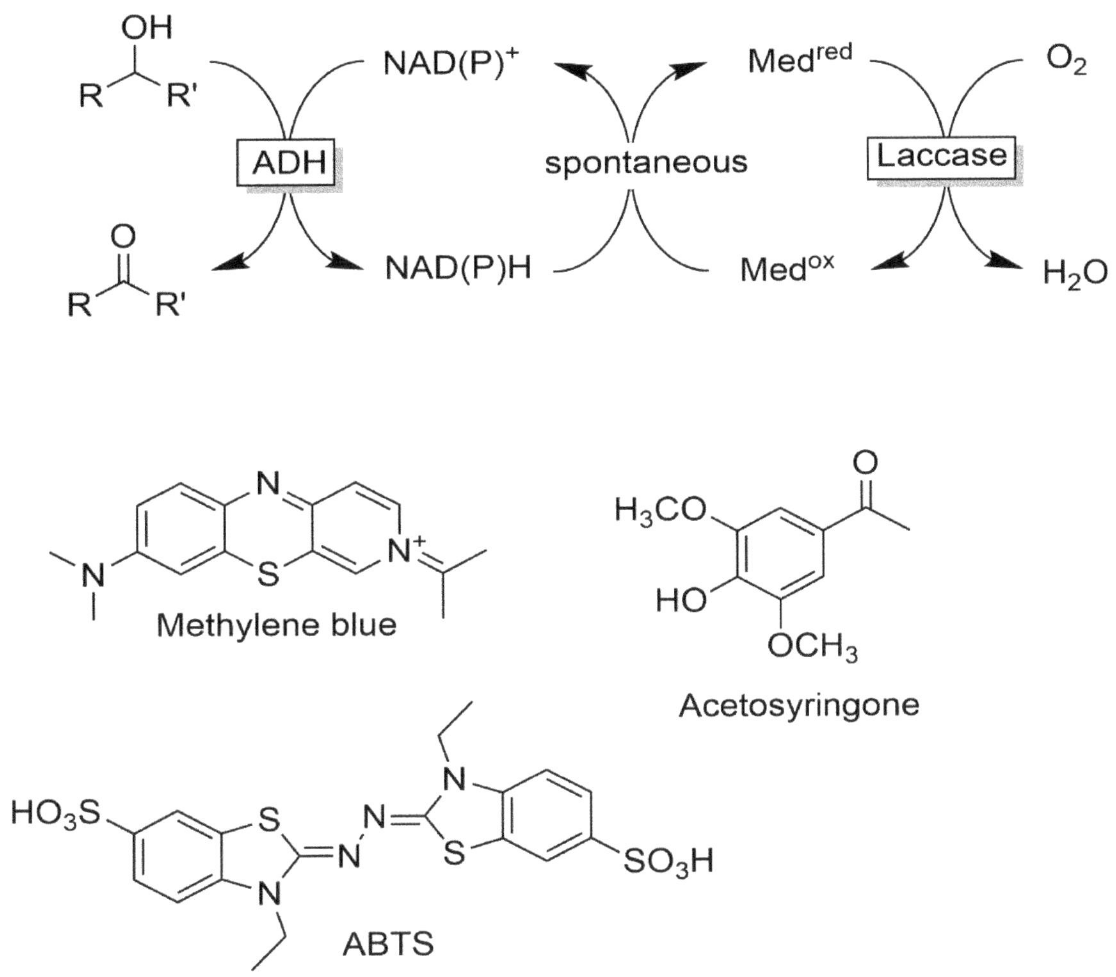

Scheme 3. Electron transport chain of a Laccase-mediator-system-promoted alcohol oxidation. The ADH catalyzes the $NAD(P)^+$-dependent oxidation reaction, yielding (NAD(P)H). The latter is spontaneously oxidized by a chemical mediator (a selection is shown on the bottom). The oxidized form of the mediator itself is re-formed by laccase-mediated aerobic oxidation.

Using molecular oxygen as a terminal electron acceptor is attractive from a thermodynamic driving force point of view as well as an environmental point of view, as only water is formed as a by-product. However, a major challenge of this approach originates from the very poor solubility of O_2 in aqueous media (under ambient conditions approximately 0.2–0.25 mM). As a consequence, the O_2 pool available is consumed, rapidly necessitating external provision with O_2 to drive the oxidation reaction [78,83,87,88]. Oxygen supply by bubbling air or O_2 through the reaction mixture represents a straightforward solution. The oxygen transfer rate is proportional to the gas–liquid interface area, which means that in principle, heavy sparking with small bubbles should be advantageous. However, many enzymes are inactivated at the liquid–gas interface [89–91]. A generally applicable solution is still elusive. It is also worth mentioning that strict regulations apply e.g., in the preparation of products used in natural foodstuffs and consumer care, excluding pressurized reactor setups.

Alcohol oxidases represent the second preparatively relevant class of enzymes useful for the oxidation of alcohols. In contrast to ADHs, AlcOXs do not rely on the nicotinamide cofactors but transfer the reducing equivalents liberated in the course of the alcohol oxidation reaction to molecular oxygen, yielding H_2O_2 as a stoichiometric by-product. Two main classes of AlcOxs are predominantly investigated: flavin-dependent AlcOxs and Cu^{2+}-dependent AlcOxs. As shown in Schemes 4 and 5, they differ significantly with respect to the oxidation mechanism.

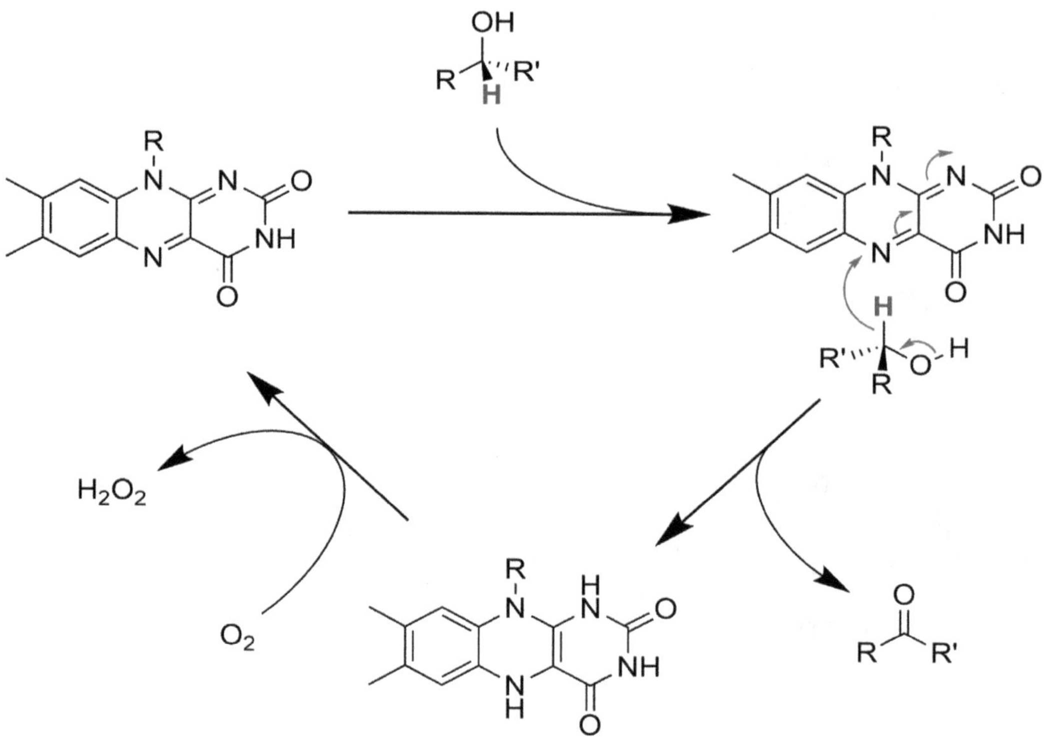

Scheme 4. Simplified mechanism of flavin-dependent oxidase-catalyzed oxidation of alcohols. The alcohol starting material binds to the enzyme active site and is oxidized by deprotonation/hydride transfer to the oxidized flavin cofactor. The resulting reduced flavin cofactor re-oxidized by molecular oxygen is a cascade of single electron transfer steps.

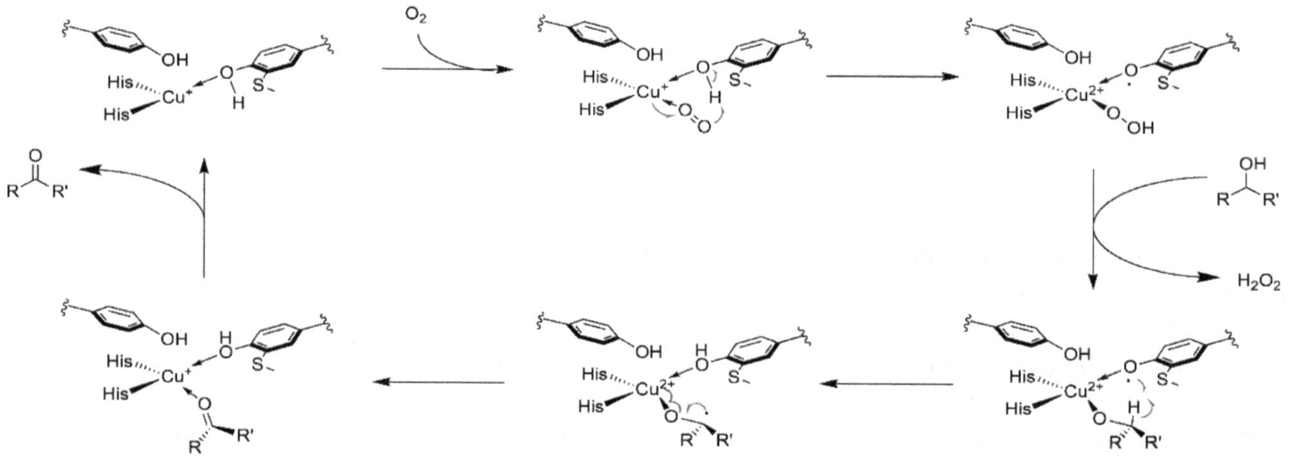

Scheme 5. Simplified mechanism of Cu-dependent oxidases. O_2 binds to the (Cu^+) resting state of the enzyme, resulting in a double oxidation by simultaneous electron transfer from Cu^+ and H-atom abstraction from a phenolic active site amino acid. Next, the Cu^+-bound H_2O_2 is substituted by the alcohol starting material which undergoes H-atom abstraction (reforming the phenolic amino acid) and electron transfer to the Cu^{2+} site, resulting in a Cu^+-coordinated carbonyl product that diffuses out of the enzyme active site to close the catalytic cycle.

In flavin-dependent oxidases, an oxidized flavin-cofactor abstracts the hydride from the alcohol C–H bond, yielding a reduced flavin. The oxidized cofactor is regenerated by re-oxidation with O_2 via a complex sequence of electron transfer steps, eventually yielding H_2O_2 as the by-product [92,93].

Cu-dependent oxidases follow a mechanism wherein a Cu^{2+}/phenoxy radical pair binds the alcohol starting material followed by a hydrogen atom abstraction step (to the phenoxy radical), yielding

Cu-dependent oxidases follow a mechanism wherein a Cu^{2+}/phenoxy radical pair binds the alcohol starting material followed by a hydrogen atom abstraction step (to the phenoxy radical), yielding the coordinated carbonyl product [94,95]. After dissociation of the product from the enzyme-active site, the reactive form is restored via O_2 reduction to H_2O_2.

Laccase-mediator systems essentially are organocatalytic alcohol oxidation reactions using oxammonium species as an oxidant. The stable *N*-oxide radical TEMPO (or its analogues) is oxidized by the blue-copper enzyme laccase (at the expense of O_2 being reduced to H_2O) to the reactive oxammonium special, forming a covalent adduct with the alcohol starting material. After the oxidation step, a hydroxylamine is formed that synproportionates with another oxammonium molecule, forming the TEMPO catalyst (Scheme 6) [96–99].

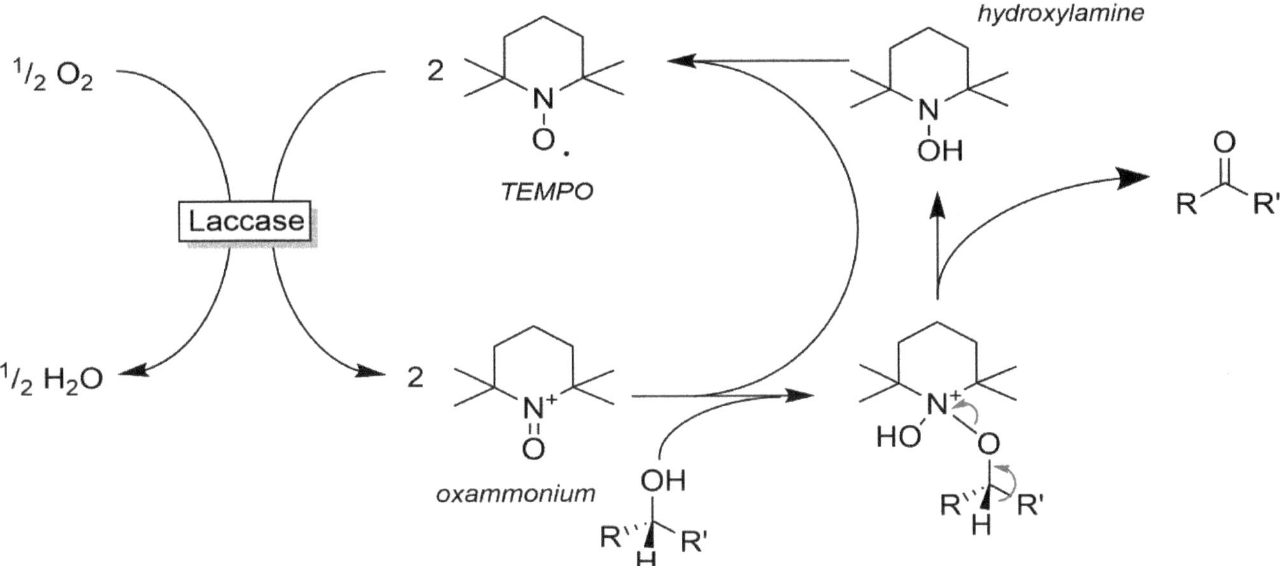

Scheme 6. Simplified mechanism of the LMS-mediated oxidation of alcohols.

2. Oxidation of Primary Alcohols

2.1. Oxidation of Primary Alcohols to Aldehydes

The selective oxidation of primary alcohols to the aldehyde level is generally not an issue using isolated enzymes. The catalytic mechanisms of ADHs and oxidases imply a hydride abstraction step, which precludes the ADH- or oxidase-catalyzed oxidation of aldehydes. However, if whole cell preparations are used, the presence of endogenous aldehyde dehydrogenases (*vide infra*) may impair the chemoselectivity of the oxidation reaction. In this case, the two liquid phase system (2LPS, Scheme 7) approach represents an elegant solution. A hydrophobic organic phase serves as a substrate reservoir (also enabling high reagent payloads) and simultaneously as a sink for the hydrophobic aldehyde product, extracting it from the aqueous, biocatalysts-containing phase and thereby protecting it from further oxidation to the acid.

For example, Schmid and coworkers used the two liquid phase system (2LPS) approach to control the multi-step oxidation of pseudocumene to 3, 4-dimethyl benzaldehyde [100,101]. As catalyst, recombinant *E. coli* overexpressing the xylose monooxygenase (XylM) from *Pseudomonas putida* was used [102,103]. In aqueous media, this catalyst performs through oxidation to the carboxylic acid; however, in the presence of dioctyl phthalate, the hydrophobic aldehyde intermediate preferentially partitions into the organic layer and thereby is removed from the catalyst, preventing the undesired final oxidation step.

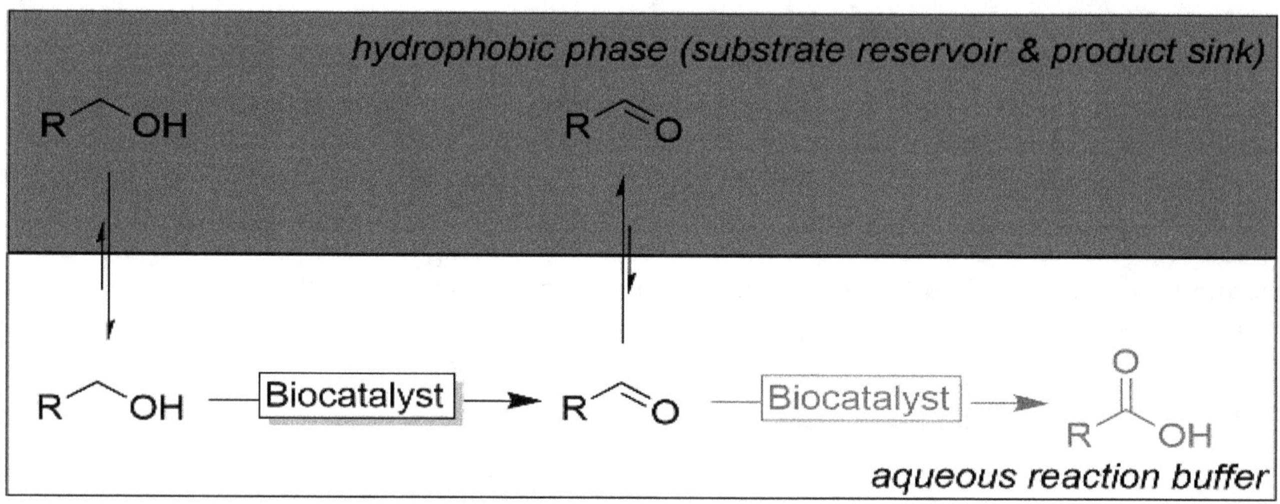

Scheme 7. The two liquid phase system (2LPS) approach for selective oxidation of alcohols to aldehydes.

Similarly, Molinari and coworkers have shown that the selectivity of acetic acid bacteria-catalyzed oxidation of primary alcohols can be controlled to either the aldehyde or acid level by performing the oxidation either in the presence or absence of isooctane as a hydrophobic organic phase (Table 2) [104–107].

Table 2. Controlling the selectivity of the whole-cell catalyzed oxidation of primary alcohols.

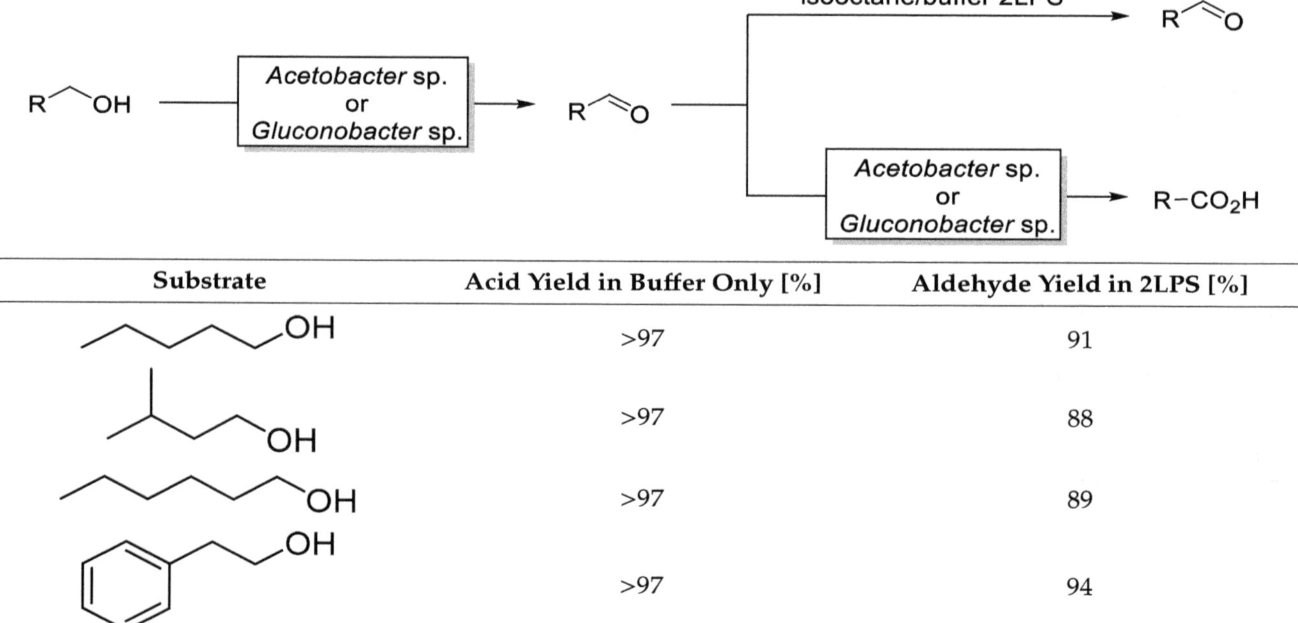

Substrate	Acid Yield in Buffer Only [%]	Aldehyde Yield in 2LPS [%]
	>97	91
	>97	88
	>97	89
	>97	94

The ADH-catalyzed oxidation of primary alcohols selectively to the aldehyde stage is rather scarce, even though the reaction is rather commonly used to assess ADH activity. AlcOxs are more frequently used as catalysts for the selective oxidation of primary alcohols to aldehydes.

Recently, we applied the 2LPS concept for the selective oxidation of (2*E*)-hex-2-enal into the corresponding aldehyde using the aryl alcohol oxidase from *Pleurotus eryngii* [108,109]. Product concentrations of up to 2.5 M in the organic phase and turnover numbers of the biocatalyst of more than 2 million could be achieved [110,111].

The selective oxidation of ethylene glycol to glycolaldehyde (Scheme 8) was achieved using oxidases from *Pichia pastoris* or *Aspergillus japonicus* [112,113]. Using the enzymes co-immobilized with catalase (to eliminate H_2O_2), molar product concentrations were achieved. The selectivity of the reaction was very high with less than 1% of the acid overoxidation product being formed.

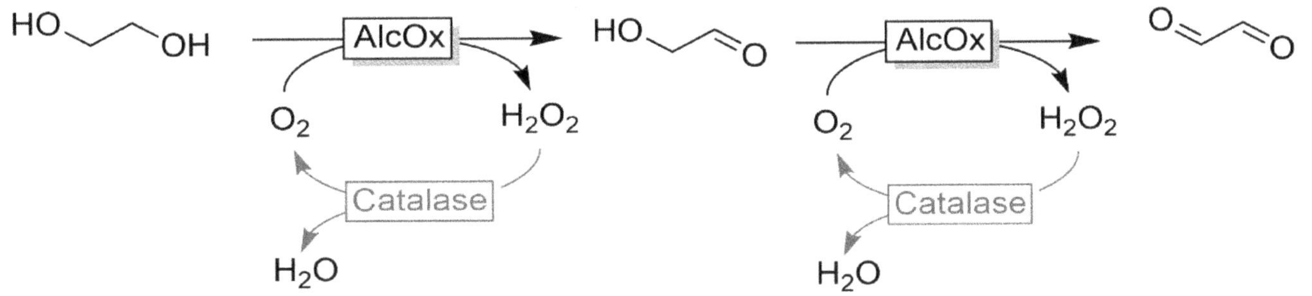

Scheme 8. Selective double oxidation of ethylene glycol to glycolaldehyde using oxidases.

In addition, the above-mentioned laccase-mediator system enjoys some popularity for the oxidation of primary alcohols [114–118]. Particularly, TEMPO is a commonly used organocatalyst enabling efficient oxidation of the alcohol starting materials. On the downside, so far, TEMPO is used in relatively high loadings (0.5–10 mol%), impairing the economic and environmental attractiveness of LMS oxidation systems.

Aldehydes are valuable reactive building blocks for further transformations. Therefore, a range of catalytic cascade reactions have been reported in which the in situ generated aldehyde is further transformed into a valuable product. Such one-pot cascades bear the advantage of circumventing at least one downstream processing, product isolation, and product purification step. This not only saves time but also resources and therefore is attractive not only from an economic but also from an environmental point of view [54].

For example, DiCosimo and coworkers reported the oxidation of glycolic acid to the corresponding aldehyde (glyoxylic acid) using methylotrophic yeasts such as *Hansenula polymorpha* or *Pichia pastoris* overexpressing spinach glycolate oxidase (GlycOx) in the presence of aminomethyl phosphonic acid (Scheme 9) [119]. The aldehyde spontaneously underwent imine formation with aminomethyl phosphonic acid, yielding N-(phosphonomethyl)glycine (glyphosate) after catalytic hydrogenation. The product was isolated and purified by simple acid precipitation and recrystallization.

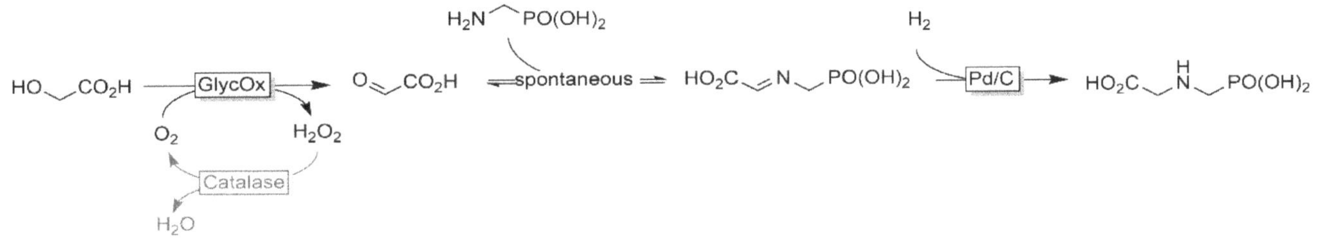

Scheme 9. Chemoenzymatic route to produce glyphosate.

Aldehydes are also attractive reagents for further C–C bond formation reactions. For example, Siebum and coworkers combined the oxidation of pent-4-en-1-ol using a commercial AlcOx and aldolase-catalyzed aldol reaction with acetone in a one-pot two-step procedure (Scheme 10) [120]. In their proof-of-concept study, the desired β-hydroxy ketone was obtained in 30% isolated yield and 70% optical purity.

Wong and coworkers reported the combination of galactose oxidase (GalOx) with rhamnulose-1-phosphate aldolase (RhaD) to generate fructose (3 chiral centers) from simple glycerol and dihydroxy acetone phosphate (Scheme 11) [121]. An optimized reaction procedure comprising heat inactivation of the oxidase prior to aldolase addition and pH adjustment between the aldolase and the dephosphorylation step gave a respectable overall yield of 55%.

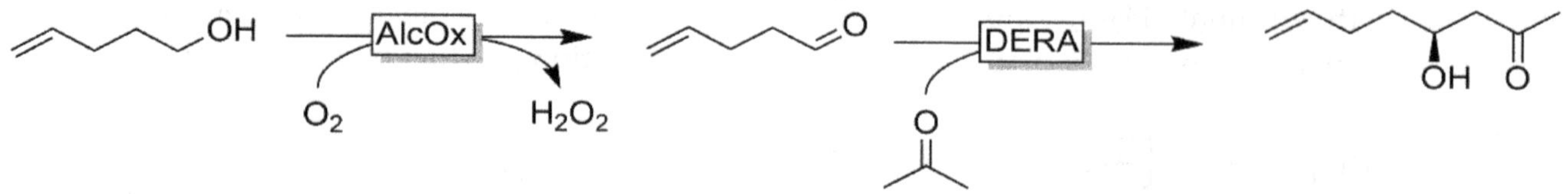

Scheme 10. Bienzymatic one-pot two-step cascade combining alcohol oxidase (AlcOx)-catalyzed oxidation of Pent-4-en-1-ol with an aldolase (2-desoxyribo-5-phosphate aldolase, DERA)-catalyzed aldol reaction.

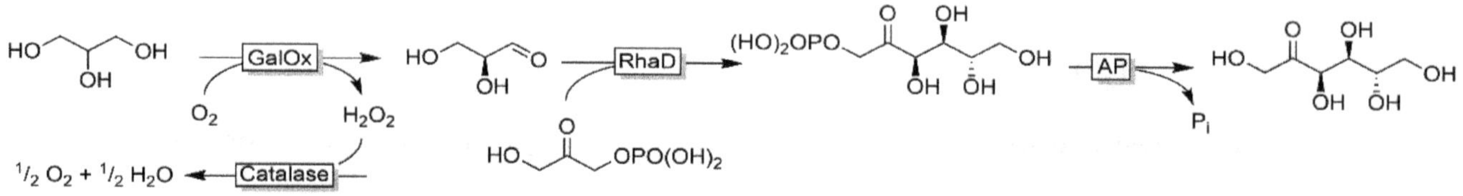

Scheme 11. Multi-enzyme cascade combining galactose oxidase (GalOx), rhamnulose-1-phosphate aldolase, and alkaline phosphatase to synthesize fructose from glycerol and dihydroxy acetone phosphate.

More recently, Turner and coworkers used a similar cascade of an engineered galactose oxidase and rabbit muscle aldolase to produce amino sugars [122].

Acyloins become available from simple alcohols if the oxidation step is coupled to a lyase-catalyzed benzoin condensation [123–126]. The in situ generation of reactive aldehydes such as formaldehyde alleviated the toxic effect of the aldehyde on the lyase (Scheme 12) [125,126].

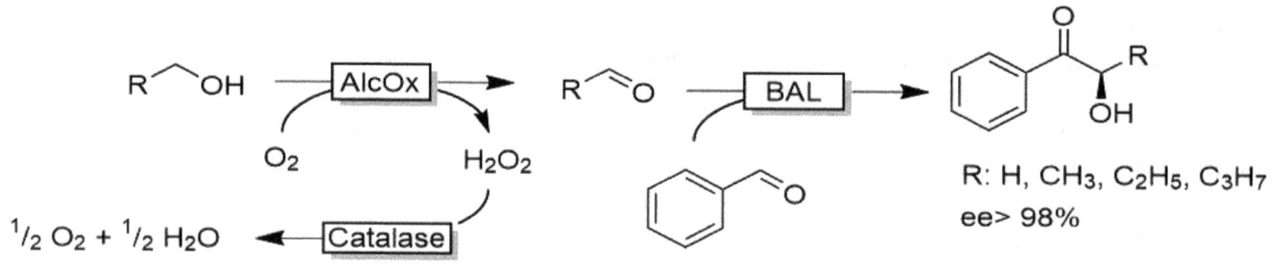

Scheme 12. Enantioselective benzoin condensation using benzaldehyde lyase (BAL) and in situ generated aldehydes.

Aldehydes can also undergo reductive amination to the corresponding amines [127,128] (Scheme 13).

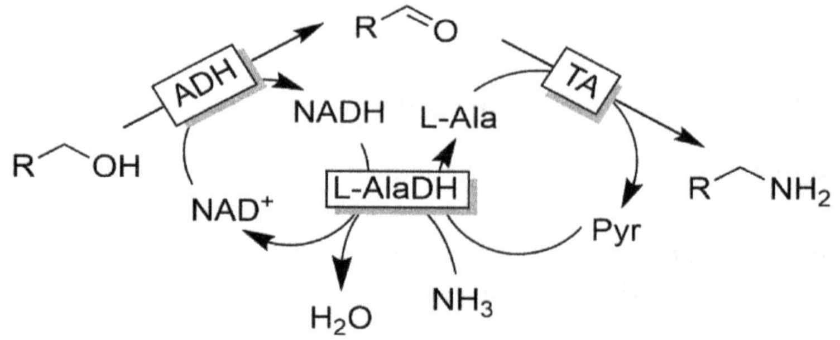

Scheme 13. Redox-neutral cascade to transform primary alcohols into primary amines.

Finally, a recently established combination of biocatalytic (ADH- or AlcOx-catalyzed) alcohol oxidation with organocatalytic (amino acid-catalyzed) aldol condensation is worth mentioning (Scheme 14) [129–131]. Starting e.g., from 1-butanol, 2-ethyl hexenal can be obtained representing an interesting approach to upgrade bio-based alcohols.

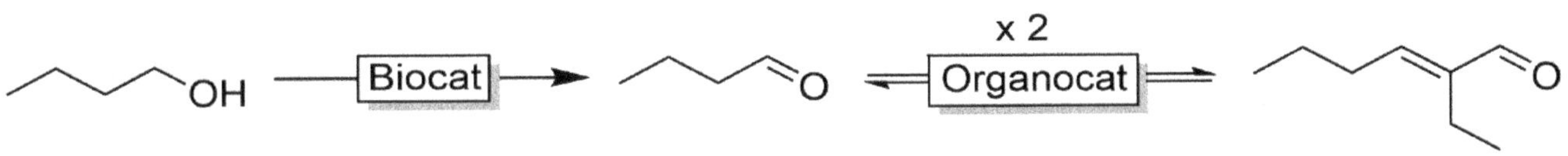

OH — Biocat → O ⇌ Organocat (x 2) ⇌ O

Scheme 14. Upgrading of alcohols (e.g., butane-1-ol to (2*E*)-2-ethylhex-2-enal) by combining biocatalytic oxidation and organocatalytic aldol condensation.

2.2. *Oxidation of Primary Alcohols to Acids*

As mentioned above, whole cell preparations often suffer from poor selectivity if oxidation of primary alcohols to the aldehyde stage is desired. However, since through oxidation to the carboxylic acid is desired, whole cell preparations have been used from an early stage onwards. Table 3 gives a representative overview over some reported through oxidations.

Table 3. Selection of microbial through oxidations of primary alcohols.

R–OH — Biocatalyst → R(C=O)OH

Catalyst	Product	Product Titer/Remarks	Ref.
Acetobacter	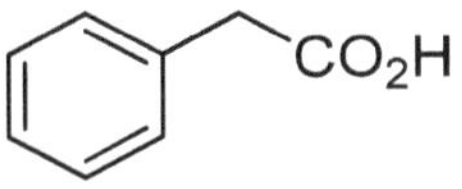 CO_2H	23 gL^{-1}/alginate immobilized cells	[105]
	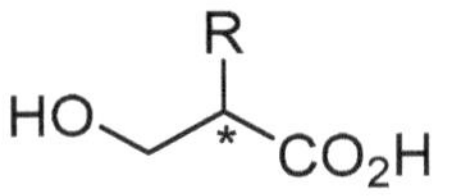 HO, R, *, CO_2H; R = CH_3 - C_5H_{11}	24 mM/dynamic kinetic resolution in a flow chemistry setup	[132–135]
	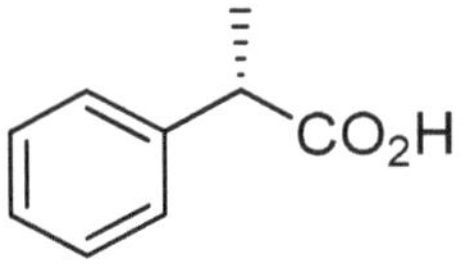 CO_2H	-	-
	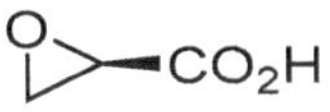 O, CO_2H	Kinetic resolution	[136]
	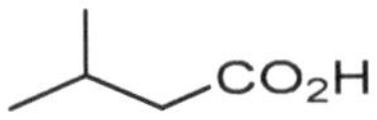 CO_2H	Up to 80 g L^{-1}	[137]
Gluconobacter	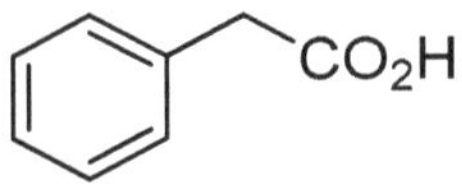 CO_2H	8 gL^{-1}	[138]
	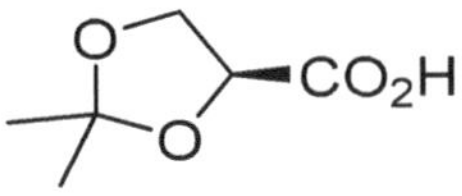 O, O, CO_2H	Kinetic resolution	[136]
	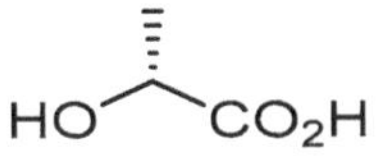 HO, CO_2H	20 g L^{-1} Kinetic resolution	[139]
Corynebacterium	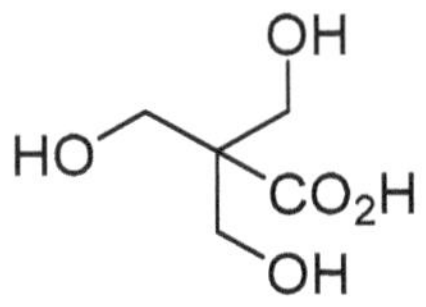 OH, HO, CO_2H, OH	25 g L^{-1}	[140,141]
Norcardia	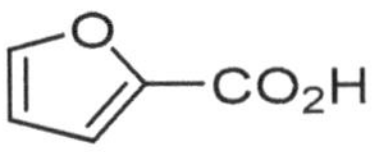 O, CO_2H	9 g L^{-1}	[142]

As mentioned above, ADHs and AlcOxs are generally not capable of oxidizing aldehydes, because the aldehyde proton is not abstractable as a hydride. This mechanistic limitation can be solved by nucleophilic attack to the carbonyl group transiently turning it into an alcohol containing a hydridically abstractable proton. Aldehyde dehydrogenases utilize this approach via a cysteine moiety in the enzyme active site (Scheme 15) [143].

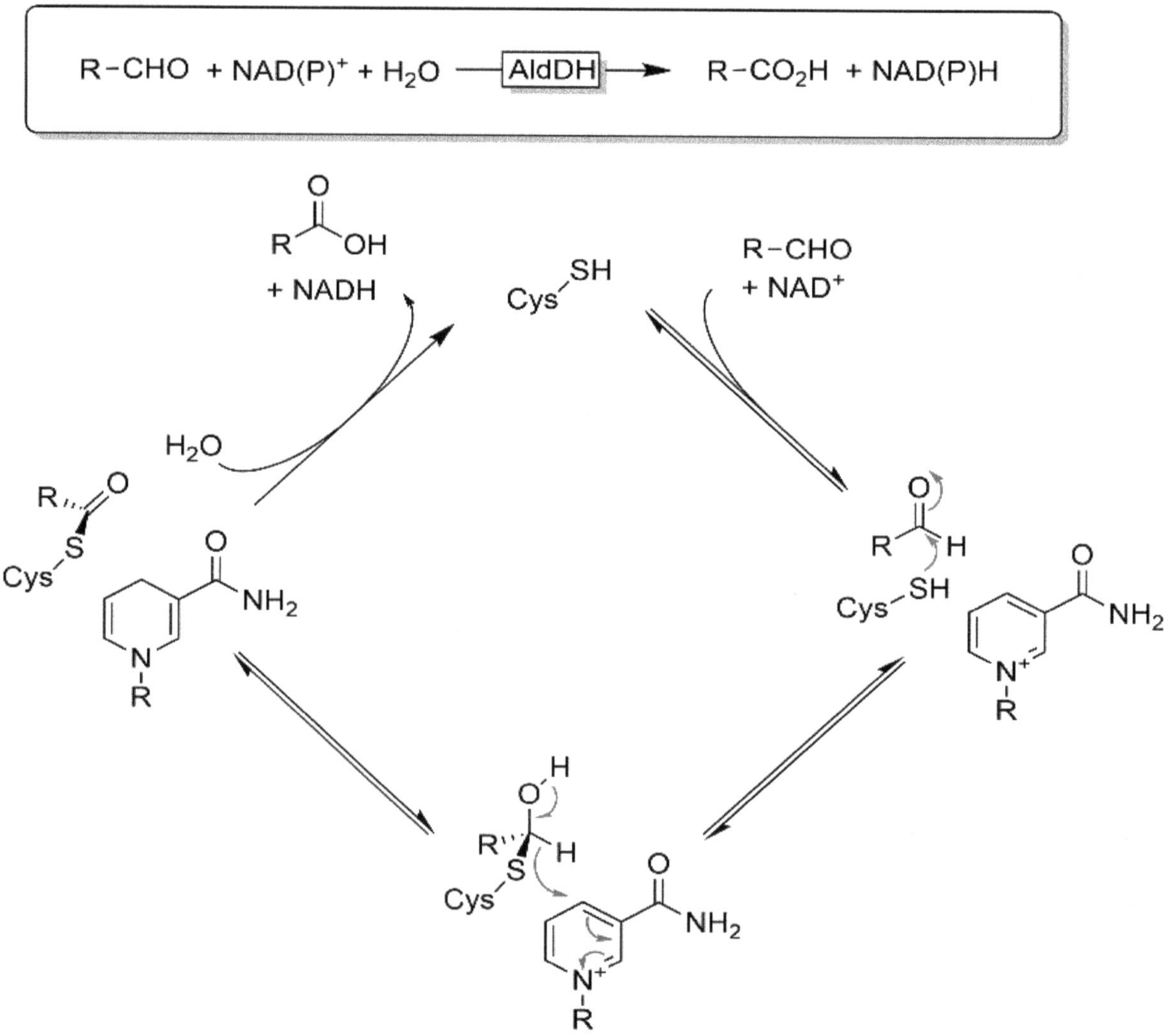

$$R{-}CHO + NAD(P)^+ + H_2O \xrightarrow{AldDH} R{-}CO_2H + NAD(P)H$$

Scheme 15. Simplified mechanism of aldehyde dehydrogenases (AldDHs). A cysteine within the active site nucleophilically attacks the aldehyde group. The resulting hemithioacetal can transfer a hydride to the enzyme-bound oxidized nicotinamide cofactor yielding a thioester, which upon hydrolysis releases the acid product.

Preparative applications of AldDHs have been reported by several groups recently [144–147]. An early example for the through oxidation of alcohols to carboxylic acids was reported by Wong and coworkers, who combined an ADH with an AldDH for this purpose (Scheme 16) [148].

Activated aldehydes, due to a favorable aldehyde-*gem* diol equilibrium, can also be converted quite efficiently by ADHs and AlcOxs to the corresponding acids [80,149–151].

Next to water, further nucleophiles have been reported such as alcohols or amines. Especially γ- and δ-diols form hemiacetals upon aldehyde formation, which can be further oxidized to the corresponding lactones (Scheme 17) [63,68,75–77,151–157]. The hemiacetal formation is kinetically and thermodynamically favored.

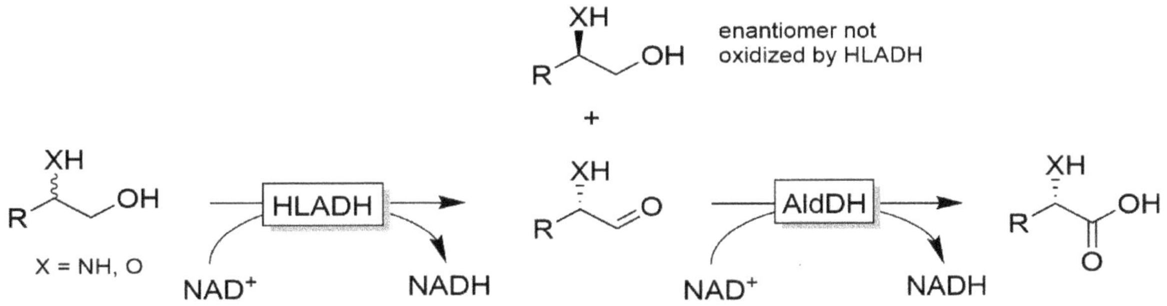

Scheme 16. Bienzymatic cascade to transform racemic 1,2-diols or 1,2-aminoalcohols into hydroxyl- or amino acids. Due to the stereoselectivity of the ADH used, the first oxidation step proceeds as kinetic resolution.

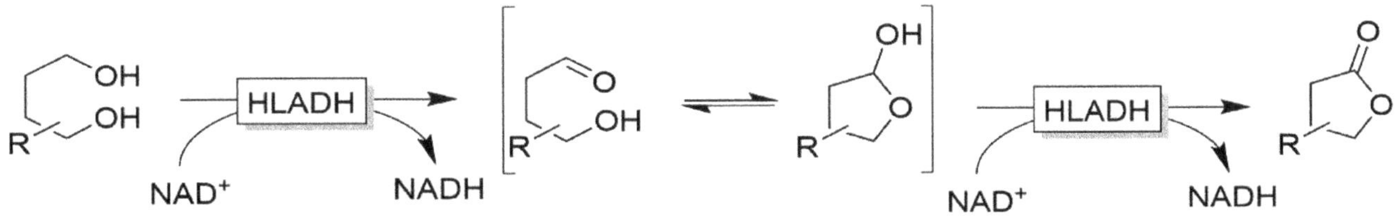

Scheme 17. Oxidative lactonization of diols using the ADH from horse liver (HLADH). The intermediate aldehyde undergoes spontaneous hemiacetal formation, yielding an oxidizable hemiacetal and finally the lactone product.

Recently, also amines have caught researchers' attention as nucleophiles. For example, Turner and coworkers reported a bienzymatic cascade to transform amino alcohols into lactames via a spontaneously formed cyclic imine (Scheme 18) [122,158]. The reaction was highly pH-responsive giving higher yields at more alkaline values, which probably reflects the protonation state of the amine functionality and its tendency to nucleophilically attack the intermediate aldehyde. Similar observations have been made in case of the ADH-catalyzed oxidative lactamization [63].

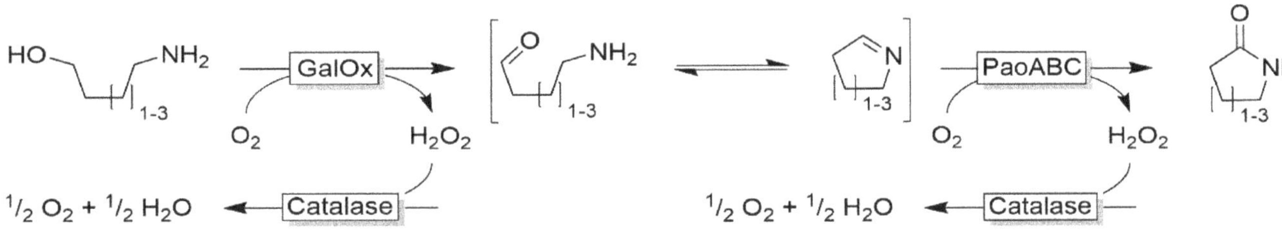

Scheme 18. Turning amino alcohols into lactames using a cascade of Galactose oxidase (mutants) (GalOx) and periplasmic aldehyde oxidase (PaoABC).

A very interesting further development of this concept was reported recently by Mutti and coworkers [150]. By performing the GalOx-catalyzed oxidation of benzylic alcohols in the presence of ammonium buffers, they were able to obtain the corresponding nitriles in satisfactory to high yields (Scheme 19). Although a fairly broad range of alcohols could be converted in decent yields, the catalyst turnover numbers are still moderate, calling for improvement; also, the catalytic mechanism remains to be elucidated. Overall, we are convinced that this interesting reaction (and possible further cascades) will gain more attention in the near future.

The oxidation of hydroxymethyl furfural (HMF) to furan dicarboxylic acid (FDCA) has been receiving particular attention in the past years. FDCA is a potential bio-based (HMF can be obtained from glucose/fructose) substitute for terephthalic acid as building block for polyesters. Therefore, significant research efforts have been devoted to the development of biocatalytic routes to oxidize HMF to FDCA (Scheme 20). In principle, all steps of this cascade can be performed by a single oxidase catalyst [159]. However, seemingly, the last oxidation step appears to be particularly difficult, which is why enzyme cascades are most promising (now) to attain an economically feasible full oxidation of HMF to FDCA [160,161].

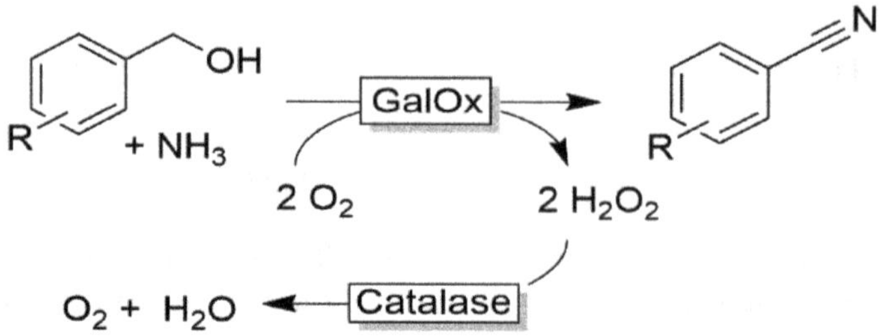

Scheme 19. Direct conversion of primary alcohols into nitriles.

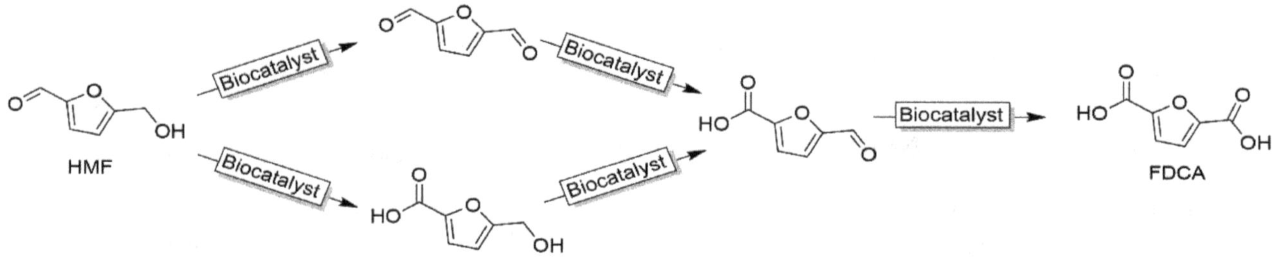

Scheme 20. Biocatalytic conversion of hydroxymethyl furfural (HMF) to furan dicarboxylic acid (FDCA). For reasons of simplicity, the various enzymes reported have been denominated generically as biocatalysts, and cosubstrates/coproducts have been omitted.

3. Oxidation of Secondary Alcohols

Compared to the reverse reaction (i.e., reduction of ketones), biocatalytic oxidations of secondary alcohols are far less common. This can be attributed to the destruction of chirality while transforming sp^3-hybridized alcohols into sp^2-hybridized carbonyl groups. Hence, value-added chiral alcohols are converted into (mostly less valuable) ketones. Nevertheless, some preparative applications of biocatalytic oxidations of secondary alcohols are known and will be discussed in the following sections.

3.1. Complete Oxidation of Racemic Secondary Alcohols

The complete oxidation of racemic alcohols necessitates non-stereoselective catalysts. However, non-stereoselectivity is a property seldom strived for in biocatalysis. As a consequence, identifying a suitable enzyme for the complete oxidation of racemic secondary alcohols can be a challenge.

Whole cells containing various enantiocomplementary ADHs are one option for the complete oxidation of racemic alcohols. For example, baker's yeast is principally capable of oxidizing both enantiomers of 2-heptanol [162] using two different ADHs. The expression level of both enzymes (depending on the growth phase) influenced the enantioselectivity of the *S. cerevisiae*-catalyzed oxidation and thereby makes a reproducible application difficult.

Another possibility for the complete oxidation of racemic alcohols would be to apply two enantiocomplementary biocatalysts; however, this will complicate the reaction scheme.

Ideally, non-selective ADHs would close the gap for the complete oxidation of racemic alcohols. Unfortunately, reports here are scarce (probably also because generally high enantioselectivity is desired, hence, seemingly negative results are not communicated clearly). One exception is the ADH from *Sphingobium yanoikuyae* (*Sy*ADH) reported by Kroutil and coworkers [55]. These authors purposely screened natural diversity for the non-selective oxidation of a range of racemic alcohols identifying *Sy*ADH (Scheme 21). In addition, this ADH also exhibited very high substrate tolerance, making it a very promising candidate for preparative-scale oxidations of a broad range of racemic alcohols.

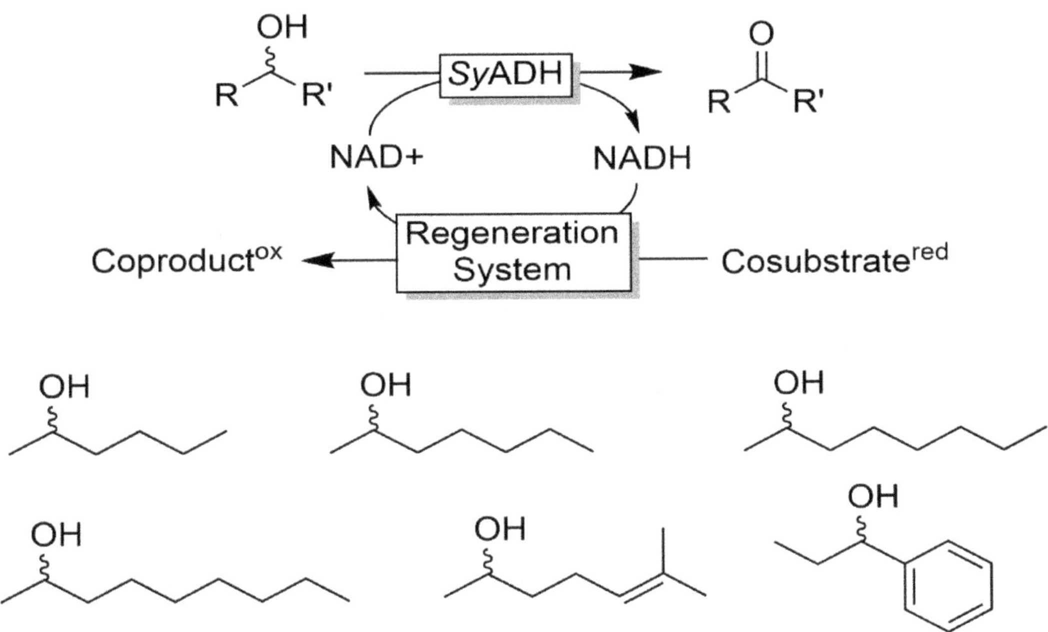

Scheme 21. Non-stereoselective oxidation of racemic alcohols using the ADH from Sphingobium yanoikuyae (SyADH).

In addition, an ADH from *Thermus thermophilius* may be an interesting candidate for the complete oxidation of racemic alcohols [163].

Finally, also the laccase-TEMPO system is worth mentioning here as the organocatalytic nature of the actual oxidation agent (2,2,6,6-Tetramethylpiperidin-1-yl)oxyl, TEMPO) also implies non-enantioselectivity and therefore is well-suited for the full oxidation of racemic alcohols (Scheme 22) [96,164–171].

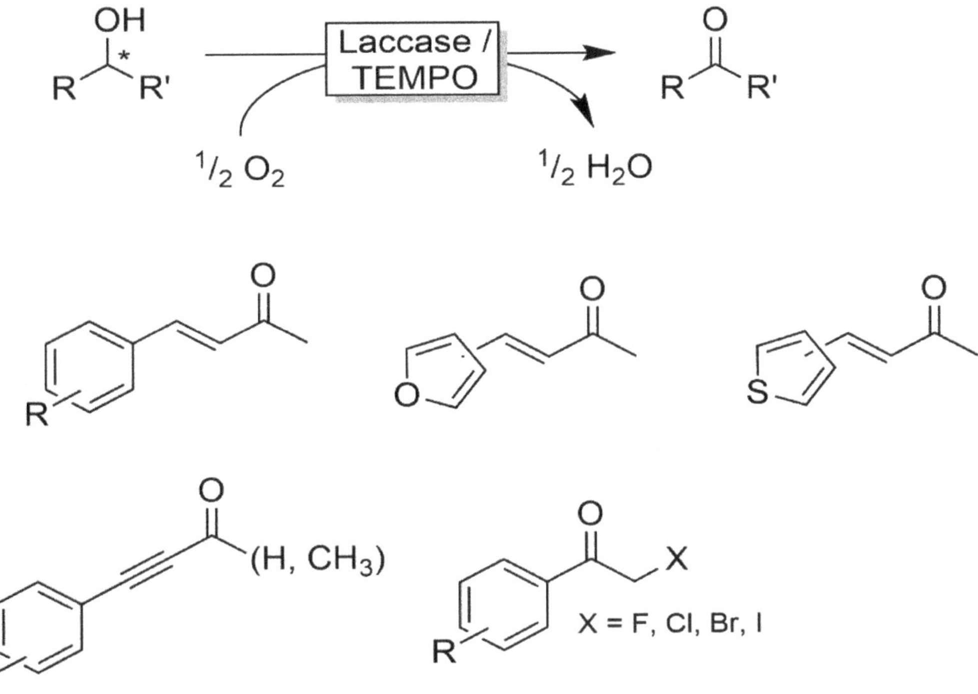

Scheme 22. Laccase-TEMPO (2,2,6,6-Tetramethylpiperidin-1-yl)oxyl) system for the full oxidation of racemic alcohols.

Especially activated benzylic, allylic, or propargylic alcohols are readily converted into the corresponding ketones. Provided, the still rather low turnover numbers of the oxidation catalyst (TEMPO, ranging below 100) have been improved, this method bears some potential for the full oxidation of racemic alcohols.

3.2. Regioselective Oxidation of Polyols

In addition to stereoselectivity, (oxidative) enzymes also frequently exhibit regioselectivity, enabling them to perform selective transformations on poly-functionalized starting materials. Such regioselectivity is particularly interesting in the case of carbohydrate oxidations; here, a selective oxidation catalyst can avoid extensive protection and deprotection chemistry. A range of oxidases and dehydrogenases catalyzing highly regioselective oxidations of polyols are known today [172,173]. Table 4 displays some representative examples.

Table 4. Selection of biocatalytic oxidations of secondary alcohols.

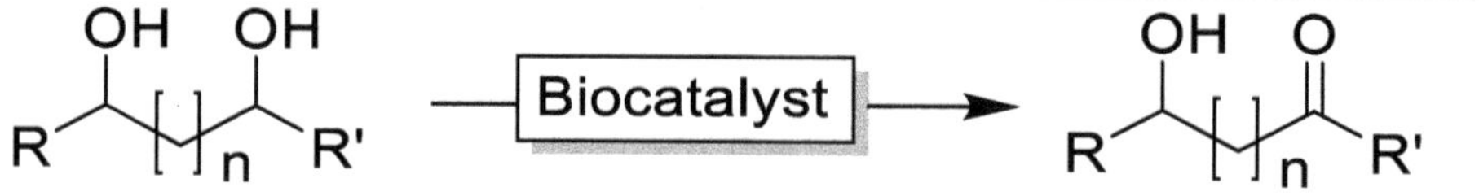

OH OH R n R' — Biocatalyst → OH O R n R'

Product	Biocatalyst	Yield [%]	Reference
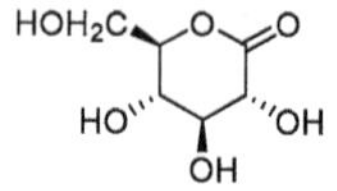	GluOx	Up to >99	[174]
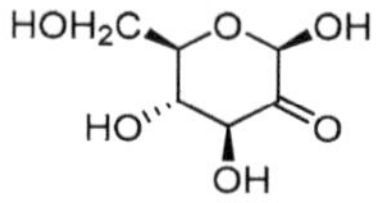	P2O	Up to >99	[175–178]
	CBOx	Up to >99	[179–183]
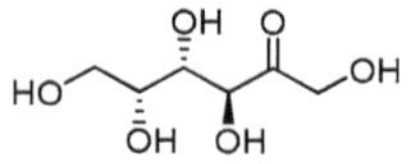	POlDH	>99	[184,185]
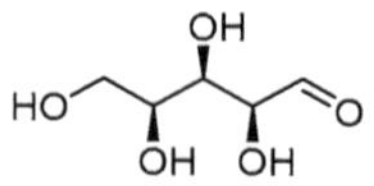	AldO	>99 (10 mM)	[186–188]

GluOx: glucose oxidase; P2O: pyranose-2-oxidase; CBOx: cellobiose oxidase; POlDH: polyol dehydrogenase; ADH-A: ADH from *Rhodococcus ruber*; ADH-9: commercial ADH; AldO: alditol oxidase from *Streptomyces coelicolor.*

Although the scope of these enzymes still is rather limited today, they exhibit a significant potential particularly in carbohydrate chemistry for protection-group independent functionalization reactions.

In the context of regioselective carbohydrate oxidation, the Reichstein process from 1934 (originally from Hoffmann–La Roche) for the transformation of glucose to ascorbic acid (vitamin C) is worth mentioning, as it still is used industrially (Scheme 23) [189].

Gluconobacter oxydans has also been investigated intensively for the oxidation of glycerol to dihydroxy acetone [190] to valorize the by-product from biodiesel synthesis into a building block for further chemical syntheses.

Other examples of regioselective oxidation deal with the conversion of steroids using selective hydroxysteroid dehydrogenases [191].

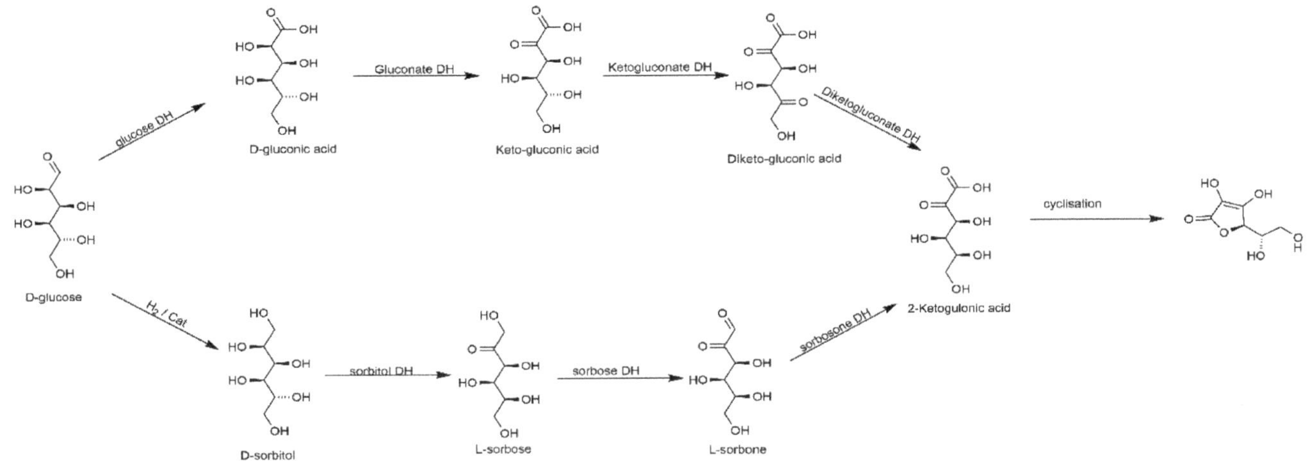

Scheme 23. Industrial pathways for the transformation of glucose to ascorbic acid.

3.3. Kinetic Oxidative Resolution and Deracemization of Racemic Secondary Alcohols

The stereoselective oxidation of only one alcohol enantiomer is a possibility for obtaining enantiomerically pure alcohols from racemic alcohols. To mention just one example, Kroutil and coworkers used the stereoselective ADH from *Rhodococcus ruber* (ADH-A, Scheme 24) for the kinetic resolution of a broad range of racemic alcohols [192].

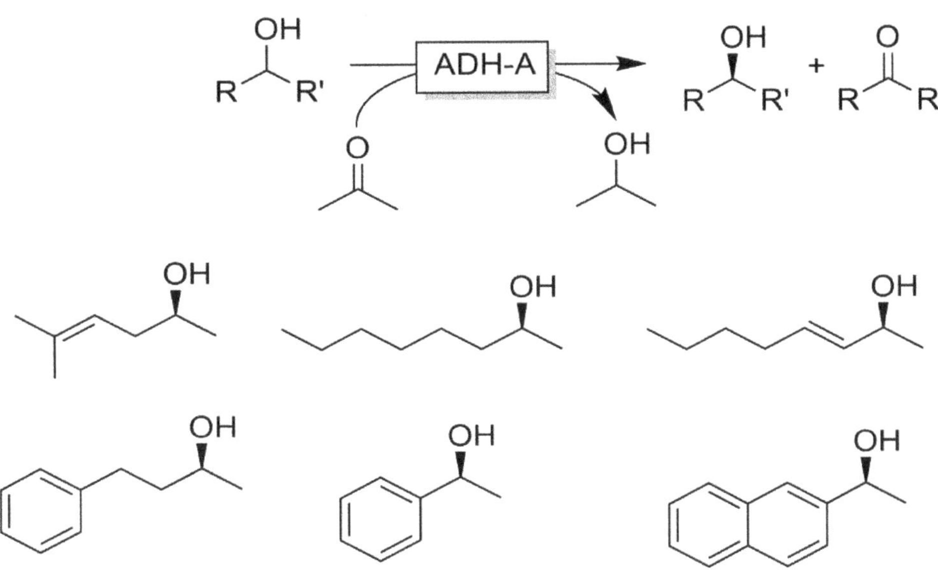

Scheme 24. Using the ADH from *Rhodococcus ruber* (ADH-A) for the oxidative kinetic resolution of racemic alcohols.

However, kinetic resolution reactions are hampered by their intrinsic maximal yield of 50%. Deracemization reactions circumvent this drawback by recycling the unwanted ketone product back into the starting alcohol [193]. Early examples used chemical reductants such as $NaBH_3CN$ yielding racemic alcohol from the ketone, which underwent further cycles of enzymatic kinetic resolution (Scheme 25a). More elegantly, Kroutil and coworkers introduced a bienzymatic reaction concept combining two enantiocomplementary ADHs wherein the first ADH catalyzes the kinetic resolution and the second ADH reduces the intermediate ketone into the desired alcohol enantiomer (Scheme 25b) [194,195].

a) deracemization via steroselecive oxidation and non-stereoselective re-reduction

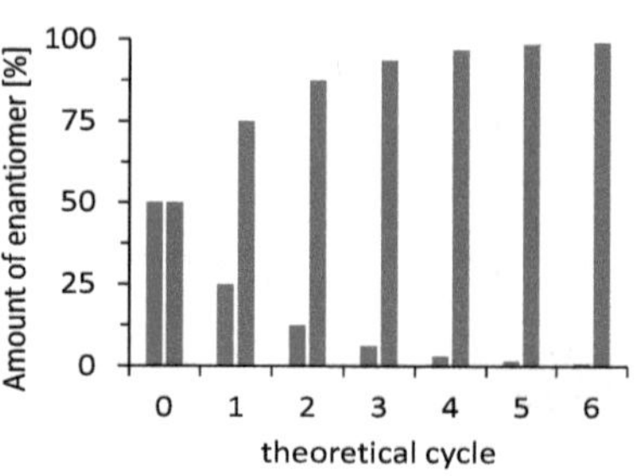

b) deracemization with stereoselective oxidation and stereoselective re-reduction

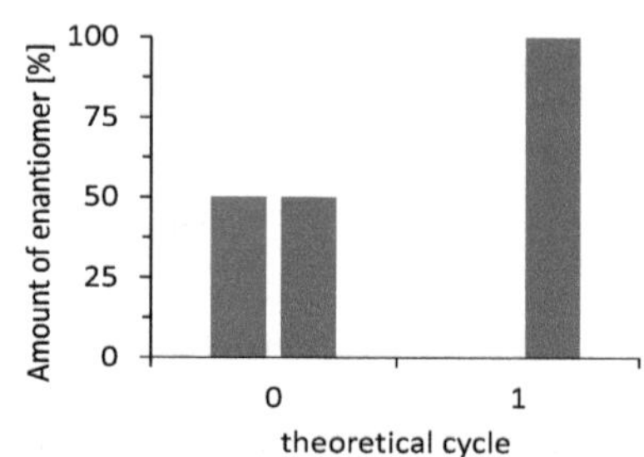

Scheme 25. Deracemization of alcohols combining (**a**) stereoselective kinetic resolution of the alcohol and non-selective re-reduction back into the racemate and (**b**) stereoselective oxidation combined with stereoselective re-reduction.

This principle is also applied for the stereoinversion of steroid alcohols [191,196]. The epimerization of e.g., cholic acid to chenodeoxycholic acid was possible using two enantiocomplementary hydroxysteroid dehydrogenases (Scheme 26) [196]. In contrast to the deracemization reactions described above, this reaction proceeded smoothly to almost full conversion even in the absence of any cofactor regeneration system, which was attributed to a lower energy content of the product and thereby resulting in a shifted equilibrium of the overall reaction.

Scheme 26. Bienzymatic epimerization of cholic acid into chenodeoxycholic acid.

4. Concluding Remarks

Biocatalysis offers manifold practical solutions for the oxidation of alcohols. Compared to many traditional chemical alternatives, selectivity is certainly the main feature of interest. Biocatalytic oxidation remains a very active field of research that has already solved issues such as the limited substrate scope of cofactor regeneration issues. Various promising approaches have been brought forward to solve the current issue of low substrate loadings. Hence, we are convinced that the importance of biocatalysis in alcohol oxidation will grow in the near future.

Acknowledgments: This research was funded by the European Research Commission (ERC consolidator grant, No. 648026) and the Netherlands Organization for Scientific Research (VICI grant No. 724.014.003). Open Access Funding by TU Delft.

References

1. Noyori, R.; Aoki, M.; Sato, K. Green oxidation with aqueous hydrogen peroxide. *Chem. Commun.* **2003**, *16*, 1977–1986. [CrossRef] [PubMed]
2. Sheldon, R.A.; Arends, I.W.C.E.; Hanefeld, U. *Green Chemistry and Catalysis*; Wiley-VCH: Weinheim, Germany, 2007.
3. Schoemaker, H.E.; Mink, D.; Wubbolts, M.G. Dispelling the myths—Biocatalysis in industrial synthesis. *Science* **2003**, *299*, 1694–1697. [CrossRef] [PubMed]
4. Wong, J.T.F.; Williams, G.R. The mechanism of liver alcohol dehydrogenase. *Arch. Biochem. Biophys.* **1968**, *124*, 344–348. [CrossRef]
5. Tsai, C.S. Relative reactivities of primary alcohols as substrates of liver alcohol dehydrogenase. *Can. J. Biochem.* **1968**, *46*, 381–385. [CrossRef]
6. Rosano, G.L.; Ceccarelli, E.A. Recombinant protein expression in *Escherichia coli*: Advances and challenges. *Front. Microbiol.* **2014**, *5*, 172. [CrossRef]
7. Baghban, R.; Farajnia, S.; Rajabibazl, M.; Ghasemi, Y.; Mafi, A.; Hoseinpoor, R.; Rahbarnia, L.; Aria, M. Yeast expression systems: Overview and recent advances. *Mol. Biotechnol.* **2019**, *61*, 365–384. [CrossRef]
8. Quaglia, D.; Irwin, J.A.; Paradisi, F. Horse liver alcohol dehydrogenase: New perspectives for an old enzyme. *Mol. Biotechnol.* **2012**, *52*, 244–250. [CrossRef]
9. Zhao, Z.Q.; Wang, H.L.; Zhang, Y.P.; Chen, L.F.; Wu, K.; Wei, D.Z. Cloning and characterization of three ketoreductases from soil metagenome for preparing optically active alcohols. *Biotechnol. Lett.* **2016**, *38*, 1799–1808. [CrossRef]
10. Berini, F.; Casciello, C.; Marcone, G.L.; Marinelli, F. Metagenomics: Novel enzymes from non-culturable microbes. *FEMS Microbiol. Lett.* **2017**, *364*, 19. [CrossRef]
11. Ausec, L.; Berini, F.; Casciello, C.; Cretoiu, M.S.; van Elsas, J.D.; Marinelli, F.; Mandic-Mulec, I. The first acidobacterial laccase-like multicopper oxidase revealed by metagenomics shows high salt and thermo-tolerance. *Appl. Microbiol. Biotechnol.* **2017**, *101*, 6261–6276. [CrossRef] [PubMed]
12. Ufarté, L.; Potocki-Veronese, G.; Cecchini, D.; Tauzin, A.S.; Rizzo, A.; Morgavi, D.P.; Cathala, B.; Moreau, C.; Cleret, M.; Robe, P.; et al. Highly promiscuous oxidases discovered in the bovine rumen microbiome. *Front. Microbiol.* **2018**, *9*, 861. [CrossRef] [PubMed]
13. Radianingtyas, H.; Wright, P.C. Alcohol dehydrogenases from thermophilic and hyperthermophilic archaea and bacteria. *FEMS Microbiol. Rev.* **2003**, *27*, 593–616. [CrossRef]
14. Feller, G. Psychrophilic enzymes: From folding to function and biotechnology. *Scientifica* **2013**, *2013*, 512840. [CrossRef] [PubMed]
15. Petratos, K.; Gessmann, R.; Daskalakis, V.; Papadovasilaki, M.; Papanikolau, Y.; Tsigos, I.; Bouriotis, V. Structure and dynamics of a thermostable alcohol dehydrogenase from the antarctic psychrophile *Moraxella* sp. Tae123. *ACS Omega* **2020**, *5*, 14523–14534. [CrossRef]
16. Liu, Y.; Pan, J.; Wei, P.; Zhu, J.; Huang, L.; Cai, J.; Xu, Z. Efficient expression and purification of recombinant alcohol oxidase in *Pichia pastoris*. *Biotechnol. Bioproc. Eng.* **2012**, *17*, 693–702. [CrossRef]
17. Hamnevik, E.; Maurer, D.; Enugala, T.R.; Chu, T.; Löfgren, R.; Dobritzsch, D.; Widersten, M. Directed evolution of alcohol dehydrogenase for improved stereoselective redox transformations of 1-phenylethane-1,2-diol and its corresponding acyloin. *Biochemistry* **2018**, *57*, 1059–1062.
18. Qu, G.; Liu, B.; Jiang, Y.; Nie, Y.; Yu, H.; Sun, Z. Laboratory evolution of an alcohol dehydrogenase towards enantioselective reduction of difficult-to-reduce ketones. *Biores. Bioproc.* **2019**, *6*, 18. [CrossRef]
19. Li, G.; Reetz, M.T. Learning lessons from directed evolution of stereoselective enzymes. *Org. Chem. Front.* **2016**, *3*, 1350–1358. [CrossRef]
20. Viña-Gonzalez, J.; Jimenez-Lalana, D.; Sancho, F.; Serrano, A.; Martinez, A.T.; Guallar, V.; Alcalde, M. Structure-guided evolution of aryl alcohol oxidase from *Pleurotus eryngii* for the selective oxidation of secondary benzyl alcohols. *Adv. Synth. Catal.* **2019**, *361*, 2514–2525. [CrossRef]
21. Viña-Gonzalez, J.; Gonzalez-Perez, D.; Ferreira, P.; Martinez, A.T.; Alcalde, M. Focused directed evolution of aryl-alcohol oxidase in *Saccharomyces cerevisiae* by using chimeric signal peptides. *Appl. Environ. Microbiol.* **2015**, *81*, 6451–6462. [CrossRef]

22. Molina-Espeja, P.; Garcia-Ruiz, E.; Gonzalez-Perez, D.; Ullrich, R.; Hofrichter, M.; Alcalde, M. Directed evolution of unspecific peroxygenase from *Agrocybe aegerita*. *Appl. Environ. Microbiol.* **2014**, *80*, 3496–3507. [CrossRef] [PubMed]
23. Garcia-Ruiz, E.; Gonzalez-Perez, D.; Ruiz-Duenas, F.J.; Martinez, A.T.; Alcalde, M. Directed evolution of a temperature-, peroxide- and alkaline pH-tolerant versatile peroxidase. *Biochem. J.* **2012**, *441*, 487–498. [CrossRef] [PubMed]
24. Qu, G.; Li, A.; Acevedo-Rocha, C.G.; Sun, Z.; Reetz, M.T. The crucial role of methodology development in directed evolution of selective enzymes. *Angew. Chem. Int. Ed.* **2020**, *59*, 13204–13231. [CrossRef]
25. Reetz, M.T. Directed evolution as a means to engineer enantioselective enzymes. In *Asymmetric Organic Synthesis with Enzymes*; Gotor, V., Alfonso, I., García-Urdiales, E., Eds.; Wiley-VCH: Hoboken, NJ, USA, 2008; pp. 21–63.
26. Arnold, F.H. Directed evolution: Bringing new chemistry to life. *Angew. Chem. Int. Ed.* **2018**, *57*, 4143–4148. [CrossRef] [PubMed]
27. Bornscheuer, U.T.; Huisman, G.W.; Kazlauskas, R.J.; Lutz, S.; Moore, J.C.; Robins, K. Engineering the third wave of biocatalysis. *Nature* **2012**, *485*, 185–194. [CrossRef] [PubMed]
28. Roiban, G.D.; Reetz, M.T. Enzyme promiscuity: Using a P450 enzyme as a carbene transfer catalyst. *Angew. Chem. Int. Ed.* **2013**, *52*, 5439–5440. [CrossRef]
29. Wang, M.; Si, T.; Zhao, H.M. Biocatalyst development by directed evolution. *Biores. Technol.* **2012**, *115*, 117–125. [CrossRef]
30. Zhou, J.Y.; Wang, Y.; Xu, G.C.; Wu, L.; Han, R.Z.; Schwaneberg, U.; Rao, Y.J.; Zhao, Y.L.; Zhou, J.H.; Ni, Y. Structural insight into enantioselective inversion of an alcohol dehydrogenase reveals a "polar gate" in stereorecognition of diaryl ketones. *J. Am. Chem. Soc.* **2018**, *140*, 12645–12654. [CrossRef]
31. Maria-Solano, M.A.; Romero-Rivera, A.; Osuna, S. Exploring the reversal of enantioselectivity. *Org. Biomol. Chem.* **2017**, *15*, 4122–4129. [CrossRef]
32. Aalbers, F.S.; Fürst, M.J.L.J.; Rovida, S.; Trajkovic, M.; Gómez Castellanos, J.R.; Bartsch, S.; Vogel, A.; Mattevi, A.; Fraaije, M.W. Approaching boiling point stability of an alcohol dehydrogenase through computationally-guided enzyme engineering. *eLife* **2020**, *9*, e54639. [CrossRef]
33. Pickl, M.; Winkler, C.K.; Glueck, S.M.; Fraaije, M.W.; Faber, K. Rational engineering of a flavoprotein oxidase for improved direct oxidation of alcohols to carboxylic acids. *Molecules* **2017**, *22*, 2205. [CrossRef] [PubMed]
34. Van den Heuvel, R.H.H.; Fraaije, M.W.; Mattevi, A.; Laane, C.; van Berkel, W.J.H. Vanillyl-alcohol oxidase, a tasteful biocatalyst. *J. Mol. Catal. B Enzym.* **2001**, *11*, 185–188. [CrossRef]
35. Van den Heuvel, R.H.H.; Fraaije, M.W.; van Berkel, W.J.H. Direction of the reactivity of vanillyl-alcohol oxidase with 4-alkylphenols. *FEBS Lett.* **2000**, *481*, 109–112. [CrossRef]
36. Van den Heuvel, R.H.H.; Fraaije, M.W.; Ferrer, M.; Mattevi, A.; van Berkel, W.J.H. Inversion of stereospecificity of vanillyl-alcohol oxidase. *Proc. Natl. Acad. Sci. USA* **2000**, *97*, 9455–9460. [CrossRef]
37. Drijfhout, F.P.; Fraaije, M.W.; Jongejan, H.; van Berkel, W.J.H.; Franssen, M.C.R. Enantioselective hydroxylation of 4-alkylphenols by vanillyl alcohol oxidase. *Biotechnol. Bioeng.* **1998**, *59*, 171–177. [CrossRef]
38. Fraaije, M.W.; van Berkel, W.J.H. Catalytic mechanism of the oxidative demethylation of 4-(methoxymethyl)phenol by vanillyl-alcohol oxidase - evidence for formation of a p-quinone methide intermediate. *J. Biol. Chem.* **1997**, *272*, 18111–18116. [CrossRef]
39. Vina-Gonzalez, J.; Alcalde, M. Directed evolution of the aryl-alcohol oxidase: Beyond the lab bench. *Comp. Struct. Biotechnol. J.* **2020**, *18*, 1800–1810. [CrossRef]
40. Vina-Gonzalez, J.; Elbl, K.; Ponte, X.; Valero, F.; Alcalde, M. Functional expression of aryl-alcohol oxidase in *Saccharomyces cerevisiae* and *Pichia pastoris* by directed evolution. *Biotechnol. Bioeng.* **2018**, *115*, 1666–1674. [CrossRef]
41. Serrano, A.; Sancho, F.; Vina-Gonzalez, J.; Carro, J.; Alcalde, M.; Guallar, V.; Martinez, A.T. Switching the substrate preference of fungal aryl-alcohol oxidase: Towards stereoselective oxidation of secondary benzyl alcohols. *Catal. Sci. Technol.* **2019**, *9*, 833–841. [CrossRef]
42. Machielsen, R.; Leferink, N.G.H.; Hendriks, A.; Brouns, S.J.J.; Hennemann, H.-G.; Dauβmann, T.; van der Oost, J. Laboratory evolution of *Pyrococcus furiosus* alcohol dehydrogenase to improve the production of (2S,5S)-hexanediol at moderate temperatures. *Extremophiles* **2008**, *12*, 587–594. [CrossRef]

43. Pennacchio, A.; Pucci, B.; Secundo, F.; La Cara, F.; Rossi, M.; Raia, C.A. Purification and characterization of a novel recombinant highly enantioselective short-chain NAD(H)-dependent alcohol dehydrogenase from *Thermus thermophilus*. *Appl. Environ. Microbiol.* **2008**, *74*, 3949–3958. [CrossRef] [PubMed]
44. Höllrigl, V.; Hollmann, F.; Kleeb, A.; Buehler, K.; Schmid, A. TADH, the thermostable alcohol dehydrogenase from *Thermus* sp. ATN1: A versatile new biocatalyst for organic synthesis. *Appl. Microbiol. Biotechnol.* **2008**, *81*, 263–273. [CrossRef] [PubMed]
45. Raia, C.A.; Giordano, A.; Rossi, M.; Adams, M.W.W.; Kelly, R.M. Alcohol dehydrogenase from *Sulfolobus solfataricus*. In *Methods Enzymol.*; Academic Press: Cambridge, MA, USA, 2001; Volume 331, pp. 176–195.
46. Tufvesson, P.; Lima-Ramos, J.; Nordblad, M.; Woodley, J.M. Guidelines and cost analysis for catalyst production in biocatalytic processes. *Org. Proc. Res. Dev.* **2010**, *15*, 266–274. [CrossRef]
47. Moa, S.; Himo, F. Quantum chemical study of mechanism and stereoselectivity of secondary alcohol dehydrogenase. *J. Inorg. Biochem.* **2017**, *175*, 259–266. [CrossRef] [PubMed]
48. Hollmann, F.; Opperman, D.J.; Paul, C.E. Enzymatic reductions - a chemist's perspective. *Angew. Chem. Int. Ed.* **2020**. [CrossRef]
49. Musa, M.M.; Ziegelmann-Fjeld, K.I.; Vieille, C.; Phillips, R.S. Activity and selectivity of w110a secondary alcohol dehydrogenase from *Thermoanaerobacter ethanolicus* in organic solvents and ionic liquids: Mono-and biphasic media. *Org. Biomol. Chem.* **2008**, *6*, 887–892. [CrossRef]
50. Fossati, E.; Polentini, F.; Carrea, G.; Riva, S. Exploitation of the alcohol dehydrogenase-acetone nadp-regeneration system for the enzymatic preparative-scale production of 12-ketochenodeoxycholic acid. *Biotechnol. Bioeng.* **2006**, *93*, 1216–1220. [CrossRef]
51. Kosjek, B.; Stampfer, W.; Pogorevc, M.; Goessler, W.; Faber, K.; Kroutil, W. Purification and characterization of a chemotolerant alcohol dehydrogenase applicable to coupled redox reactions. *Biotechnol. Bioeng.* **2004**, *86*, 55–62. [CrossRef]
52. Orbegozo, T.; Vries, J.G.d.; Kroutil, W. Biooxidation of primary alcohols to aldehydes through hydrogen transfer employing *Janibacter terrae*. *Eur. J. Org. Chem.* **2010**, *2010*, 3445–3448. [CrossRef]
53. Chenault, H.; Whitesides, G. Regeneration of nicotinamide cofactors for use in organic synthesis. *App. Biochem. Biotechnol.* **1987**, *14*, 147–197. [CrossRef]
54. Ni, Y.; Holtmann, D.; Hollmann, F. How green is biocatalysis? To calculate is to know. *ChemCatChem* **2014**, *6*, 930–943. [CrossRef]
55. Lavandera, I.; Kern, A.; Resch, V.; Ferreira-Silva, B.; Glieder, A.; Fabian, W.M.F.; de Wildeman, S.; Kroutil, W. One-way biohydrogen transfer for oxidation of sec-alcohols. *Org. Lett.* **2008**, *10*, 2155–2158. [CrossRef] [PubMed]
56. Riebel, B.R.; Gibbs, P.R.; Wellborn, W.B.; Bommarius, A.S. Cofactor regeneration of NAD^+ from NADH: Novel water-forming NADH oxidases. *Adv. Synth. Catal.* **2002**, *344*, 1156–1168. [CrossRef]
57. Riebel, B.; Gibbs, P.; Wellborn, W.; Bommarius, A. Cofactor regeneration of both NAD^+ from NADH and $NADP^+$ from NADPH:NADH oxidase from *Lactobacillus sanfranciscensis*. *Adv. Synth. Catal.* **2003**, *345*, 707–712. [CrossRef]
58. Hummel, W.; Riebel, B. Isolation and biochemical characterization of a new NADH oxidase from lactobacillus brevis. *Biotechnol. Lett.* **2003**, *25*, 51–54. [CrossRef]
59. Hummel, W.; Kuzu, M.; Geueke, B. An efficient and selective enzymatic oxidation system for the synthesis of enantiomerically pure d-tert-leucine. *Org. Lett.* **2003**, *5*, 3649–3650. [CrossRef]
60. Geueke, B.; Riebel, B.; Hummel, W. NADH oxidase from *Lactobacillus brevis*: A new catalyst for the regeneration of NAD. *Enzym. Microb. Technol.* **2003**, *32*, 205–211. [CrossRef]
61. Jiang, R.; Bommarius, A.S. Hydrogen peroxide-producing NADH oxidase (nox-1) from *Lactococcus lactis*. *Tetrahedron Asymm.* **2004**, *15*, 2939–2944. [CrossRef]
62. Hirano, J.; Miyamoto, K.; Ohta, H. Purification and characterization of thermostable H2O2-forming NADH oxidase from 2-phenylethanol-assimilating *Brevibacterium* sp. KU1309. *Appl. Microbiol. Biotechnol.* **2008**, *80*, 71–78. [CrossRef]
63. Huang, L.; Sayoga, G.V.; Hollmann, F.; Kara, S. Horse liver alcohol dehydrogenase-catalyzed oxidative lactamization of amino alcohols. *ACS Catal.* **2018**, *8*, 8680–8684. [CrossRef]
64. Nowak, C.; Beer, B.; Pick, A.; Roth, T.; Lommes, P.; Sieber, V. A water-forming NADH oxidase from *Lactobacillus pentosus* suitable for the regeneration of synthetic biomimetic cofactors. *Front. Microbiol.* **2015**, *6*, 957. [CrossRef] [PubMed]

65. Pham, N.H.; Hollmann, F.; Kracher, D.; Preims, M.; Haltrich, D.; Ludwig, R. Engineering an enzymatic regeneration system for NAD(P)H oxidation. *J. Mol. Catal. B Enzym.* **2015**, *120*, 38–46. [CrossRef]
66. Kochius, S.; Ni, Y.; Kara, S.; Gargiulo, S.; Schrader, J.; Holtmann, D.; Hollmann, F. Light-accelerated biocatalytic oxidation reactions. *ChemPlusChem* **2014**, *79*, 1554–1557. [CrossRef]
67. Könst, P.; Kara, S.; Kochius, S.; Holtmann, D.; Arends, I.W.C.E.; Ludwig, R.; Hollmann, F. Expanding the scope of laccase-mediator systems. *ChemCatChem* **2013**, *5*, 3027–3032. [CrossRef]
68. Kara, S.; Spickermann, D.; Schrittwieser, J.H.; Weckbecker, A.; Leggewie, C.; Arends, I.W.C.E.; Hollmann, F. Access to lactone building blocks via horse liver alcohol dehydrogenase-catalyzed oxidative lactonization. *ACS Catal.* **2013**, *3*, 2436–2439. [CrossRef]
69. Aksu, S.; Arends, I.W.C.E.; Hollmann, F. A new regeneration system for oxidized nicotinamide cofactors. *Adv. Synth. Catal.* **2009**, *351*, 1211–1216. [CrossRef]
70. Ferrandi, E.E.; Monti, D.; Patel, I.; Kittl, R.; Haltrich, D.; Riva, S.; Ludwig, R. Exploitation of a laccase/meldola's blue system for nad+ regeneration in preparative scale hydroxysteroid dehydrogenase-catalyzed oxidations. *Adv. Synth. Catal.* **2012**, *354*, 2821–2828. [CrossRef]
71. Tonin, F.; Martì, E.; Arends, I.W.C.E.; Hanefeld, U. Laccase did it again: A scalable and clean regeneration system for NAD+ and its application in the synthesis of 12-oxo-hydroxysteroids. *Catalysts* **2020**, *10*, 677. [CrossRef]
72. Zhu, C.; Li, Q.; Pu, L.; Tan, Z.; Guo, K.; Ying, H.; Ouyang, P. Nonenzymatic and metal-free organocatalysis for in situ regeneration of oxidized cofactors by activation and reduction of molecular oxygen. *ACS Catal.* **2016**, *6*, 4989–4994. [CrossRef]
73. Rauch, M.; Schmidt, S.; Arends, I.W.C.E.; Oppelt, K.; Kara, S.; Hollmann, F. Photobiocatalytic alcohol oxidation using led light sources. *Green Chem.* **2017**, *19*, 376–379. [CrossRef]
74. Gargiulo, S.; Arends, I.W.C.E.; Hollmann, F. A photoenzymatic system for alcohol oxidation. *ChemCatChem* **2011**, *3*, 338–342. [CrossRef]
75. Boratyński, F.; Dancewicz, K.; Paprocka, M.; Gabryś, B.; Wawrzeńcz, C. Chemo-enzymatic synthesis of optically active γ- and δ-decalactones and their effect on aphid probing, feeding and settling behavior. *PLoS ONE* **2016**, *11*, e0146160. [CrossRef] [PubMed]
76. Boratynski, F.; Kielbowicz, G.; Wawrzenczyk, C. Lactones 34. Application of alcohol dehydrogenase from horse liver (hladh) in enantioselective synthesis of δ- and ε-lactones. *J. Mol. Catal. B Enzym.* **2010**, *65*, 30–36. [CrossRef]
77. Irwin, A.J.; Jones, J.B. Asymmetric syntheses via enantiotopically selective horse liver alcohol dehydrogenase catalyzed oxidations of diols containing a prochiral center. *J. Am. Chem. Soc.* **1977**, *99*, 556–561. [CrossRef]
78. Dias Gomes, M.; Bommarius, B.R.; Anderson, S.R.; Feske, B.D.; Woodley, J.M.; Bommarius, A.S. Bubble column enables higher reaction rate for deracemization of (R,S)-1-phenylethanol with coupled alcohol dehydrogenase/NADH oxidase system. *Adv. Synth. Catal.* **2019**, *361*, 2574–2581. [CrossRef]
79. Presecki, A.V.; Vasic-Racki, Ð. Mathematical modelling of the dehydrogenase catalyzed hexanol oxidation with coenzyme regeneration by NADH oxidase. *Proc. Biochem.* **2009**, *44*, 54–61. [CrossRef]
80. Könst, P.; Merkens, H.; Kara, S.; Kochius, S.; Vogel, A.; Zuhse, R.; Holtmann, D.; Arends, I.W.C.E.; Hollmann, F. Oxidation von aldehyden mit alkoholdehydrogenasen. *Angew. Chem. Int. Ed.* **2012**, *51*, 9914–9917. [CrossRef]
81. Zhang, J.D.; Cui, Z.M.; Fan, X.J.; Wu, H.L.; Chang, H.H. Cloning and characterization of two distinct water-forming NADH oxidases from *Lactobacillus pentosus* for the regeneration of nad. *Bioproc. Biosys. Eng.* **2016**, *39*, 603–611. [CrossRef]
82. Holec, C.; Neufeld, K.; Pietruszka, J. P450 bm3 monooxygenase as an efficient NAD(P)H-oxidase for regeneration of nicotinamide cofactors in ADH-catalysed preparative scale biotransformations. *Adv. Synth. Catal.* **2016**, *358*, 1810–1819. [CrossRef]
83. Rehn, G.; Pedersen, A.T.; Woodley, J.M. Application of nad(p)h oxidase for cofactor regeneration in dehydrogenase catalyzed oxidations. *J. Mol. Catal. B Enzym.* **2016**, *134*, 331–339. [CrossRef]
84. Ni, Y.; Fernández-Fueyo, E.; Baraibar, A.G.; Ullrich, R.; Hofrichter, M.; Yanase, H.; Alcalde, M.; van Berkel, W.J.H.; Hollmann, F. Peroxygenase-catalyzed oxyfunctionalization reactions promoted by the complete oxidation of methanol. *Angew. Chem. Int. Ed.* **2016**, *55*, 798–801. [CrossRef] [PubMed]
85. Jia, H.-Y.; Zong, M.-H.; Zheng, G.-W.; Li, N. Myoglobin-catalyzed efficient in situ regeneration of nad(p)+ and their synthetic biomimetic for dehydrogenase-mediated oxidations. *ACS Catal.* **2019**, *9*, 2196–2202. [CrossRef]

86. Jia, H.-Y.; Zong, M.-H.; Zheng, G.-W.; Li, N. One-pot enzyme cascade for controlled synthesis of furancarboxylic acids from 5-hydroxymethylfurfural by H2O2 internal recycling. *ChemSusChem* **2019**, *12*, 4764–4768. [CrossRef] [PubMed]
87. Ramesh, H.; Mayr, T.; Hobisch, M.; Borisov, S.; Klimant, I.; Krühne, U.; Woodley, J.M. Measurement of oxygen transfer from air into organic solvents. *J. Chem. Technol. Biotechnol.* **2016**, *91*, 832–836. [CrossRef]
88. Pedersen, A.T.; Birmingham, W.R.; Rehn, G.; Charnock, S.J.; Turner, N.J.; Woodley, J.M. Process requirements of galactose oxidase catalyzed oxidation of alcohols. *Org. Proc. Res. Dev.* **2015**, *19*, 1580–1589. [CrossRef]
89. Bommarius, A.S.; Karau, A. Deactivation of formate dehydrogenase (FDH) in solution and at gas-liquid interfaces. *Biotechnol. Prog.* **2005**, *21*, 1663–1672. [CrossRef]
90. Thomas, C.R.; Geer, D. Effects of shear on proteins in solution. *Biotechnol. Lett.* **2011**, *33*, 443–456. [CrossRef]
91. Van Hecke, W.; Ludwig, R.; Dewulf, J.; Auly, M.; Messiaen, T.; Haltrich, D.; Van Langenhove, H. Bubble-free oxygenation of a bi-enzymatic system: Effect on biocatalyst stability. *Biotechnol. Bioeng.* **2009**, *102*, 122–131. [CrossRef]
92. Massey, V. The chemical and biological versatility of riboflavin. *Biochem. Soc. Trans.* **2000**, *28*, 283–296. [CrossRef]
93. Massey, V. Activation of molecular oxygen by flavins and flavoproteins. *J. Biol. Chem.* **1994**, *269*, 22459–22462.
94. Whittaker, J.W. The radical chemistry of galactose oxidase. *Arch. Biochem. Biophys.* **2005**, *433*, 227–239. [CrossRef] [PubMed]
95. Whittaker, J.W. Free radical catalysis by galactose oxidase. *Chem. Rev.* **2003**, *103*, 2347–2363. [CrossRef] [PubMed]
96. Díaz-Rodríguez, A.; Lavandera, I.; Kanbak-Aksu, S.; Sheldon, R.A.; Gotor, V.; Gotor-Fernández, V. From diols to lactones under aerobic conditions using a laccase/TEMPO catalytic system in aqueous medium. *Adv. Synth. Catal.* **2012**, *18*, 3405–3408. [CrossRef]
97. Tromp, S.; Matijošytė, I.; Sheldon, R.; Arends, I.; Mul, G.; Kreutzer, M.; Moulijn, J.; de Vries, S. Mechanism of laccase–TEMPO-catalyzed oxidation of benzyl alcohol. *ChemCatChem* **2010**, *2*, 827–833. [CrossRef]
98. Arends, I.; Li, Y.X.; Ausan, R.; Sheldon, R.A. Comparison of tempo and its derivatives as mediators in laccase catalysed oxidation of alcohols. *Tetrahedron* **2006**, *62*, 6659–6665. [CrossRef]
99. Sheldon, R.A.; Arends, I.W.C.E. Organocatalytic oxidations mediated by nitroxyl radicals. *Adv. Synth. Catal.* **2004**, *346*, 1051–1071. [CrossRef]
100. Bühler, B.; Bollhalder, I.; Hauer, B.; Witholt, B.; Schmid, A. Chemical biotechnology for the specific oxyfunctionalization of hydrocarbons on a technical scale. *Biotechnol. Bioeng.* **2003**, *82*, 833–842. [CrossRef] [PubMed]
101. Bühler, B.; Bollhalder, I.; Hauer, B.; Witholt, B.; Schmid, A. Use of the two-liquid phase concept to exploit kinetically controlled multistep biocatalysis. *Biotechnol. Bioeng.* **2003**, *81*, 683–694. [CrossRef] [PubMed]
102. Bühler, B.; Witholt, B.; Hauer, B.; Schmid, A. Characterization and application of xylene monooxygenase for multistep biocatalysis. *Appl. Environ. Microbiol.* **2002**, *68*, 560–568. [CrossRef] [PubMed]
103. Bühler, B.; Schmid, A.; Hauer, B.; Witholt, B. Xylene monooxygenase catalyzes the multistep oxygenation of toluene and pseudocumene to corresponding alcohols, aldehydes, and acids in *Escherichia coli* jm101. *J. Biol. Chem.* **2000**, *275*, 10085–10092. [CrossRef]
104. Molinari, F.; Gandolfi, R.; Aragozzini, F.; Leon, R.; Prazeres, D.M.F. Biotransformations in two-liquid-phase systems: Production of phenylacetaldehyde by oxidation of 2-phenylethanol with acetic acid bacteria. *Enzym. Microb. Technol.* **1999**, *25*, 729–735. [CrossRef]
105. Gandolfi, R.; Cavenago, K.; Gualandris, R.; Sinisterra Gago, J.V.; Molinari, F. Production of 2-phenylacetic acid and phenylacetaldehyde by oxidation of 2-phenylethanol with free immobilized cells of *Acetobacter aceti*. *Proc. Biochem.* **2004**, *39*, 749–753. [CrossRef]
106. Villa, R.; Romano, A.; Gandolfi, R.; Sinisterra Gago, J.V.; Molinari, F. Chemoselective oxidation of primary alcohols to aldehydes with *Gluconobacter oxydans*. *Tetrahedron Lett.* **2002**, *43*, 6059–6061. [CrossRef]
107. Gandolfi, R.; Ferrara, N.; Molinari, F. An easy and efficient method for the production of carboxylic acids and aldehydes by microbial oxidation of primary alcohols. *Tetrahedron Lett.* **2001**, *42*, 513–514. [CrossRef]
108. Ferreira, P.; Medina, M.; Guillén, F.; Martínez, M.J.; Van Berkel, W.J.H.; Martínez, Á.T. Spectral and catalytic properties of aryl-alcohol oxidase, a fungal flavoenzyme acting on polyunsaturated alcohols. *Biochem. J.* **2005**, *389*, 731–738. [CrossRef]

109. Francisco, G.; Angel, T.M.; Maria Jesús, M. Substrate specificity and properties of the aryl-alcohol oxidase from the ligninolytic fungus *Pleurotus eryngii*. *Eur. J. Biochem.* **1992**, *209*, 603–611.
110. De Almeida, T.P.; van Schie, M.; Tieves, F.; Ma, A.; Younes, S.; Fernandez Fueyo, E.; Arends, I.; Riul, A.; Hollmann, F. Efficient aerobic oxidation of *trans*-2-hexen-1-ol using the aryl alcohol oxidase from *Pleurotus eryngii*. *Adv. Synth. Catal.* **2019**, *361*, 2668–2672. [CrossRef]
111. Van Schie, M.M.C.H.; Pedroso de Almeida, T.; Laudadio, G.; Tieves, F.; Fernández-Fueyo, E.; Noël, T.; Arends, I.W.C.E.; Hollmann, F. Biocatalytic synthesis of the green note trans-2-hexenal in a continuous-flow microreactor. *Beilstein J. Org. Chem.* **2018**, *14*, 697–703. [CrossRef]
112. Ukeda, H.; Ishii, T.; Sawamura, M.; Isobe, K. Glycolaldehyde production from ethylene glycol with immobilized alcohol oxidase and catalase. *Biosci. Biotech. Biochem.* **1998**, *62*, 1589–1591. [CrossRef]
113. Isobe, K.; Nishise, H. A new enzymatic method for glycolaldehyde production from ethylene glycol. *J. Mol. Catal. B Enzym.* **1995**, *1*, 37–43. [CrossRef]
114. Matijosyte, I.; Arends, I.W.C.E.; de Vries, S.; Sheldon, R.A. Preparation and use of cross-linked enzyme aggregates (cleas) of laccases. *J. Mol. Catal. B Enzym.* **2010**, *62*, 142–148. [CrossRef]
115. Baratto, L.; Candido, A.; Marzorati, M.; Sagui, F.; Riva, S.; Danieli, B. Laccase-mediated oxidation of natural glycosides. *J. Mol. Catal. B Enzym.* **2006**, *39*, 3–8. [CrossRef]
116. Barreca, A.M.; Fabbrini, M.; Galli, C.; Gentili, P.; Ljunggren, S. Laccase/mediated oxidation of a lignin model for improved delignification procedures. *J. Mol. Catal. B Enzym.* **2003**, *26*, 105–110. [CrossRef]
117. Baiocco, P.; Barreca, A.M.; Fabbrini, M.; Galli, C.; Gentili, P. Promoting laccase activity towards non-phenolic substrates: A mechanistic investigation with some laccase-mediator systems. *Org. Biomol. Chem.* **2003**, *1*, 191–197. [CrossRef]
118. Fabbrini, M.; Galli, C.; Gentili, P.; Macchitella, D. An oxidation of alcohols by oxygen with the enzyme laccase and mediation by tempo. *Tetrahedron Lett.* **2001**, *42*, 7551–7553. [CrossRef]
119. Gavagan, J.E.; Fager, S.K.; Seip, J.E.; Clark, D.S.; Payne, M.S.; Anton, D.L.; DiCosimo, R. Chemoenzymic synthesis of n-(phosphonomethyl)glycine. *J. Org. Chem.* **1997**, *62*, 5419–5427. [CrossRef]
120. Siebum, A.; van Wijk, A.; Schoevaart, R.; Kieboom, T. Galactose oxidase and alcohol oxidase: Scope and limitations for the enzymatic synthesis of aldehydes. *J. Mol. Catal. B Enzym.* **2006**, *41*, 141–145. [CrossRef]
121. Franke, D.; Machajewski, T.; Hsu, C.-C.; Wong, C.-H. One-pot synthesis of l-fructose using coupled multienzyme systems based on rhamnulose-1-phosphate aldolase. *J. Org. Chem.* **2003**, *68*, 6828–6831. [CrossRef]
122. Herter, S.; McKenna, S.M.; Frazer, A.R.; Leimkuhler, S.; Carnell, A.J.; Turner, N.J. Galactose oxidase variants for the oxidation of amino alcohols in enzyme cascade synthesis. *ChemCatChem* **2015**, *7*, 2313–2317. [CrossRef]
123. Zhang, W.; Fernandez Fueyo, E.; Hollmann, F.; Leemans Martin, L.; Pesic, M.; Wardenga, R.; Höhne, M.; Schmidt, S. Combining photo-organo redox- and enzyme catalysis facilitates asymmetric C-H bond functionalization. *Eur. J. Org. Chem.* **2019**, *10*, 80–84. [CrossRef]
124. Schmidt, S.; Pedroso de Almeida, T.; Rother, D.; Hollmann, F. Towards environmentally acceptable synthesis of chiral α-hydroxy ketones via oxidase-lyase cascades. *Green Chem.* **2017**, *19*, 1226–1229. [CrossRef]
125. Pérez-Sánchez, M.; Müller, C.R.; Domínguez de María, P. Multistep oxidase–lyase reactions: Synthesis of optically active 2-hydroxyketones by using biobased aliphatic alcohols. *ChemCatChem* **2013**, *5*, 2512–2516. [CrossRef]
126. Shanmuganathan, S.; Natalia, D.; Greiner, L.; Dominguez de Maria, P. Oxidation-hydroxymethylation-reduction: A one-pot three-step biocatalytic synthesis of optically active α-aryl vicinal diols. *Green Chem.* **2012**, *14*, 94–97. [CrossRef]
127. Sattler, J.H.; Fuchs, M.; Tauber, K.; Mutti, F.G.; Faber, K.; Pfeffer, J.; Haas, T.; Kroutil, W. Redox self-sufficient biocatalyst network for the amination of primary alcohols. *Angew. Chem. Int. Ed.* **2012**, *51*, 9156–9159. [CrossRef] [PubMed]
128. Velasco-Lozano, S.; Santiago-Arcos, J.; Mayoral, J.A.; López-Gallego, F. Co-immobilization and colocalization of multi-enzyme systems for the cell-free biosynthesis of aminoalcohols. *ChemCatChem* **2020**, *12*, 3030–3041. [CrossRef]
129. Stewart, K.N.; Hicks, E.G.; Domaille, D.W. Merger of whole cell biocatalysis with organocatalysis upgrades alcohol feedstocks in a mild, aqueous, one-pot process. *ACS Sustain. Chem. Eng.* **2020**, *8*, 4114–4119. [CrossRef]

130. Hafenstine, G.R.; Ma, K.; Harris, A.W.; Yehezkeli, O.; Park, E.; Domaille, D.W.; Cha, J.N.; Goodwin, A.P. Multicatalytic, light-driven upgrading of butanol to 2-ethylhexenal and hydrogen under mild aqueous conditions. *ACS Catal.* **2017**, *7*, 568–572. [CrossRef]
131. Hafenstine, G.R.; Harris, A.W.; Ma, K.; Cha, J.N.; Goodwin, A.P. Conversion of ethanol to 2-ethylhexenal at ambient conditions using tandem, biphasic catalysis. *ACS Sustain. Chem. Eng.* **2017**, *5*, 10483–10489. [CrossRef]
132. De Vitis, V.; Dall'oglio, F.; Tentori, F.; Contente, M.L.; Romano, D.; Brenna, E.; Tamborini, L.; Molinari, F. Bioprocess intensification using flow reactors: Stereoselective oxidation of achiral 1,3-diols with immobilized *Acetobacter aceti*. *Catalysts* **2019**, *9*, 208. [CrossRef]
133. De Vitis, V.; Dall'Oglio, F.; Pinto, A.; De Micheli, C.; Molinari, F.; Conti, P.; Romano, D.; Tamborini, L. Chemoenzymatic synthesis in flow reactors: A rapid and convenient preparation of captopril. *ChemistryOpen* **2017**, *6*, 668–673. [CrossRef]
134. Brenna, E.; Cannavale, F.; Crotti, M.; De Vitis, V.; Gatti, F.G.; Migliazza, G.; Molinari, F.; Parmeggiani, F.; Romano, D.; Santangelo, S. Synthesis of enantiomerically enriched 2-hydroxymethylalkanoic acids by oxidative desymmetrisation of achiral 1,3-diols mediated by *Acetobacter aceti*. *ChemCatChem* **2016**, *8*, 3796–3803. [CrossRef]
135. Gandolfi, R.; Borrometi, A.; Romano, A.; Sinisterra Gago, J.V.; Molinari, F. Enantioselective oxidation of (±)-2-phenyl-1-propanol to (S)-2-phenyl-1-propionic acid with *Acetobacter aceti*: Influence of medium engineering and immobilization. *Tetrahedron Asymm.* **2002**, *13*, 2345–2349. [CrossRef]
136. Geerlof, A.; Vantol, J.B.A.; Jongejan, J.A.; Duine, J.A. Enantioselective conversions of the racemic C-3-alcohol synthons, glycidol (2,3-epoxy-1-propanol), and solketal (2,2-dimethyl-4-(hydroxymethyl)-1,3-dioxolane) by quinohemoprotein alcohol dehydrogenases and bacteria containing such enzymes. *Biosci. Biotechnol. Biochem.* **1994**, *58*, 1028–1036. [CrossRef]
137. Molinari, F.; Villa, R.; Aragozzini, F.; Cabella, P.; Barbeni, M.; Squarcia, F. Multigram-scale production of aliphatic carboxylic acids by oxidation of alcohols with *Acetobacter pasteurianus* NCIMB 11664. *J. Chem. Technol. Biotechnol.* **1997**, *70*, 294–298. [CrossRef]
138. Mihal', M.; Červeňanský, I.; Markoš, J.; Rebroš, M. Bioproduction of phenylacetic acid in airlift reactor by immobilized *Gluconobacter oxydans*. *Chem. Pap.* **2017**, *71*, 103–118. [CrossRef]
139. Su, W.; Chang, Z.; Gao, K.; Wei, D. Enantioselective oxidation of racemic 1,2-propanediol to d-(-)-lactic acid by gluconobacter oxydans. *Tetrahedron Asymm.* **2004**, *15*, 1275–1277. [CrossRef]
140. Sayed, M.; Dishisha, T.; Sayed, W.F.; Salem, W.M.; Temerk, H.M.; Pyo, S.-H. Enhanced selective oxidation of trimethylolpropane to 2,2-bis(hydroxymethyl)butyric acid using *Corynebacterium* sp. Atcc 21245. *Proc. Biochem.* **2017**, *63*, 1–7. [CrossRef]
141. Sayed, M.; Dishisha, T.; Sayed, W.F.; Salem, W.M.; Temerk, H.A.; Pyo, S.-H. Selective oxidation of trimethylolpropane to 2,2-bis(hydroxymethyl)butyric acid using growing cells of *Corynebacterium* sp. ATCC 21245. *J. Biotechnol.* **2016**, *221*, 62–69. [CrossRef] [PubMed]
142. Perez, H.I.; Manjarrez, N.; Solis, A.; Luna, H.; Ramirez, M.A.; Cassani, J. Microbial biocatalytic preparation of 2-furoic acid by oxidation of 2-furfuryl alcohol and 2-furanaldehyde with *Nocardia corallina*. *Afr. J. Biotechnol.* **2009**, *8*, 2279–2282.
143. Imber, M.; Pietrzyk-Brzezinska, A.J.; Antelmann, H. Redox regulation by reversible protein S-thiolation in gram-positive bacteria. *Redox Biol.* **2019**, *20*, 130–145. [CrossRef] [PubMed]
144. Knaus, T.; Tseliou, V.; Humphreys, L.D.; Scrutton, N.S.; Mutti, F.G. A biocatalytic method for the chemoselective aerobic oxidation of aldehydes to carboxylic acids. *Green Chem.* **2018**, *20*, 3931–3943. [CrossRef]
145. Knaus, T.; Mutti, F.G.; Humphreys, L.D.; Turner, N.J.; Scrutton, N.S. Systematic methodology for the development of biocatalytic hydrogen-borrowing cascades: Application to the synthesis of chiral α-substituted carboxylic acids from α-substituted α,β-unsaturated aldehydes. *Org. Biomol. Chem.* **2015**, *13*, 223–233. [CrossRef] [PubMed]
146. Winkler, T.; Gröger, H.; Hummel, W. Enantioselective rearrangement coupled with water addition: Direct synthesis of enantiomerically pure saturated carboxylic acids from α,β-unsaturated aldehydes. *ChemCatChem* **2014**, *6*, 961–964. [CrossRef]
147. Wu, S.; Zhou, Y.; Seet, D.; Li, Z. Regio- and stereoselective oxidation of styrene derivatives to arylalkanoic acids via one-pot cascade biotransformations. *Adv. Synth. Catal.* **2017**, *359*, 2132–2141. [CrossRef]

148. Wong, C.H.; Matos, J.R. Enantioselective oxidation of 1,2-diols to l-α-hydroxy acids using co-immobilized alcohol and aldehyde dehydrogenases as catalysts. *J. Org. Chem.* **1985**, *50*, 1992–1994. [CrossRef]
149. Birmingham, W.R.; Turner, N.J. A single enzyme oxidative "cascade" via a dual-functional galactose oxidase. *ACS Catal.* **2018**, *8*, 4025–4032. [CrossRef]
150. Vilím, J.; Knaus, T.; Mutti, F.G. Catalytic promiscuity of galactose oxidase: A mild synthesis of nitriles from alcohols, air, and ammonia. *Angew. Chem. Int. Ed.* **2018**, *57*, 14240–14244. [CrossRef]
151. Martin, C.; Trajkovic, M.; Fraaije, M.W. Production of hydroxy acids: Selective double oxidation of diols by flavoprotein alcohol oxidase. *Angew. Chem. Int. Ed.* **2020**, *59*, 4869–4872. [CrossRef]
152. Irwin, A.J.; Lok, K.P.; Huang, K.W.-C.; Jones, J.B. Enzymes in organic synthesis. Influence of substrate structure on rates of horse liver alcohol dehydrogenase-catalysed oxidoreductions. *J. Chem. Soc. Perkin* **1978**, *1*, 1636–1642. [CrossRef]
153. Irwin, A.J.; Jones, J.B. Regiospecific and enantioselective horse liver alcohol dehydrogenase catalyzed oxidations of some hydroxycyclopentanes. *J. Am. Chem. Soc.* **1977**, *99*, 1625–1630. [CrossRef]
154. Schröder, I.; Steckhan, E.; Liese, A. In situ NAD+ regeneration using 2,2′-azinobis(3-ethylbenzothiazoline-6-sulfonate) as an electron transfer mediator. *J. Electroanal. Chem.* **2003**, *541*, 109–115. [CrossRef]
155. Hilt, G.; Lewall, B.; Montero, G.; Utley, J.H.P.; Steckhan, E. Efficient in-situ redox catalytic NAD(P)+ regeneration in enzymatic synthesis using transition-metal complexes of 1,10-phenanthroline-5,6-dione in its n-monomethylated derivate as catalysts. *Liebigs Ann.* **1997**, *1997*, 2289–2296. [CrossRef]
156. Boratyński, F.; Janik-Polanowicz, A.; Szczepańska, E.; Olejniczak, T. Microbial synthesis of a useful optically active (+)-isomer of lactone with bicyclo[4.3.0]nonane structure. *Sci. Rep.* **2018**, *8*, 468. [CrossRef] [PubMed]
157. Boratynski, F.; Smuga, M.; Wawrzenczyk, C. Lactones 42. Stereoselective enzymatic/microbial synthesis of optically active isomers of whisky lactone. *Food Chem.* **2013**, *141*, 419–427. [CrossRef]
158. Bechi, B.; Herter, S.; McKenna, S.; Riley, C.; Leimkuhler, S.; Turner, N.J.; Carnell, A.J. Catalytic bio-chemo and bio-bio tandem oxidation reactions for amide and carboxylic acid synthesis. *Green Chem.* **2014**, *16*, 4524–4529. [CrossRef]
159. Dijkman, W.P.; Groothuis, D.E.; Fraaije, M.W. Enzyme-catalyzed oxidation of 5-hydroxymethylfurfural to furan-2,5-dicarboxylic acid. *Angew. Chem. Int. Ed.* **2014**, *53*, 6515–6518. [CrossRef]
160. Carro, J.; Ferreira, P.; Rodriguez, L.; Prieto, A.; Serrano, A.; Balcells, B.; Arda, A.; Jimenez-Barbero, J.; Gutierrez, A.; Ullrich, R.; et al. 5-hydroxymethylfurfural conversion by fungal aryl-alcohol oxidase and unspecific peroxygenase. *FEBS J.* **2015**, *282*, 3218–3229. [CrossRef]
161. McKenna, S.M.; Mines, P.; Law, P.; Kovacs-Schreiner, K.; Birmingham, W.R.; Turner, N.J.; Leimkühler, S.; Carnell, A.J. The continuous oxidation of HMF to FDCA and the immobilisation and stabilisation of periplasmic aldehyde oxidase (paoabc). *Green Chem.* **2017**, *19*, 4660–4665. [CrossRef]
162. Cappaert, L.; Larroche, C. Oxidation of a mixture of 2-(r) and 2-(s)-heptanol to 2-heptanone by saccharomyces cerevisiae in a biphasic system. *Biocatal. Biotransf.* **2004**, *22*, 291–296. [CrossRef]
163. Velasco-Lozano, S.; Rocha-Martin, J.; Favela-Torres, E.; Calvo, J.; Berenguer, J.; Guisán, J.M.; López-Gallego, F. Hydrolysis and oxidation of racemic esters into prochiral ketones catalyzed by a consortium of immobilized enzymes. *Biochem. Eng. J.* **2016**, *112*, 136–142. [CrossRef]
164. Albarrán-Velo, J.; Gotor-Fernández, V.; Lavandera, I. One-pot two-step chemoenzymatic deracemization of allylic alcohols using laccases and alcohol dehydrogenases. *Mol. Catal.* **2020**, *493*, 111087. [CrossRef]
165. Gonzalez-Granda, S.; Mendez-Sanchez, D.; Lavandera, I.; Gotor-Fernandez, V. Laccase-mediated oxidations of propargylic alcohols. Application in the deracemization of 1-arylprop-2-yn-1-ols in combination with alcohol dehydrogenases. *ChemCatChem* **2020**, *12*, 520–527. [CrossRef]
166. Albarran-Velo, J.; Lavandera, I.; Gotor-Fernandez, V. Sequential two-step stereoselective amination of allylic alcohols through the combination of laccases and amine transaminases. *ChemBioChem* **2020**, *21*, 200–211. [CrossRef] [PubMed]
167. Martinez-Montero, L.; Gotor, V.; Gotor-Fernandez, V.; Lavandera, I. Stereoselective amination of racemic sec-alcohols through sequential application of laccases and transaminases. *Green Chem.* **2017**, *19*, 474–480. [CrossRef]
168. Mendez-Sanchez, D.; Mangas-Sanchez, J.; Lavandera, I.; Gotor, V.; Gotor-Fernandez, V. Chemoenzymatic deracemization of secondary alcohols by using a TEMPO-iodine-alcohol dehydrogenase system. *ChemCatChem* **2015**, *7*, 4016–4020. [CrossRef]

169. Diaz-Rodriguez, A.; Rios-Lombardia, N.; Sattler, J.H.; Lavandera, I.; Gotor-Fernandez, V.; Kroutil, W.; Gotor, V. Deracemisation of profenol core by combining laccase/TEMPO-mediated oxidation and alcohol dehydrogenase-catalysed dynamic kinetic resolution. *Catal. Sci. Technol.* **2015**, *5*, 1443–1446. [CrossRef]
170. Kedziora, K.; Diaz-Rodriguez, A.; Lavandera, I.; Gotor-Fernandez, V.; Gotor, V. Laccase/TEMPO-mediated system for the thermodynamically disfavored oxidation of 2,2-dihalo-1-phenylethanol derivatives. *Green Chem.* **2014**, *16*, 2448–2453. [CrossRef]
171. Diaz-Rodriguez, A.; Martinez-Montero, L.; Lavandera, I.; Gotor, V.; Gotor-Fernandez, V. Laccase/2,2,6,6-tetramethylpiperidinoxyl radical (TEMPO): An efficient catalytic system for selective oxidations of primary hydroxy and amino groups in aqueous and biphasic media. *Adv. Synth. Catal.* **2014**, *356*, 2321–2329. [CrossRef]
172. Winter, R.T.; Fraaije, M.W. Applications of flavoprotein oxidases in organic synthesis: Novel reactivities that go beyond amine and alcohol oxidations. *Curr. Org. Chem.* **2012**, *16*, 2542–2550. [CrossRef]
173. Turner, N.J. Enantioselective oxidation of c-o and c-n bonds using oxidases. *Chem. Rev.* **2011**, *111*, 4073–4087. [CrossRef]
174. Bankar, S.B.; Bule, M.V.; Singhal, R.S.; Ananthanarayan, L. Glucose oxidase—An overview. *Biotechnol. Adv.* **2009**, *27*, 489–501. [CrossRef]
175. Spadiut, O.; Pisanelli, I.; Maischberger, T.; Peterbauer, C.; Gorton, L.; Chaiyen, P.; Haltrich, D. Engineering of pyranose 2-oxidase: Improvement for biofuel cell and food applications through semi-rational protein design. *J. Biotechnol.* **2009**, *139*, 250–257. [CrossRef]
176. Pisanelli, I.; Kujawa, M.; Spadiut, O.; Kittl, R.; Halada, P.; Volc, J.; Mozuch, M.D.; Kersten, P.; Haltrich, D.; Peterbauer, C. Pyranose 2-oxidase from *Phanerochaete chrysosporium*-expression in *E. coli* and biochemical characterization. *J. Biotechnol.* **2009**, *142*, 97–106. [CrossRef] [PubMed]
177. Giffhorn, F. Fungal pyranose oxidases: Occurrence, properties and biotechnical applications in carbohydrate chemistry. *Appl. Microbiol. Biotechnol.* **2000**, *54*, 727–740. [CrossRef]
178. Freimund, S.; Huwig, A.; Giffhorn, F.; Köpper, S. Rare keto-aldoses from enzymatic oxidation: Substrates and oxidation products of pyranose 2-oxidase. *Chem. Eur. J.* **1998**, *4*, 2442–2455. [CrossRef]
179. Tan, T.-C.; Kracher, D.; Gandini, R.; Sygmund, C.; Kittl, R.; Haltrich, D.; Hallberg, B.M.; Ludwig, R.; Divne, C. Structural basis for cellobiose dehydrogenase action during oxidative cellulose degradation. *Nat. Commun.* **2015**, *6*, 7542. [CrossRef] [PubMed]
180. Fujita, K.; Nakamura, N.; Igarashi, K.; Samejima, M.; Ohno, H. Biocatalytic oxidation of cellobiose in an hydrated ionic liquid. *Green Chem.* **2009**, *11*, 351–354. [CrossRef]
181. Maischberger, T.; Nguyen, T.-H.; Sukyai, P.; Kittl, R.; Riva, S.; Ludwig, R.; Haltrich, D. Production of lactose-free galacto-oligosaccharide mixtures: Comparison of two cellobiose dehydrogenases for the selective oxidation of lactose to lactobionic acid. *Carbohydr. Res.* **2008**, *343*, 2140–2147. [CrossRef]
182. Ludwig, R.; Ozga, M.; Zamocky, M.; Peterbauer, C.; Kulbe, K.D.; Haltrich, D. Continuous enzymatic regeneration of electron acceptors used by flavoenzymes: Cellobiose dehydrogenase-catalyzed production of lactobionic acid as an example. *Biocatal. Biotransf.* **2004**, *22*, 97–104. [CrossRef]
183. Henriksson, G.; Johansson, G.; Pettersson, G. A critical review of cellobiose dehydrogenases. *J. Biotechnol.* **2000**, *78*, 93–113. [CrossRef]
184. Sha, F.; Zheng, Y.; Chen, J.; Chen, K.; Cao, F.; Yan, M.; Ouyang, P. D-tagatose manufacture through bio-oxidation of galactitol derived from waste xylose mother liquor. *Green Chem.* **2018**, *20*, 2382–2391. [CrossRef]
185. Matsushita, K.; Fujii, Y.; Ano, Y.; Toyama, H.; Shinjoh, M.; Tomiyama, N.; Miyazaki, T.; Sugisawa, T.; Hoshino, T.; Adachi, O. 5-keto-d-gluconate production is catalyzed by a quinoprotein glycerol dehydrogenase, major polyol dehydrogenase, in *Gluconobacter* species. *Appl. Environ. Microbiol.* **2003**, *69*, 1959–1966. [CrossRef] [PubMed]
186. Van Hellemond, E.W.; Vermote, L.; Koolen, W.; Sonke, T.; Zandvoort, E.; Heuts, D.P.; Janssen, D.B.; Fraaije, M.W. Exploring the biocatalytic scope of alditol oxidase from *Streptomyces coelicolor*. *Adv. Synth. Catal.* **2009**, *351*, 1523–1530. [CrossRef]
187. Leferink, N.G.H.; Heuts, D.P.H.M.; Fraaije, M.W.; van Berkel, W.J.H. The growing vao flavoprotein family. *Arch. Biochem. Biophys.* **2008**, *474*, 292–301. [CrossRef] [PubMed]

188. Forneris, F.; Heuts, D.P.H.M.; Delvecchio, M.; Rovida, S.; Fraaije, M.W.; Mattevi, A. Structural analysis of the catalytic mechanism and stereoselectivity in *Streptomyces coelicolor* alditol oxidase. *Biochemistry* **2008**, *47*, 978–985.
189. Hancock, R.D.; Viola, R. Biotechnological approaches for l-ascorbic acid production. *Trends Biotechnol.* **2002**, *20*, 299–305. [CrossRef]
190. Dikshit, P.K.; Moholkar, V.S. Optimization of 1,3-dihydroxyacetone production from crude glycerol by immobilized *Gluconobacter oxydans* MTCC 904. *Biores. Technol.* **2016**, *216*, 1058–1065. [CrossRef]
191. Ferrandi, E.E.; Bertuletti, S.; Monti, D.; Riva, S. Hydroxysteroid dehydrogenases: An ongoing story. *Eur. J. Org. Chem.* **2020**, *2020*, 4463–4473. [CrossRef]
192. Stampfer, W.; Kosjek, B.; Moitzi, C.; Kroutil, W.; Faber, K. Biocatalytic asymmetric hydrogen transfer. *Angew. Chem. Int. Ed.* **2002**, *41*, 1014–1017. [CrossRef]
193. Musa, M.M.; Hollmann, F.; Mutti, F.G. Synthesis of enantiomerically pure alcohols and amines via biocatalytic deracemisation methods. *Catal. Sci. Technol.* **2019**, *9*, 5487–5503. [CrossRef]
194. Voss, C.V.; Gruber, C.C.; Kroutil, W. Deracemization of secondary alcohols through a concurrent tandem biocatalytic oxidation and reduction. *Angew. Chem. Int. Ed.* **2008**, *47*, 741–745. [CrossRef] [PubMed]
195. Voss, C.V.; Gruber, C.C.; Faber, K.; Knaus, T.; Macheroux, P.; Kroutil, W. Orchestration of concurrent oxidation and reduction cycles for stereoinversion and deracemisation of sec-alcohols. *J. Am. Chem. Soc.* **2008**, *130*, 13969–13972. [CrossRef] [PubMed]
196. Tonin, F.; Otten, L.G.; Arends, I. NAD(+)-dependent enzymatic route for the epimerization of hydroxysteroids. *ChemSusChem* **2019**, *12*, 3192–3203. [CrossRef] [PubMed]

Screening of Biocatalysts for Synthesis of the Wieland–Miescher Ketone

Mitul P. Patel [1], Nathaneal T. Green [1], Jacob K. Burch [1], Kimberly A. Kew [2] and Robert M. Hughes [1,*]

[1] Department of Chemistry, East Carolina University, Greenville, NC 27858, USA; patelmi14@students.ecu.edu (M.P.P.); greenn15@students.ecu.edu (N.T.G.); burchj13@students.ecu.edu (J.K.B.)
[2] Department of Biochemistry and Molecular Biology, Brody School of Medicine, East Carolina University, Greenville, NC 27834, USA; kimykew@gmail.com
* Correspondence: hughesr16@ecu.edu

Abstract: Lipases, a versatile class of biocatalysts, have been shown to function in non-aqueous media/organic solvents and to possess "promiscuous" catalytic activity for a wide range of organic transformations. In this study, we explored the biocatalytic properties of a library of commercially available lipases by screening them for catalysis of a one-pot synthesis of Wieland–Miescher ketone, an important intermediate in the synthesis of biologically active compounds such as steroids and terpenoids, from methyl vinyl ketone and 2-methyl-1,3-cyclohexanedione. As a direct outgrowth of this screen, we created an optimized procedure for Wieland–Miescher ketone (WMK) synthesis using crude lipase preparations, characterizing both reaction yield and enantiomeric excess. We also identified principal components of the crude lipase mixture through proteomics and present evidence for a non-lipolytic origin of the observed catalysis. Finally, using the optimized conditions developed in this study, we propose a general absorbance-based screening methodology for assessing biocatalytic potential of crude enzyme preparations for synthesis of WMK.

Keywords: lipase; biocatalysis; Wieland-Miescher ketone; biocatalyst screening; amylase; peptidase; enantiomeric excess; Robinson annulation

---

1. Introduction

Metal-based catalysts have revolutionized synthetic chemistry, enabling challenging chemical reactions with exquisite stereochemical control [1,2]. Disadvantages of their frequent use, however, include high cost and potentially adverse environmental effects. As a result, chemists have explored biologically derived catalysts as less expensive and less toxic alternatives [1–4]. In particular, the pharmaceutical industry has adapted biocatalysts to "green" large-scale synthetic pathways [3]. Numerous enzymes, for example, have been evolved to synthesize pharmaceutical compounds, or their intermediates, with high yields and enantiomeric excess, improving overall process efficiency and environmental compatibility [4].

Lipases, a versatile class of biocatalysts, have undergone extensive study over the past two decades [5]. They have been reported to function in non-aqueous media/organic solvents, possess catalytic activity towards a wide range of organic reactions, and provide recyclability via immobilization [1,5–7]. In their native (biological) setting, lipases catalyze the hydrolysis of esters from triacylglycerides. However, when placed under minimal or no-water conditions, they can catalyze various types of organic transformations such as Michael additions and aldol additions [6,8]. Catalysis of these non-native transformations by enzymes has been previously deemed "promiscuous" [9,10], a general descriptor for proteins that catalyze non-native transformations. Hult and Berglund previously defined three types of promiscuity: conditional (due to change in solvent, temperature, etc.); substrate (adaptable substrate specificity); and catalytic (accommodation of mechanistically distinct reactions) [9]. The catalytic promiscuity of lipases has been the foundation on which various studies have been conducted to demonstrate a broad spectrum of lipase catalyzed organic transformations including the Robinson annulation, the Knoevenagel condensation, and the Morita–Baylis–Hillman reaction [7,11,12]. Lipases derived from plant, animal, and fungal sources, or recombinantly expressed from a variety of host organisms, are readily commercially available and thus are an attractive source of new catalytic moieties.

Despite intensive investigation of lipase biocatalysis, unanswered questions remain regarding fundamental aspects of their propensity for promiscuous catalysis [13]. Some studies have identified crude lipase preparations with biocatalytic properties presuming, but not demonstrating, that lipase, in particular the lipase active site responsible for the in vivo hydrolysis of fatty acids (Figure 1A,B), is also responsible for promiscuous biocatalytic behavior observed in organic or aqueous/organic solvent mixtures (Figure 1C) [12,14]. While in many instances the lipase activity present in crude lipase preparations is directly responsible for promiscuous catalysis, this is not always the case [15,16]. In the absence of a more thorough characterization of these catalytic moieties, it is challenging to improve on the initial findings through the now-standard means of rational design and/or directed evolution.

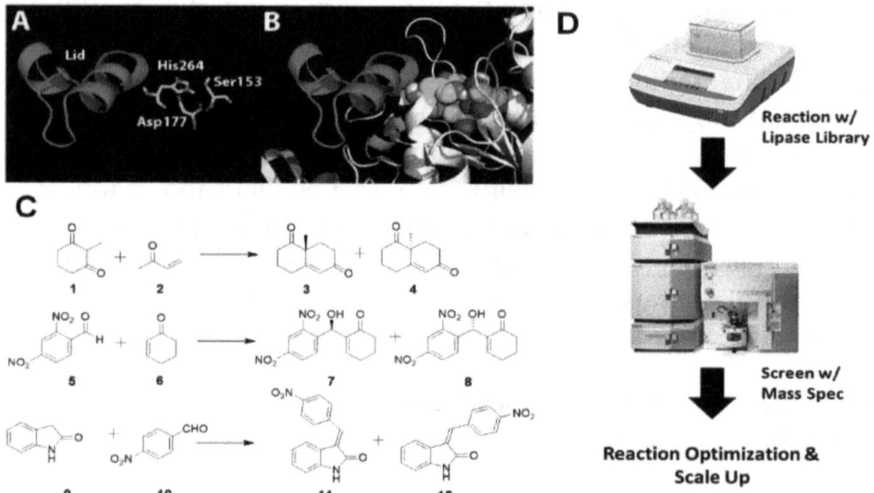

Figure 1. Porcine pancreatic lipase (PPL) is a model lipase for biocatalysis of C-C bond forming reactions. The active site ((**A**) colored spheres) of PPL is presumed to catalyze the C-C bond forming reactions via carbonyl binding and activation in an oxy-anion hole. The lid domain ((**A**) red ribbon) is presumed to undergo interfacial activation (assuming an open conformation in more hydrophobic solvents). (**B**) Both lid domain and active shown in context of 3D protein ribbon structure (PDB ID 1ETH). (**C**) Examples of carbon-carbon bond forming reactions catalyzed by lipases (Robinson annulation: reaction of cyclohexanedione **1** and methyl vinyl ketone **2** provides bicyclic Wieland–Miescher ketone products **3** and **4**; Morita–Baylis–Hillman: reaction of dinitrobenzaldehyde **5** and cyclohexenone **6** provides MBH products **7** and **8**; and Knoevenagel condensation: reaction of 2-oxindole **9** and nitrobenzaldehyde **10** provides KC products **11** and **12**). (**D**) Overview of screening methodology. Biocatalytic reactions were scaled down but otherwise conducted according to literature procedures. Mass spectrometry methodologies (LC/MS and MALDI) were used to screen the resulting biocatalytic reactions for product formation. Commercially available compounds were purchased for use as standards in mass spec calibration. Screening was then followed by reaction optimization and measurement of yield and stereoselectivity.

Our approach to the characterization of biocatalysts involved screening crude preparations for biocatalytic activity with mass spectrometry followed by reaction optimization (Figure 1D). In addition, we utilized mass spectrometry for identification of the primary components of biocatalytic lipase preparations. This opens the possibility of improving reproducibility and further development of biocatalytic methods. In this work, we describe methods to screen lipase preparations for biocatalytic activity towards a synthetically useful target, the Wieland–Miescher ketone (**3** and **4**, Figure 1C), an important intermediate for the synthesis of biologically active compounds such as steroids, terpenoids, and taxol, via a one-pot Robinson annulation reaction. In the Robinson annulation, a Michael addition between a ketone and a methyl vinyl ketone is followed by an intramolecular aldol condensation to form a six-membered ring. We also describe how initial results from a mass spectrometry-based screen can be improved through a process of reaction optimization. We took lead catalysts from the screen and further characterized their proteinaceous components, identifying non-lipase proteins that are potentially responsible for the observed catalytic activity. The methodology and fundamental characterization of biocatalysts described in this report have general utility in the screening and identification of biocatalysts in numerous organic transformations.

2. Results

We began by assembling a 14-member library of commercially available lipases preparations, selecting lipases previously reported to catalyze C-C bond forming reactions (porcine pancreatic lipase, *Candida antarctica* lipase B) [11,12,14,17], in addition to lipase preparations with unknown biocatalytic properties in promiscuous biocatalysis (Wheat germ lipase). Biocatalytic potential of the lipase library was then assessed for synthesis of the Wieland–Miescher ketone from methyl vinyl ketone and 2-methyl-1,3-cyclohexanedione using recently reported literature conditions [12]. To rapidly assess biocatalytic potential of the library, an LC/MS method was developed (Materials and Methods) that detected only the main fragmentation pattern Wieland–Miescher ketone product (Figure 2).

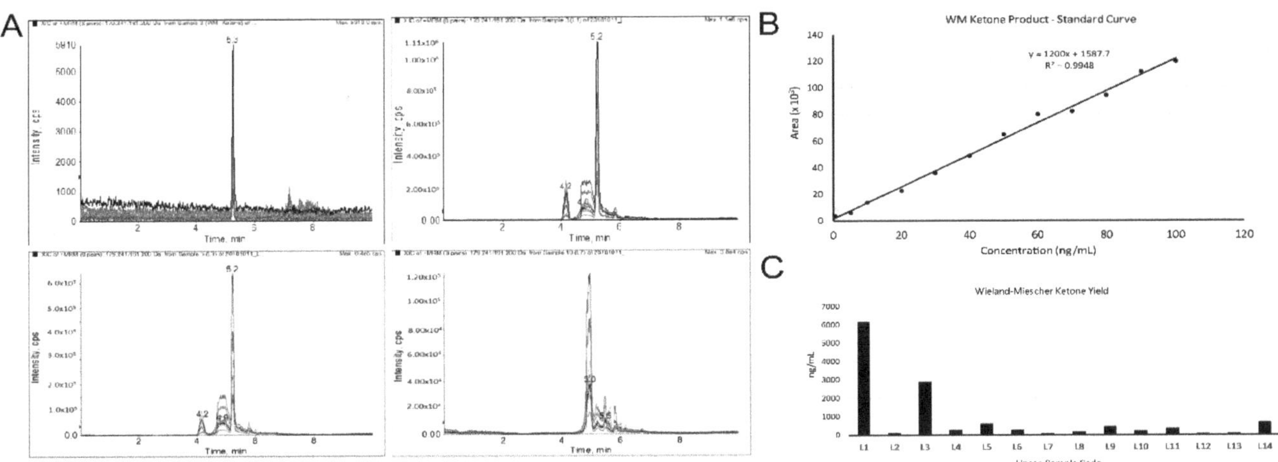

Figure 2. Development of an LC/MS method for biocatalyst screening. (**A**) LC/MS data for standard (upper left) and lipases L1, L3, and L7 (clockwise from upper right). (**B**) Standard curve for Robinson annulation product (**3** and **4**) generated with LC/MS method. (**C**) Quantification of data for lipase library using standard curve from (**B**). Reaction conditions: The following reagents/reactants were mixed in a centrifuge tube for synthesis of Wieland–Miescher ketone: methanol (1 mL), methyl vinyl ketone (81 μL), 2-methyl-1,3-cyclohexanedione (23 mg), deionized water (100 μL), and lipase (5 mg). After reactant assembly, centrifuge tubes were placed in a shaker (30 °C, 250 rpm) for 48 h. Target mass (m/z) = 178.10.

Reaction yields were determined for each lipase using a standard curve (Figure 2C and Table 1). While we were able to identify lipases (Entries 1, 3, and 5 in Table 1) with biocatalytic activity, the yields were less than 1%, in contrast to a literature report using these conditions [12]. As a result, we undertook optimization studies using our top hit from the screen (L1-PPL) to attempt better reaction yields.

Table 1. Screening of lipase library for synthesis of Wieland–Miescher ketone.

Entry	Lipase Code	Lipase Name	Yield(ng/μL) [a]
1	L1	Porcine Pancreatic Lipase (PPL)	6140
2	L2	Rhizopus Oryzae	92
3	L3	Wheat Germ Lipase	2880
4	L4	Candida Rugosa	269
5	L5	Aspergillus Niger	591
6	L6	Aspergillus Oryzae	242
7	L7	Pseudomonas Cepacia	59
8	L8	Candida Sp.	148
9	L9	Rhizopus Niveus	457
10	L10	Mucor Miehei	211
11	L11	Mucor Javanicus	331
12	L12	Candida Antarctica "B"	15
13	L13	Candida Antarctica "A"	77
14	L14	Pseudomonas Fluorescens	648

[a] Yields determined by LC/MS.

To optimize this reaction, we turned to a previously described one-pot synthesis of WMK using L-proline in DMSO [18]. Therefore, in these initial optimization studies, we switched to DMSO as the primary solvent and included L-proline as a positive control. In the optimization experiments, reaction yields were assessed using HPLC (Figure 3C). We found that the PPL-catalyzed reaction in DMSO produced the triketone product **13** (Figure 3A), whereas the L-proline reaction formed the expected WMK products **3** and **4** (Figure 3A). We hypothesized that L-proline plays dual roles as a biocatalyst: one in which the amino acid promotes the necessary transition state for conversion of **13** to **14** through an aldol addition, and another in which the α-amino sidechain functions as a base to convert **14** to a mixture of products **3** and **4** through dehydration. Therefore, we reasoned that addition of a basic co-catalyst such as imidazole might better mimic the bi-functional L-Pro catalyst. We found that a combination of PPL and imidazole was able to produce a modest yield of **3** and **4** (Figure 3 and Table 2). By contrast, in reactions in which only PPL was present, or in which only imidazole was present, no WMK product was detected via HPLC (Figure 3B and Table 2). Interestingly, the combination of BSA and imidazole also resulted in an appreciable yield of product (Table 2).

Table 2. Addition of imidazole as cocatalyst for reaction optimization.

Entry	Biocatalyst	Cocatalyst (Imidazole)	Yield (%) [a]
1	-	-	n.r.
2	-	+	n.r.
3	PPL (5 mg)	-	2.6
4	PPL (5 mg)	+	14.4
5	BSA (5 mg)	-	n.r.
6	BSA (5 mg)	+	13.4
7	L-Pro (6.4 mg)	-	68.8

[a] Yields were determined by HPLC; n.r., no reaction/product not detected by HPLC.

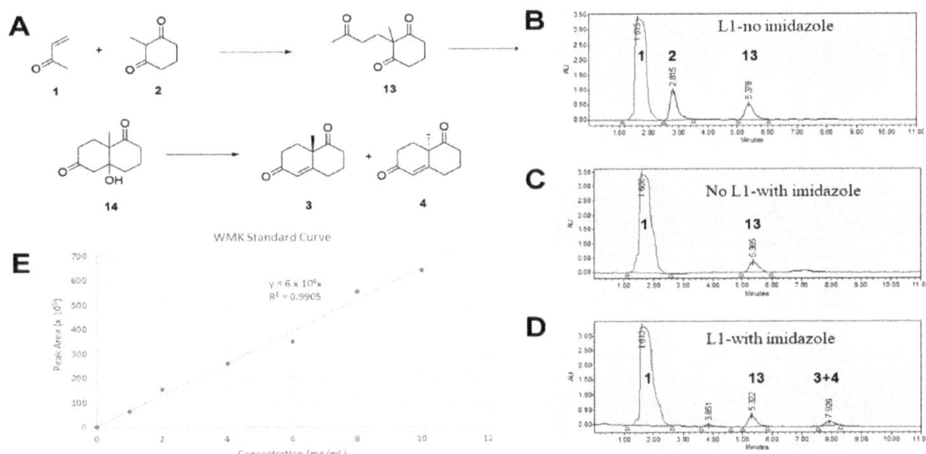

Figure 3. Investigation of reaction products in the presence and absence of imidazole and lipase preparation. (**A**) Starting materials 2-methyl-1,3-cyclohexanedione **1** and methyl vinyl ketone **2** produce triketone intermediate **13**, which proceeds through bicyclic species **14** to WMK products **3** and **4**. (**B**) HPLC chromatogram of reaction sample with L1 but no imidazole. (**C**) HPLC chromatogram of reaction sample with imidazole but no L1. (**D**) HPLC chromatogram of reaction sample with both L1 and imidazole. (**E**) Standard curve generated using HPLC data obtained from commercially available Wieland–Miescher ketone. Reaction conditions taken from [18].

2.1. Reaction Optimization: Catalyst Loading, Solvent Effects, Enantiomeric Excess

Having identified the components necessary for forming modest amounts of WMK from protein and co-catalyst, we sought to further optimize the reaction by focusing on catalyst loading and solvent system. Setting up a series of reactions in DMSO, we varied catalyst loading from 5 to 35 mg/mL, finding that reaction yield increased up to 30 mg of catalyst per 1 mL reaction volume, but decreased with amounts in excess of 30 mg (Table 3 and Figure 4A). The observed decrease in yield with higher catalyst loading is largely due to an increased amount of insoluble, aggregated material from the crude lipase preparation in the reaction vials, which interferes with efficient mixing of the lipase and substrates. Note that mechanical stirring, rather than shaking, might reverse the observed trend. We then investigated reaction yields in solvents other than DMSO. Reactions in methanol produced similar yields to DMSO, yet reactions with increasingly longer chain alcohols (ethanol, propanol, butanol, and pentanol) steadily decreased reaction yields (Table 3). Enantiomeric excess of the Wieland–Miescher ketone products produced with the lipase/imidazole system in DMSO were substantially lower than those produced with L-Proline (Figure 4), whereas performing the PPL reaction in MeOH slightly improved enantiomeric excess. Reactions performed in longer chain alcohols also resulted in progressively lower enantiomeric excess (ee) values (Table 3, Entries 6–9).

Table 3. Determination of optical purity for reactions with PPL and imidazole.

Entry	Biocatalyst	Solvent	Yield (%) [a]	Ee (%) [b]
1	L-Proline	DMSO	68.8	70.1
2	PPL-5 mg	DMSO	14.4	9.8
3	PPL-35 mg	DMSO	51.9	17.7
4	PPL-5 mg	Methanol	7.3	24.1
5	PPL-30 mg	Methanol	55.1	21.7
6	PPL-30 mg	Ethanol	48.3	10.3
7	PPL-30 mg	Propanol	35.3	8.3
8	PPL-30 mg	Butanol	28.7	6.1
9	PPL-30 mg	Pentanol	29.6	3.6

[a] Yields were determined by HPLC. [b] Enantiomeric excess (ee) determined by chiral HPLC; *S*-configuration.

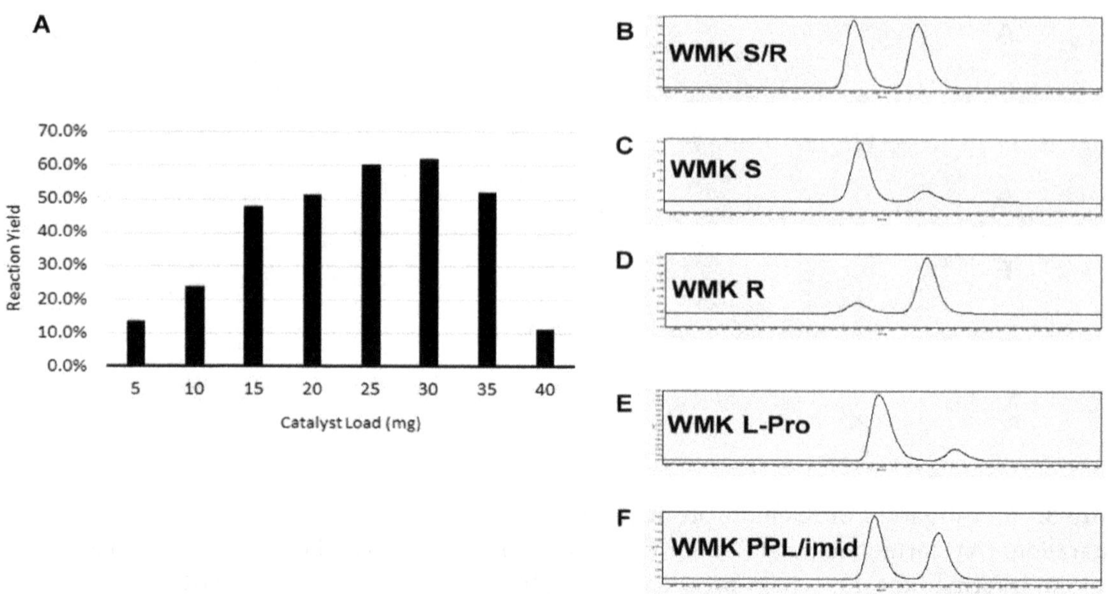

Figure 4. Effect of catalyst loading on reaction yield and determination of enantiomeric excess in L-Proline and PPL/imidazole catalyzed WMK reactions. (**A**) PPL/imidazole reactions in DMSO with varying amounts of PPL catalyst. Chiral HPLC chromatograms (Chiralcel OD-H column; mobile phase: isopropanol/heptane mixture (20/80, *v*/*v*) containing 0.1% formic acid) are shown for: (**B**) racemic WMK standard; (**C**) S-WMK enriched standard; (**D**) R-WMK enriched standard; (**E**) purified L-proline catalyzed WMK product; and (**F**) purified PPL/imidazole catalyzed WMK product.

2.2. Reaction Optimization: Exploration of Co-Catalysts

We next investigated the effects of co-catalysts other than imidazole on reaction yield. Dimethylaminopyridine (DMAP), which is more nucleophilic but less basic than imidazole, gave a substantially lower yield (Table 4, Entry 4). Alternately, ionic base co-catalysts sodium hydroxide and sodium bicarbonate gave less significantly reduced yields of WMK

Table 4. Screening optimal PPL loading with different cocatalyst and solvents.

Entry	Cocatalyst	Solvent	Yield(%) [a]
1	Sodium Hydroxide	Methanol	43.0
2	Sodium Bicarbonate	Methanol	46.5
3	Imidazole	Methanol	55.1
4	DMAP	Methanol	24.5

[a] Yields were determined by HPLC.

2.3. Determination of Primary Components of Biocatalytic Preparations

Previous reports have demonstrated the presence of multiple biocatalytic entities present in commonly used lipase preparations [19]. Having optimized our reaction conditions with crude lipase preparations, we sought to more precisely define their composition. We separated the crude PPL mixture on a size exclusion column, isolating two primary components (Figure 5A). Bands were excised and subjected to proteolytic digest and subsequent characterization by mass spectrometry (see Section 4). The higher molecular weight species was identified as porcine alpha-amylase, whereas the lower molecular weight species was identified as a porcine carboxypeptidase (both carboxypeptidases A1 and B were present in the analysis). Surprisingly, neither of the principle components of this commonly used lipase preparation was porcine pancreatic lipase. This is significant, since numerous reports use this preparation of porcine pancreatic lipase (Type II), and others without further purification [12,14,20,21]. A second lipase from our library with biocatalytic activity for WMK production (L3; wheat germ lipase) was also characterized by this method (Figure 5B and Supplementary Materials Figure S1). Again, none of the principle components of this preparation was identified as wheat germ lipase.

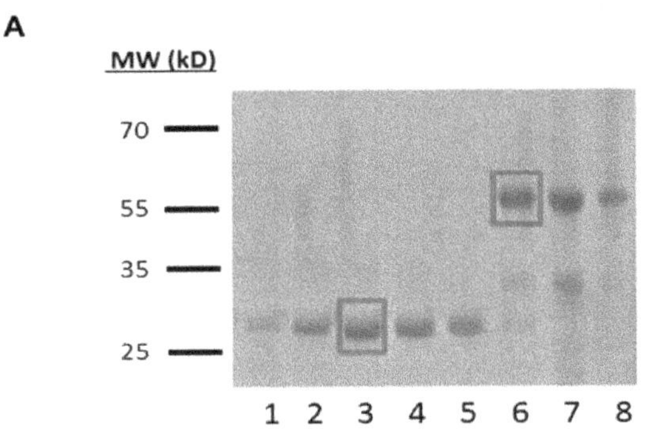

B

Sample(Band/Spot)	Protein Name	Species	MW(Da)
PPL - High MW	Alpha-amylase	Sus scrofa	57094
PPL - Low MW	Carboxypeptidase A1	Sus scrofa	47206
PPL - Low MW	Carboxypeptidase B	Sus scrofa	47351
WGL – High MW	HSP70	Triticum aestivum	70986
WGL – Low MW	Phosphoglycerate kinase	Triticum aestivum	39331

Figure 5. Identification of components of lipase L1 (Porcine Pancreatic Lipase) and L3 (Wheat Germ Lipase) preparations. (**A**) PPL lipase preparation was purified using size exclusion chromatography. The fractions were also analyzed using SDS-PAGE and two distinct bands were found to be present in the crude preparation of PPL and were subsequently characterized by mass spectrometry fingerprinting analysis. (**B**) Peptides identified in the fingerprint analysis indicated that the band with lower molecular weight (PPL, Low MW; blue rectangle, Lane 3 in (**A**)) contained two proteins Carboxypeptidase A1 and Carboxypeptidase B) and the band with higher molecular weight (PPL, High MW; blue rectangle, Lane 6 in (**A**)) contained the protein Alpha-amylase. Electrophoretic mobility of the Low MW band corresponds to the mature (truncated) forms of Carboxypeptidase A1 and B. An identical analysis was conducted for the Wheat Germ Lipase (WGL) protein preparation, resulting in the identification of HSP70 and Phosphoglycerate kinase (only two of many *T. aestivum* proteins present in this preparation, see Figure S1). The namesake "porcine pancreatic lipase" or "wheat germ lipase" were not identified as principal components in either analysis.

2.4. Reassessment of Biocatalytic Preparations with Optimized Reaction Conditions

Having identified optimized solvent, catalyst loading, and basic cofactor, we then rescreened our lipase library for biocatalytic activity towards WMK. In addition, as an outcome of this screen, we discovered that absorbance, in addition to mass spec, could be an effective means of screening for biocatalytic potential towards the WMK synthesis. In addition to select lipases from our initial screen, we also included commercially available proteins that were identified by proteomics as principal components of the crude PPL preparation (carboxypeptidase A (PA), carboxypeptidase B (PB), alpha amylase (P alphaS)) in addition to immobilized forms of *Candida antarctica* lipase B (L12-acr and L12-imo). By correlating absorbance (350 nm) of the crude reaction mixture with reaction yield (Figure 6A), we demonstrate a correlation between biocatalytic potential and yield of WMK. Furthermore, plotting absorbance (350 nm) versus percent yield reveals that applying a colorimetric cutoff of OD_{350} = 1.0 would eliminate all of the reactions with yields below 10% (Figure 6B). This indicates that colorimetric analysis could be a convenient basis for a large-scale screen of biocatalysts for WMK production.

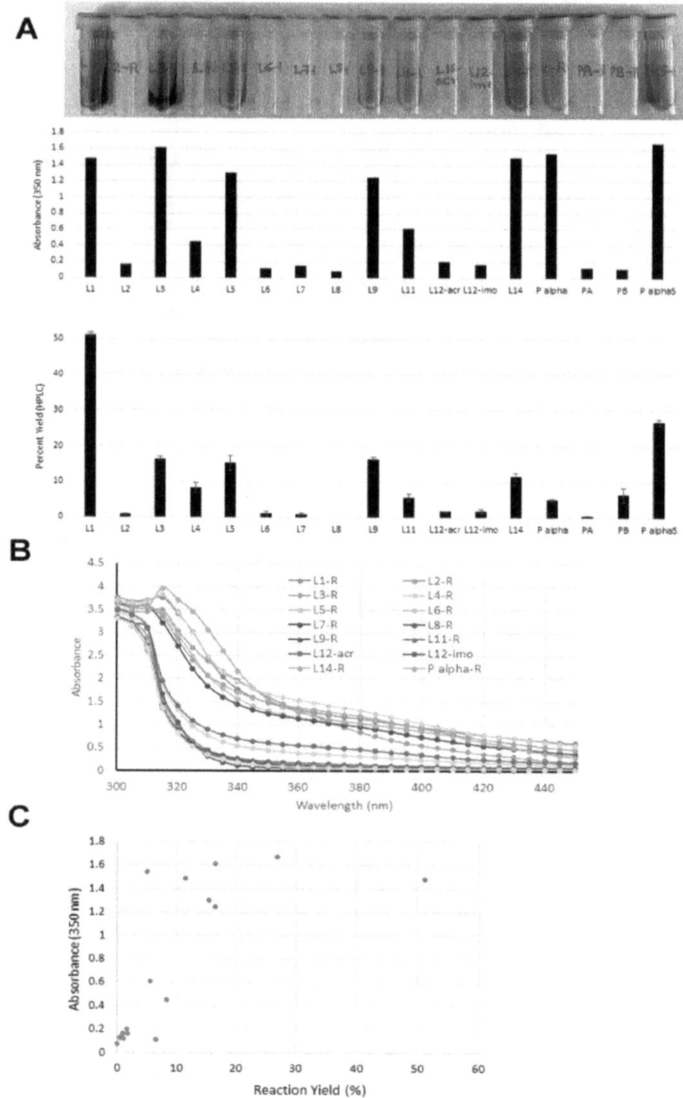

Figure 6. Rescreening of the lipase library under optimized conditions. (**A**) Images of reactions after 89 h (top); absorbance of crude reaction mixtures at 350 nm (middle); and reaction yields from rescreening of each the lipase library (bottom). Error bars determined from three replicate HPLC injections. (**B**) Wavelength scan of crude reaction absorbance. (**C**) Plot of absorbance (350 nm) versus reaction yield.

A more in-depth analysis of reaction yields from this screen reveals that the biocatalysts identified in the initial LC/MS screen (L1, L3, L5, L9, and L14) also emerged as top biocatalysts in the colorimetric screen (Figure 6 and Table 5). In addition, alpha amylase alone, which was identified as a primary component of the PPL preparation, produced a significant yield of WMK. The yield did not match that of the crude PPL preparation, indicating that multiple proteinaceous components, including porcine pancreatic lipase, carboxypeptidases, and amylases, may be involved in the biocatalytic reaction. However, based on the ability of a non-lipase protein (BSA) and an individual amino acid (L-Pro) to promote product formation, we hypothesized that the endogenous catalytic activities of the component proteins were likely not contributing to product formation. To investigate this possibility, we then conducted our biocatalytic reactions in the presence of amylase and lipase inhibitors (acarbose and orlistat, respectively; Table 6). In the presence of excess inhibitor, the yield of the PPL reaction was essentially unchanged. Similarly, the yield of the reaction with the amylase component was only modestly reduced in the presence of excess inhibitor. We also confirmed that intrinsic lipase activity of the PPL preparation was fully inhibited in the presence of orlistat by monitoring hydrolytic activity with p-nitrophenylpalmitate with and without inhibitor (Figure S2). Based upon these results, we conclude that lipase activity is not responsible for the promiscuous catalysis observed with crude preparations of PPL. Furthermore, while the reaction can be catalyzed by multiple proteinaceous entities regardless of whether they possess intrinsic enzymatic activity, not all proteins are capable of catalyzing product formation.

Table 5. Rescreening of lipase library using optimized reaction conditions.

Entry	Biocatalyst	Yield (%) [a]	ee (%) [b]
1	L1	51.9	21.7
2	L3	17.1	1.2
3	L5	13.5	19.9
4	L9	16.0	14.4
5	L14	10.9	(−)1.7
6	α-Amylase(solid)	26.4	19.3

[a] Yields were determined by HPLC; n.r., no reaction. [b] Determined by chiral HPLC (insert column information here); *S*-configuration.

Table 6. Addition of catalyst inhibitors

Entry	Biocatalyst	Inhibitor (Acarbose)	Inhibitor (Orlistat)	Yield (%) [a]
1	PPL	-	-	50.7
2	PPL	+	-	52.5
3	PPL	-	+	54.9
4	α-Amylase (solid)	-	-	31.0
5	α-Amylase (solid)	+	-	27.4
6	α-Amylase (solid)	-	+	27.6

[a] Yields were determined by HPLC.

3. Discussion

In this study, we screened lipase preparations for catalysis of a one-pot Wieland–Miescher ketone synthesis. Our initial screen detected minimal product formation in the presence of lipases alone. As a result, we turned to literature conditions used in the L-Pro catalyzed WMK synthesis for reaction optimization purposes. We found that inclusion of a basic co-catalyst was essential for promoting higher yields of WMK in the presence of a protein catalyst, in contrast to the L-Pro catalyzed reaction, which requires no co-catalyst. In addition, we used proteomics to characterize the primary components of a crude lipase mixture that has been previously used in numerous studies of biocatalysis. We anticipate that these results will inform future studies using this catalytic preparation. After further optimizing catalyst loading and solvent conditions, we then re-screened a lipase library, correlating absorbance of the crude reaction mixture with production of WMK product. This approach could easily be expanded to screen larger libraries of potential biocatalysts. Finally, we demonstrated that the innate enzymatic activities of the proteins present are not required for product formation. While the identification of the essential catalytic moieties remains incomplete, it appears that future optimization work might best be focused on alternative considerations, such as shape complementarity and surface exposed sidechains, rather than the native catalytic mechanisms of these promiscuous biocatalysts. For example, the poorer enantioselectivity of protein biocatalysts versus the individual amino acid L-Pro is intriguing, suggesting that the presence of a well-defined active-site cavity contributes little to stabilization of the transition state. However, the advantages of a protein biocatalyst over an individual amino acid catalyst (more engineerable, more adaptable to immobilization, etc.) support future engineering efforts of these nascent catalytic entities. Furthermore, as not all lipase preparations tested in this screen supported promiscuous biocatalytic activity, there are protein preparation-dependent aspects of this biocatalytic synthesis that require further investigation.

Promiscuity in biocatalysis remains an intriguing topic, often challenging chemists to think outside the comfort zone of well-established catalytic mechanisms. In this work, we demonstrated that the assumptions regarding the role of intrinsic catalytic properties of promiscuous biocatalysts should be carefully considered. In many studies, catalytic promiscuity has undoubtedly been misattributed to various properties of one enzyme or another, when the role of surface exposed amino acids, protein shape, or even influence of the dielectric environment may be more important. For promiscuous biocatalysts to fulfill their potential as inexpensive and reusable "green" reagents, it is critical that their modes of action be more precisely characterized. For example, in this study, we identified alpha amylase as a key catalytic component of a crude lipase preparation from porcine pancreas. As a result, future studies can now be conducted on the optimization of this protein for promiscuous biocatalysis through rational design, directed evolution, or exploration of immobilization methodologies. Furthermore, this study established that colorimetric screening can be useful for identifying catalysts of WMK formation. In addition to this being a general platform for identifying new biocatalysts, we anticipate that this screen can be extended to other variations on the Robinson annulation, enabling screening for substrate scope.

4. Materials and Methods

Chemicals, substrates, and solvents were purchased from commercial sources (Millipore-Sigma (St. Louis, MO, USA); TCI America (Portland, OR, USA); VWR (Radnor, PA, USA); and Fisher Scientific, Waltham, MA, USA). Lipases, amylases, and carboxypeptidases were purchased from Sigma-Aldrich (St. Louis, MO, USA). Bovine serum albumin (BSA) was purchased from Fisher Scientific (see Table 7).

LC/MS mass spectrometry method. Liquid chromatography (LC) was performed on a HPLC system (3200 Exion LC100; SCIEX Corporation, Framingham, MA, USA). Five microliters of samples were injected into the reversed-phase Gemini NX-C18 column (50 mm × 2.0 mm, 3 μm; Phenomenex, Torrance, CA, USA) by using auto sampler. The mobile phase composition for A was water/acetonitrile mixture (95/5, *v*/*v*) with 0.1% formic acid, and for B was methanol. The gradient method was as follows: 100% A for 1 min, decrease to 15% over 4 min, hold at 15% for 1 min, increase back to 100% over 1 min, and hold at 100% for 3 min. The flow rate was 0.3 mL min^{-1} and total run time was 10 min for each sample injection. The column temperature was kept at 35 °C. The LC elute was introduced into the Applied Biosystem Triple Quad API 3200, a triple-quadrupole tandem mass spectrometer (Beverly, MA, USA) equipped with a turbo spray ionization source, for quantification of compounds in positive ionization mode. Detection of the target molecule was performed in a multiple reaction monitoring (MRM) mode, and transition of *m*/*z* 179.24 to 161.20 was used for detection of WMK.

Biocatalysis general conditions. The following reagents/reactants were mixed in a microcentrifuge tube for synthesis of Wieland–Miescher ketone: solvent (900 μL), methyl vinyl ketone (19.8 μL), 2-methyl-1,3-cyclohexanedione (20 mg), deionized water (100 μL), and lipase (0–40 mg) or L-proline (6.4 mg), with or without imidazole (2.7 mg, 0.04 mmol). For reaction involving inhibitors, the biocatalyst was mixed with the inhibitor (5 mg) for 10 min at room temperature before adding the substrates and cocatalyst. After reaction assembly, capped microcentrifuge tubes were placed in a temperature-controlled shaker (New Brunswick Excella e24; 35 °C, 250 rpm) for 89 h.

Cofactor and cosolvent optimization. The following reagents/reactants were mixed in a microcentrifuge tube for synthesis of Wieland–Miescher ketone: alcohol solvent (900 μL), methyl vinyl ketone (19.8 μL), 2-methyl-1,3-cyclohexanedione (20 mg), deionized water (100 μL), lipase (L1, 30 mg), and cocatalyst (0.04 mmol). After reaction assembly, capped microcentrifuge tubes were placed in a temperature-controlled shaker (New Brunswick Excella e24; 35 °C, 250 rpm) for 89 h.

Lipase purification and proteomics. Solutions of porcine pancreatic lipase and wheat germ lipase were fractionated on a HiPrep 26/60 Sephacryl S-200 HR column connected to an Akta FPLC (GE Healthcare Life Sciences, city, State Abbr. (if has), country). Protein fractions were characterized via SDS-PAGE gels stained with SimplyBlue SafeStain (ThermoFisher). The bands of greatest intensity were excised and digested with trypsin overnight using a standard in-gel digestion protocol. The resultant tryptic peptides were desalted using a C18 Zip-Tip and applied to a MALDI target plate with α-cyano-4-hydroxycinnamic acid as the matrix. The sample was analyzed on an AB Sciex 5800 MALDI TOF/TOF mass spectrometer (Framingham, MA, USA). MS/MS spectra were searched against Uniprot databases using Mascot version 2.3 (Matrix Science) (Boston, MA, USA) within ProteinPilot software version 3.0 (AB Sciex, Framingham, MA, USA). A significance score threshold was calculated in Mascot, with ion scores above the threshold considered positive IDs ($p < 0.05$).

Table 7. Lipase library components [a].

Lipase Code	Lipase	Catalog Number
L1	Porcine Pancreatic Lipase (PPL) *	L3126
L2	Rhizopus Oryzae *	62305
L3	Wheat Germ Lipase *	L3001
L4	Candida Rugosa *	L1754
L5	Aspergillus Niger *	62301
L6	Rhizomucor miehei **	L4277
L7	Pseudomonas Cepacia *	62309
L8	Candida Sp. **	L3170
L9	Rhizopus Niveus *	62310
L10	Mucor Miehei *	62298
L11	Mucor Javanicus *	L8906
L12	Candida Antarctica B **	62288
L13	Candida Antarctica A **	62287
L14	Pseudomonas Fluorescens *	534730
L12-acr	Candida Antarctica B **	L4777
L12-imo	Candida Antarctica B **	52583
PA	Carboxypeptidase A *	C9268
PB	Carboxypeptidase B *	C9584
P alpha	Alpha amylase (suspension) *	10102814001
P alpha S	Alpha amylase (powder) *	A3176

[a] Lipase library components were purchased from Sigma-Aldrich. Catalog numbers are shown for rapid identification of lipases. * Indicates lipases extracted from native biological source; ** indicates lipase produced in a non-native host (i.e., recombinant).

Hi-Res Mass Spec analysis of triketone and WMK. Samples were analyzed with a Q Exactive HF-X (ThermoFisher, Bremen, Germany) mass spectrometer. Samples were introduced via a heated electrospray source (HESI) at a flow rate of 10 μL/min. One hundred times domain transients were averaged in the mass spectrum. HESI source conditions were set as: nebulizer temperature 100 deg C, sheath gas (nitrogen) 15 arb, auxiliary gas (nitrogen) 5 arb, sweep gas (nitrogen) 0 arb, capillary temperature 250 °C, RF voltage 100 V, and spray voltage 3.5 KV. The mass range was set to 150–2000 m/z. All measurements were recorded at a resolution setting of 120,000. Solutions were analyzed at 0.1 mg/mL or less based on responsiveness to the ESI mechanism. Xcalibur (ThermoFisher, Breman, Germany) was used to analyze the data. Molecular formula assignments were determined with Molecular Formula Calculator (v 1.2.3). All observed species were singly charged, as verified by unit m/z separation between mass spectral peaks corresponding to the ^{12}C and $^{13}C^{12}C_{c\text{-}1}$ isotope for each elemental composition.

Characterization of triketone (2-methyl-2-(3-oxobutyl)-1,3-cyclohexanedione): Hi-Res mass spec: m/z = 197.11761 (mass error = 2.0 ppm) Assigned Chemical Formula: C11H17O3 $[M+H]^+$; 1H-NMR

(400 MHz, CDCl3): 2.60–2.77 (m, 4H), 2.35 (t, 2H), 2.11 (s, 3H), 2.04–2.08 (m, 2H), 1.89–1.95 (m, 2H), 1.25 (s, 3H).

Characterization of WMK ((S,R)-(+/−)-8a-methyl-3,4,8,8a-tetrahydro-1,6(2H,7H)-naphthalenedione): Hi-Res mass spec: m/z = 179.10583 (mass error = −4.6 ppm) Assigned Chemical Formula: C11H15O2 $[M+H]^+$; 1H NMR (400 MHz): 5.85 (s, 1H), 2.67–2.76 (m, 2H), 2.38–2.52 (m, 4H), 2.09–2.19 (m, 3H), 1.65–1.78 (m, 1H), 1.45 (s, 3H).

HPLC analysis of reaction yield. Liquid chromatography was performed on a HPLC system (Waters Corporation, Milford, MA, USA) consisting of a binary pump (1525 Binary HPLC Pump), absorbance detector (2487 Dual Wavelength Absorbance Detector), and auto sampler (717-plus Autosampler). Five microliters of sample were injected into the reversed-phase Xterra MS C18 column (4.6 mm × 100 mm, 5 μm; Waters, Milford, MA, USA) using the auto sampler. The samples were analyzed using an isocratic method: 80% Solvent A (water) and 20% Solvent B (methanol). The flow rate was 1.0 mL min^{-1} and the total run time was 13 min for each sample injection. The detector wavelength was set to 210 nm and the retention time for WMK was 7.9 min.

HPLC analysis of enantiomeric excess. Liquid chromatography for enantiomeric excess was performed on a HPLC system (Waters Corporation) consisting of binary pump (1525 Binary HPLC Pump), absorbance detector (2487 Dual Wavelength Absorbance Detector), and auto sampler (717-plus Autosampler). Five microliters of sample were injected into the Chiralcel OD-H column (0.46 cm × 25 cm; Daicel, Minato, Japan) using the auto sampler. Isopropanol/heptane mixture (20/80, *v/v*) containing 0.1% formic acid was used as a mobile phase at a flow rate of 0.1 mL min^{-1}. Total run time was 111 min for each sample injection. The detector wavelength was set to 250 nm and the retention times for (S)-WMK and (R)-WMK were 43.5 and 46.3 min, respectively.

5. Conclusions

We characterized the biocatalytic potential of commercially available lipase preparations for catalysis of Wieland–Miescher ketone synthesis. Optimization experiments demonstrated the importance of a basic co-catalyst for efficient production of Wieland–Miescher ketone. Enantioselectivities were lower than a previously reported L-Pro catalyzed WMK synthesis. We were also able to identify non-lipase proteins present in a commercially available lipase preparation with biocatalytic activity and establish a colorimetric screen for identification of biocatalysts. Taken together, these methods comprise a potential platform for future engineering studies, where biocatalysts identified in this work can be optimized for better yields, enantioselectivities, and substrate scope in WMK synthesis and in related carbon–carbon bond forming reactions.

Author Contributions: Conceptualization, M.P.P. and R.M.H.; methodology, M.P.P., K.A.K., and R.M.H.; formal analysis, M.P.P., K.A.K., and R.M.H.; investigation, M.P.P., N.T.G., J.K.B., and R.M.H.; resources, R.M.H.; data curation, K.A.K.; writing—original draft preparation, M.P.P. and R.M.H.; writing—review and editing, M.P.P., K.A.K., and R.M.H.; supervision, R.M.H.; project administration, R.M.H.; and funding acquisition, R.M.H. All authors have read and agreed to the published version of the manuscript.

Acknowledgments: We thank the ECU Department of Chemistry Mass Spectrometry facility, Kim Kew, director and the UNC Michael Hooker Proteomics Core, Laura J. Herring, director, for their assistance with proteomics experiments and analysis. We also thank the University of North Carolina's Department of Chemistry Mass Spectrometry Core Laboratory, Brandie Ehrmann, director, and Diane Wallace for their assistance with mass spectrometry analysis. We also thank Brian E. Love, ECU Department of Chemistry, for helpful discussions.

References

1. Sheldon, R.A.; Woodley, J.M. Role of Biocatalysis in Sustainable Chemistry. *Chem. Rev.* **2018**, *118*, 801–838. [CrossRef] [PubMed]

2. Sheldon, R.A.; Brady, D. Broadening the Scope of Biocatalysis in Sustainable Organic Synthesis. *ChemSusChem* **2019**, *12*, 2859–2881. [CrossRef] [PubMed]
3. Wenda, S.; Illner, S.; Mell, A.; Kragl, U. Industrial biotechnology—The future of green chemistry? *Green Chem.* **2011**, *13*, 3007–3047. [CrossRef]
4. Ma, S.K.; Gruber, J.; Davis, C.; Newman, L.; Gray, D.; Wang, A.; Grate, J.; Huisman, G.W.; Sheldon, R.A. A green-by-design biocatalytic process for atorvastatin intermediate. *Green Chem.* **2010**, *12*, 81–86. [CrossRef]
5. Reetz, M.T. Biocatalysis in Organic Chemistry and Biotechnology: Past, Present, and Future. *J. Am. Chem. Soc.* **2013**, *135*, 12480–12496. [CrossRef] [PubMed]
6. Kapoor, M.; Gupta, M.N. Lipase promiscuity and its biochemical applications. *Process Biochem.* **2012**, *47*, 555–569. [CrossRef]
7. Tian, X.; Zhang, S.; Zheng, L. First Novozym 435 lipase-catalyzed Morita—Baylis—Hillman reaction in the presence of amides. *Enzyme Microb. Technol.* **2016**, *84*, 32–40. [CrossRef]
8. Miao, Y.; Rahimi, M.; Geertsema, E.M. Recent developments in enzyme promiscuity for carbon—carbon bond-forming reactions. *Curr. Opin. Chem. Biol.* **2015**, *25*, 115–123. [CrossRef]
9. Hult, K.; Berglund, P. Enzyme promiscuity: Mechanism and applications. *Trends Biotechnol.* **2007**, *25*, 231–238. [CrossRef]
10. Fu, Y.; Fan, B.; Chen, H.; Huang, H.; Hu, Y. Promiscuous enzyme-catalyzed cascade reaction: Synthesis of xanthone derivatives. *Bioorg. Chem.* **2018**, *80*, 555–559. [CrossRef]
11. Ding, Y.; Xiang, X.; Gu, M.; Xu, H.; Huang, H.; Hu, Y. Efficient lipase-catalyzed Knoevenagel condensation: Utilization of biocatalytic promiscuity for synthesis of benzylidene-indolin-2-ones. *Bioprocess Biosyst. Eng.* **2016**, *39*, 125–131. [CrossRef] [PubMed]
12. Zhang, Y.; Lai, P. One-pot synthesis of Wieland—Miescher ketone. *Res. Chem. Intermed.* **2015**, *41*, 4077–4082. [CrossRef]
13. Babtie, A.; Tokuriki, N.; Hollfelder, F. What makes an enzyme promiscuous? *Curr. Opin. Chem. Biol.* **2010**, *14*, 200–207. [CrossRef] [PubMed]
14. Zhang, H. A Novel One-Pot Multicomponent Enzymatic Synthesis of 2,4-Disubstituted Thiazoles. *Catal. Lett.* **2014**, *144*, 928–934. [CrossRef]
15. Maruyama, T.; Nakajima, M.; Kondo, H.; Kawasaki, K.; Seki, M.; Goto, M. Can lipases hydrolyze a peptide bond? *Enzyme Microb. Technol.* **2003**, *32*, 655–657. [CrossRef]
16. Evitt, A.S.; Bornscheuer, U.T. Lipase CAL-B does not catalyze a promiscuous decarboxylative aldol addition or Knoevenagel reaction. *Green Chem.* **2011**, *13*, 1141–1142. [CrossRef]
17. Svedendahl, M.; Hult, K.; Berglund, P. Fast Carbon—Carbon Bond Formation by a Promiscuous Lipase. *J. Org. Chem.* **2005**, *127*, 17988–17989. [CrossRef]
18. Bui, T.; Barbas, C.F., III. A proline-catalyzed asymmetric Robinson annulation reaction. *Tetrahedron Lett.* **2000**, *41*, 6951–6954. [CrossRef]
19. Segura, R.L.; Palomo, J.M.; Mateo, C.; Cortes, A.; Terreni, M.; Fernandez-Lafuente, R.; Guisan, J.M. Different properties of the lipases contained in porcine pancreatic lipase extracts as enantioselective biocatalysts. *Biotechnol. Prog.* **2004**, *20*, 825–829. [CrossRef]
20. Kapoor, M.; Majumder, A.B.; Gupta, M.N. Promiscuous Lipase-Catalyzed C—C Bond Formation Reactions Between 4 Nitrobenzaldehyde and 2-Cyclohexen-1-one in Biphasic Medium: Aldol and Morita—Baylis—Hillman Adduct Formations. *Catal. Lett.* **2015**, *145*, 527–532. [CrossRef]
21. Reetz, M.T.; Mondière, R.; Carballeira, J.D. Enzyme promiscuity: First protein-catalyzed Morita–Baylis–Hillman reaction. *Tetrahedron Lett.* **2007**, *48*, 1679–1681. [CrossRef]

3

Comparative Analysis and Biochemical Characterization of Two Endo-β-1,3-Glucanases from the Thermophilic Bacterium *Fervidobacterium* sp.

Christin Burkhardt [1], Christian Schäfers [1], Jörg Claren [2], Georg Schirrmacher [2] and Garabed Antranikian [1,*]

1 Institute of Technical Microbiology, Hamburg University of Technology (TUHH), Kasernenstr. 12, 21073 Hamburg, Germany; christin.burkhardt@tuhh.de (C.B.); christian.schaefers@tuhh.de (C.S.)
2 Group Biotechnology, Clariant Produkte (Deutschland) GmbH, Semmelweisstr. 1, 82152 Planegg, Germany; joerg.claren@clariant.com (J.C.); georg.schirrmacher@eitfood.eu (G.S.)
* Correspondence: antranikian@tuhh.de

Abstract: Laminarinases exhibit potential in a wide range of industrial applications including the production of biofuels and pharmaceuticals. In this study, we present the genetic and biochemical characteristics of FLamA and FLamB, two laminarinases derived from a metagenomic sample from a hot spring in the Azores. Sequence comparison revealed that both genes had high similarities to genes from *Fervidobacterium nodosum* Rt17-B1. The two proteins showed sequence similarities of 62% to each other and belong to the glycoside hydrolase (GH) family 16. For biochemical characterization, both laminarinases were heterologously produced in *Escherichia coli* and purified to homogeneity. FLamA and FLamB exhibited similar properties and both showed highest activity towards laminarin at 90 °C and pH 6.5. The two enzymes were thermostable but differed in their half-life at 80 °C with 5 h and 1 h for FLamA and FLamB, respectively. In contrast to other laminarinases, both enzymes prefer β-1,3-glucans and mixed-linked glucans as substrates. However, FLamA and FLamB differ in their catalytic efficiency towards laminarin. Structure predictions were made and showed minor differences particularly in a kink adjacent to the active site cleft. The high specific activities and resistance to elevated temperatures and various additives make both enzymes suitable candidates for application in biomass conversion.

Keywords: *Fervidobacterium*; endo-β-1,3-glucanase; laminarinase; glycoside hydrolase; thermostable; gene duplication

---

1. Introduction

β-1,3-Glucans are non-cellulosic carbohydrates that are widespread in nature. They can be found in the cell walls of fungi (pachyman), in reproductive structures of plants (callose) or as exopolysaccharide from bacteria (curdlan) [1]. β-1,3-Glycosidic linkages are also present in mixed-linked β-glucans from cereals (e.g., barley) or lichens [2]. Moreover, β-1,3-glucans are one of the most abundant carbohydrates in marine ecosystems [3]. In micro- and macroalgae, β-1,3-glucans are structurally diverse and serve as storage glucans. In brown algae, the β-1,3-glucan is named laminarin and represents up to 25% of the dry weight, depending on species, season, and growing conditions [4]. Due to the significantly higher production yields than terrestrial biomass, a high carbohydrate content, and the lack of hemicellulose and lignin, macroalgae biomass is a promising feedstock for new biorefinery concepts [5,6]. For industrial utilization of this feedstock, robust and efficient enzymes like laminarinases are required [7]. Furthermore, laminarinases could be applied for yeast extract production, as a biocontrol agent against fungal plant pathogens [8], and for partial hydrolysis of β-1,3-glucans for the production of antiviral and antitumor therapeutics [9].

For the complete enzymatic hydrolysis of β-1,3-glucans endo-acting β-1,3-glucanases (EC 3.2.1.39) and β-1,3(4)-glucanases (EC 3.2.1.6), both known as laminarinases, exo-β-1,3-glucosidases (EC 3.2.1.58) are required. These enzymes and their corresponding substrates have in common that they both are widely distributed among plants, fungi, and bacteria from many different habitats. On the basis of amino acid sequence similarities, all endo-acting laminarinases from plants can be assigned to the glycoside hydrolase family GH 17, whereas most of the bacterial laminarinases belong to GH 16 [10]. According to the Carbohydrate-Active Enzymes database (CAZy), around 40 β-1,3-glucanases from bacteria are already characterized (August 2019). Nevertheless, laminarinases with high stability at varying conditions and temperatures are desired for industrial applications [11]. To obtain highly stable and efficient enzymes for industrial application, thermophilic organisms represent an excellent resource. Moreover, these enzymes allow reactions at elevated process temperatures, which do not only reduce microbial contamination, but also increase the solubility and diffusion rates of the catalysts for complex polymeric substrates [12].

In this study, an environmental sample from an Azorean hot spring (São Miguel, Portugal) was used as a source for novel laminarinase-encoding genes. Recently, this environmental sample has been proven to be an excellent source for unique thermo-active enzymes [13]. By sequence-based screening, two putative genes were identified with sequence similarities to parts of the genomic sequence of the thermophilic bacterium *Fervidobacterium nodosum* Rt17-B1, which was completely sequenced in 2007 [14]. *Fervidobacterium nodosum* which was isolated from a hot spring in New Zealand, is able to ferment a wide range of carbohydrates [15] and is considered as a good source for novel carbohydrate degrading enzymes. So far, only one highly active and thermostable cellulase from this bacterium has been characterized [16].

Here, we describe the recombinant production and purification of the two thermoactive laminarinases FLamA and FLamB originating from *Fervidobacterium* sp. To estimate their potential relevance for industrial applications, both recombinant enzymes were characterized in detail.

2. Results

2.1. Sequence Analysis of FLamA and FLamB

By sequence-based screening, two new putative laminarinase encoding genes were identified. Metagenomic DNA from a hot spring of the Azores, which was known to contain genomic DNA of a *Fervidobacterium* strain, was used as template DNA for PCR. The primers for the amplification of the two genes encoding for FLamA and FLamB were based on two putative laminarinase genes of the complete genome sequence of *Fervidobacterium nodosum* Rt17-B1 (GenBank: CP000771). The amplified DNA fragment for *flamA* showed 99% sequence similarity to putative laminarinase genes from the genomes of *F. nodosum* Rt17-B1 and *F. pennivorans* DSM 9078 (GenBank: NC_017095). The corresponding amino acid sequence is annotated as a multispecies endo-β-1,3-glucanase found in various *Fervidobacterium* species (GenBank: WP_011994743). In comparison to that, the amplified DNA fragment for *flamB* showed 95% and 67% sequence similarity to GenBank sequences of two other putative laminarinase genes from *F. nodosum* Rt17-B1 and *F. pennivorans* DSM 9078, respectively. Due to the 99% similarity of the amino acid sequence to next hits in the database, *flamB* was annotated in GenBank with the accession number LT882624.

In the genome of *F. nodosum* Rt17-B1, the highly similar genes of *flamA* and *flamB* are embedded in two different operons. By comparing these operons, we identified putative open reading frames for proteins with the same predicted functions (Figure 1). Among both gene clusters, the related proteins revealed high sequence similarities of 62–89% on amino acid level (100% query coverage). *FlamA* and *flamB* showed 62% sequence similarity to each other. Adjacent to the putative laminarinase encoding genes, we identified genes for putative proteins involved in ABC transporter system and genes for putative alanine racemases and GH 3 proteins. Moreover, a truncated gene encoding a

Transposase_20 and a repeat region nearby one of the predicted operons indicated that the two related operons originate from a gene duplication event of a DNA cassette around 20 kb in size.

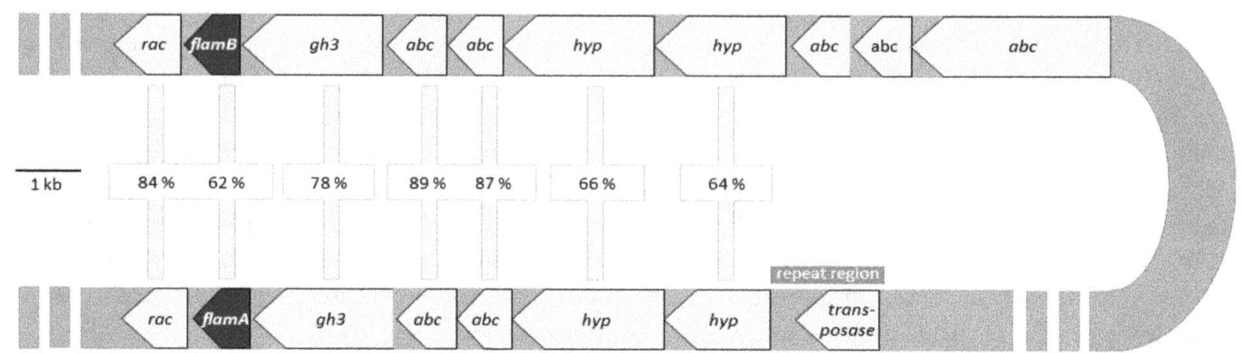

Figure 1. Predicted structural organization of the two operons containing *flamA* and *flamB* based on the highly similar genes in the genome of *Fervidobacterium nodosum*. Sequence identities of similar genes are shown in the center. Predicted genes: *rac*—putative gene for alanine racemase; *flamA* and *flamB*—genes for β-1,3-glucanases; *gh3*—putative genes for glycoside hydrolase of family GH 3; *abc*—putative genes for ABC transporter proteins; *transposase*—truncated gene of a Transposase_20 with a repeat region; *hyp*—hypothetical genes.

Structure predictions for FLamA and FLamB revealed a sandwich-like β-jelly roll fold, which is characteristic for laminarinases of class GH 16 (Figure S1). The structural prediction of both proteins was based on the catalytic residue of *Thermotoga maritima*, with 58% and 60% sequence identity to *flamA* and *flamB*, respectively. The comparison of both predictions demonstrated the high structural similarities of FLamA and FLamB with only small differences in a loop adjacent to the cleft of the active site (highlighted in color).

The sequences of both proteins showed similarities to those of laminarinases from other thermophilic bacteria. Indeed, FLamA and FLamB were similar to catalytic domains of experimentally verified β-1,3-glucanases from *T. maritima*, *T. petrophila*, and *T. neapolitana* with a 54–55% and 58–59% identity, respectively. Moreover, FLamA and FLamB were 55% and 58% similar to the catalytic domain of a β-1,3-glucanase from the hyperthermophilic Archaeon *Pyrococcus furiosus*. The phylogenetic relationship among both enzymes and other biochemically characterized GH 16 family members of bacteria is shown in Figure 2. Concerning the predicted tree based on the homologous region of the catalytic domain, FLamA and FLamB form a solid clade with the β-1,3-glucanases of the Archaeon *P. furiosus* and members of the same eubacterial order Thermotogales, which includes some of the most extremely thermophilic species currently known.

2.2. Recombinant Production of FLamA and FLamB

Both genes were expressed in *E. coli* C43(DE3) at 37 °C over 4 h of induction. The obtained proteins harboring a C-terminal 6xHis affinity tag were purified via affinity chromatography and size exclusion chromatography. Both FLamA and FLamB were purified to homogeneity with a factor of 178 and 593 and a final yield of 30% and 18%, respectively (Table S1). The SDS-PAGE revealed a molecular weight of approximately 30 kDa for both proteins, which was slightly smaller than the predicted molecular size of 34.9 kDa and 34.1 kDa for FLamA and FLamB, respectively (Figure 3). Domain prediction revealed that FLamA and FLamB consisted, in each case, of one single GH 16 domain without any further known structural elements.

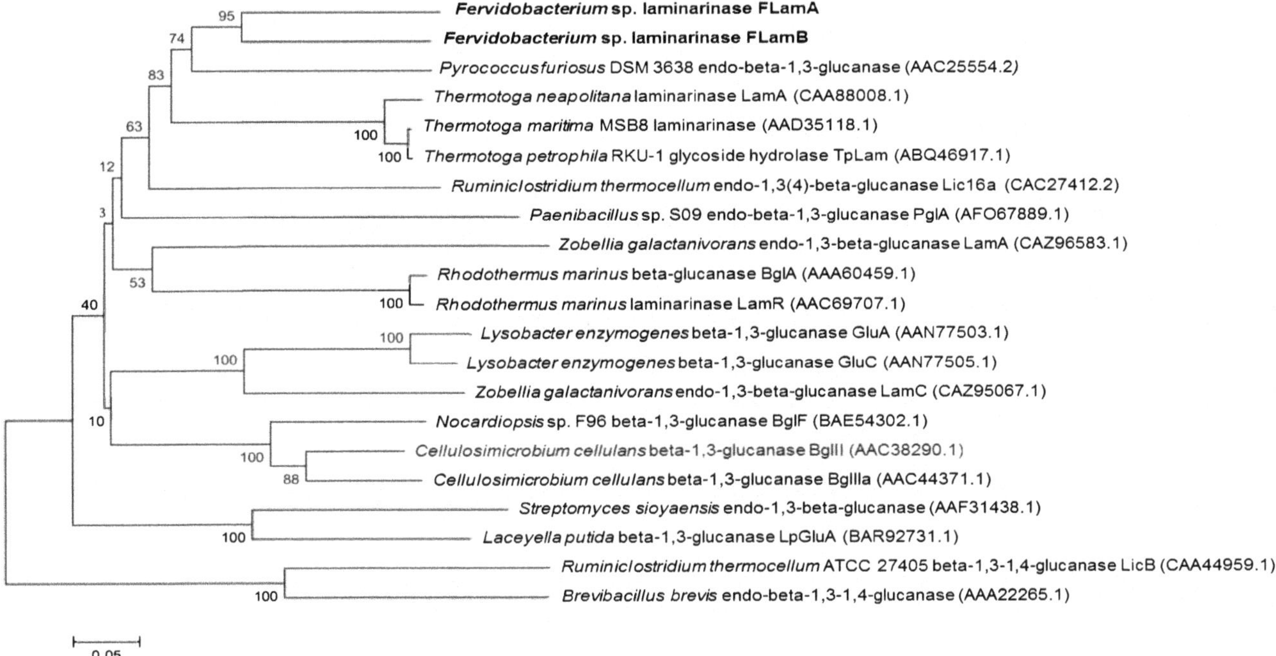

Figure 2. Phylogenetic tree of biochemically characterized GH 16 enzymes. For construction of the tree, sequences of the catalytic domains were used. GenBank accession numbers are indicated in brackets. Bootstrap values are designated on each branch of the tree.

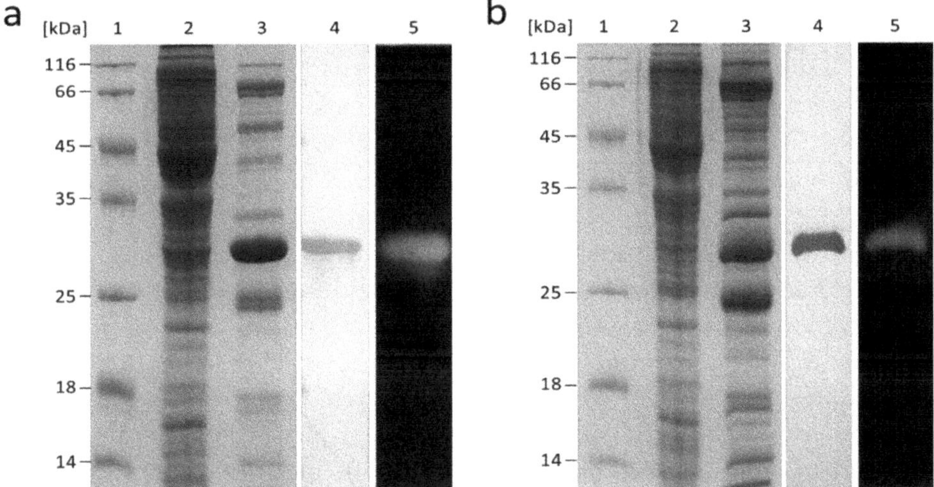

Figure 3. SDS-PAGE analysis of the His-tagged β-1,3-glucanases FLamA (**a**) and FLamB (**b**). *Line 1*, molecular weight marker; *line 2*, crude extract; *line 3*, eluate after the Ni-NTA affinity chromatography; *line 4*, purified enzymes after size exclusion chromatography; *line 5*, zymogram for activity staining of the purified enzyme.

2.3. Substrate Specificity of FLamA and FLamB

The substrate specificity of FLamA and FLamB was tested towards a number of complex carbohydrates in a range from 40 to 100 °C (Figure 4, Table 1). Significant differences between FLamA and FLamB were detected in the specific activities towards β-1,3-glucans. Highest activities of both enzymes were measured at 90 °C towards laminarin, whereby FLamB showed a higher specific activity than FLamA, with 876 U/mg and 609 U/mg, respectively. Compared to that, highest activities towards amorphous curdlan were revealed at 70 °C with 78% and 94% relative activity. For undissolved curdlan, the highest activity was determined at 80 °C, with even lower relative activities of 44% and 50% for FLamA and FLamB, respectively.

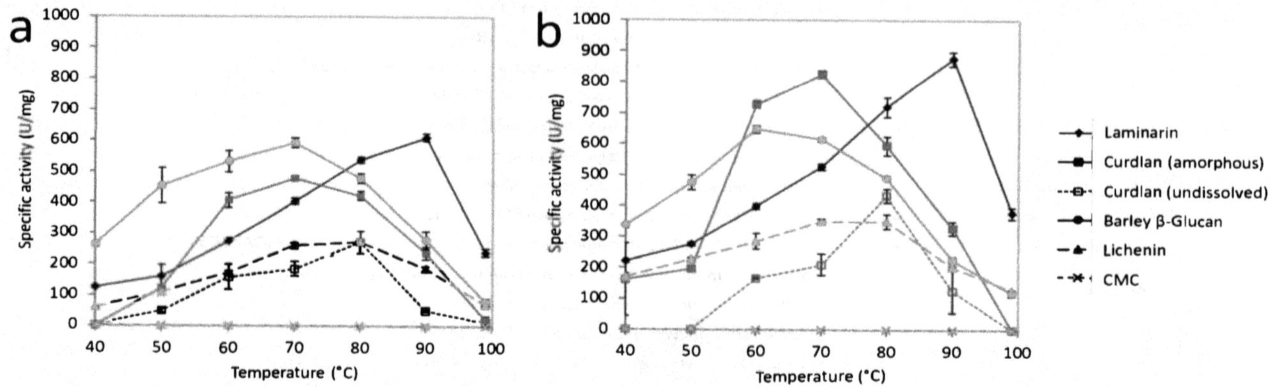

Figure 4. The temperature profiles of recombinant FLamA (**a**) and FLamB (**b**) depending on substrates laminarin, amorphous curdlan, unsolved curdlan, barley β-glucan, lichenin, and carboxymethyl cellulose (CMC).

Table 1. Substrate specificity of FLamA and FLamB at the optimal temperatures for each substrate.

Substrate	FLamA			FLamB		
	T (°C)	Specific Activity (U/mg)	Relative Activity (%)	T (°C)	Specific Activity (U/mg)	Relative Activity (%)
Laminarin	90	609 ± 12	100	90	876 ± 23	100
Curdlan *	70	478 ± 07	78	70	825 ± 13	94
Curdlan	80	270 ± 37	44	80	434 ± 22	50
Barley β-glucan	70	592 ± 16	97	60	648 ± 12	74
Lichenin	80	271 ± 5	45	70	350 ± 8	40
CMC +	-	0	0	-	0	0

* Amorphous curdlan, diluted in NaOH and neutralized; + between 40–99 °C, no activity was detectable.

In comparison to β-1,3-glucans, the specific activities towards mixed-linked glucans did not differ substantially between FLamA and FLamB. The specific activities of FLamA and FLamB towards barley β-glucan from barley were in the same range with 592 U/mg and 648 U/mg, respectively. This was determined at optimal temperatures of 70 °C (FLamA) and 60 °C (FLamB). For both enzymes, 44% and 52% residual activities towards barley β-glucan were detected at 40 °C. Moreover, both enzymes showed, at 40 °C, higher activities towards mixed-linked β-glucan (262 U/mg and 336 U/mg) than towards laminarin (124 U/mg and 220 U/mg). The specific activities towards lichenin at optimal conditions (80 °C and 70 °C) were with 271 U/mg and 350 U/mg for FLamA and FLamB, lower than towards the other tested substrates. Hydrolysis of the β-1,4-glucan CMC was not observed.

2.4. Degradation Pattern and Enzyme Kinetics

The hydrolysis products of laminarin and barley β-glucan were investigated by HPLC analysis. The FLamA and FLamB produced similar degradation patterns. Hydrolysis of laminarin (Figure 5a) mainly results in laminaribiose, glucose-mannitol units (DP2), and glucose (DP1) and to a lesser extent laminaritriose (DP3) and higher oligosaccharides. The degradation pattern of barley β-glucan was different (Figure 5b). Laminaritriose was the major product and to a lesser extent DP1 and DP2 were detected. The produced oligosaccharides indicated that FLamA and FLamB are β-1,3-glucanases with endo-acting mode. Moreover, the shifted product variation with barley β-glucan suggested that both enzymes hydrolyzed the β-1,3-glycosidic linkages in mixed-linked glucans and to a lesser extent the β-1,4-building blocks.

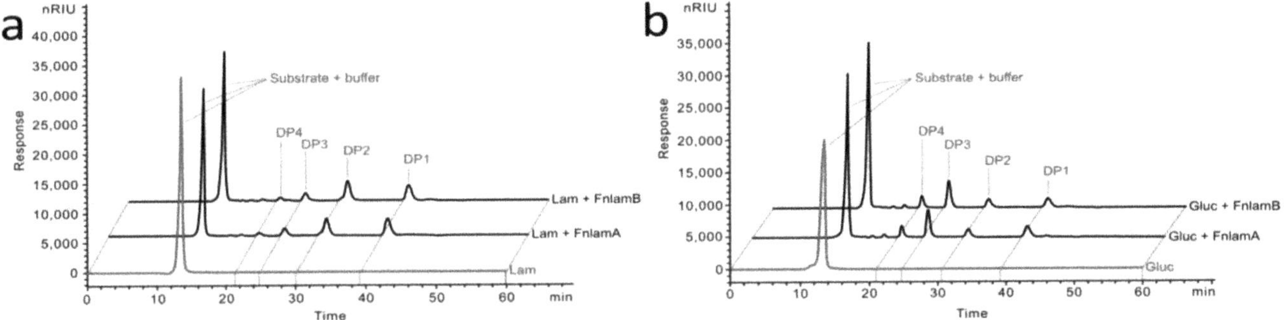

Figure 5. HPLC analysis of products from polysaccharide degradation. Hydrolysis products from laminarin (**a**) and barley β-glucan (**b**) after 18 h of incubation at 70 °C. DP: degree of polymerization. DP1: glucose. Lam: laminarin. Gluc: barley β-glucan.

The kinetic parameters were determined in the presence of laminarin. At optimal conditions (90 °C, pH 6.5) the K_m value of FLamA was 2.01 mg ml^{-1} and with that higher than that of FLamB (1.64 mg ml^{-1}). Moreover, the catalytic efficiency values (k_{cat}/K_m) were 494.3 ml s^{-1} mg^{-1} and 806.3 ml s^{-1} mg^{-1} for FLamA and FLamB, respectively. The results indicated that FLamB has a 61% higher catalytic efficiency on laminarin than FLamA, which is in agreement with the determined substrate specificities.

2.5. *Effects of pH and Temperature*

The effect of pH was analyzed in a range of pH 2 to 12. Measurements revealed similar pH spectra for FLamA and FLamB (Figure 6). Both enzymes showed activity in a range between pH 5 and 9. Optimal activity was detected at pH 6.5. Concerning the stability of FLamA, there was a decrease by approximately 8% in activity over the entire pH range after 24 h (Figure S3). In the same pH range, FLamB showed at least 82% residual activity.

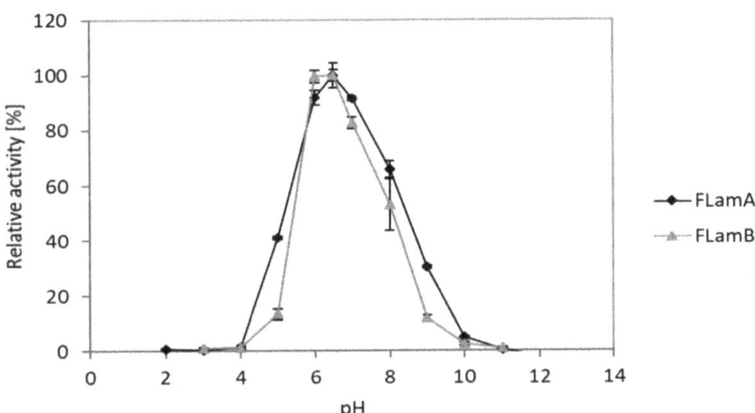

Figure 6. Influence of pH on the activity of recombinant FLamA and FLamB towards laminarin.

To examine the temperature stability of the two thermoactive enzymes, the residual activities after heat incubation at 70, 80, and 90 °C were measured (Figure 7). FLamA and FLamB exhibited more than 80% and 40% residual activity over 24 h at 70 °C, respectively. At 80 °C, the enzymes had a half-life time of 5 h (FLamA) and 1 h (FLamB). By incubating the enzymes at 90 °C after 30 min, no residual activity was detected. FLamA was, in the tested temperature range, more stable than FLamB.

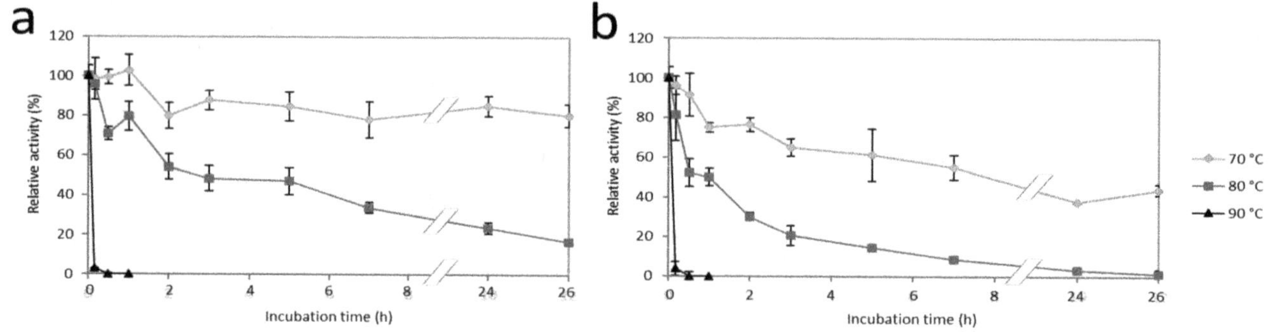

Figure 7. Influence of temperature on the stability of FLamA (**a**) and FLamB (**b**).

2.6. Effects of Metal Ions and Chemical Additives

The effect of various additives on the activities of FLamA and FLamB was investigated. For both enzymes, the examined additives had similar influences. Among the tested metal ions (5 mM), Ag^{+}, Al^{3+}, Cr^{3+}, Cu^{2+}, Zn^{2+}, and Fe^{2+} completely inhibited FLamA and FLamB (Table 2a). Both Co^{2+} and Ni^{2+} significantly reduced the activity of the enzymes. The same effect was observed by the addition of 5 mM SDS, cetyltrimethylammonium bromide (CTAB), and Pefabloc (Table 2b). In contrast, the reducing agents dithiothreitol (DTT) and β-mercaptoethanol had positive effects on enzyme activities.

Table 2. Influence of metal ions (**a**) and different reagents (**b**) on laminarinase activity of FLamA and FLamB.

a	Relative Residual Activity (%) *		b	Relative Residual Activity (%) *	
Metal ion	**FLamA**	**FLamB**	**Reagent**	**FLamA**	**FLamB**
$AgNO_3$	0.65 ± 0.00	3.29 ± 0.58	CHAPS	78.95 ± 3.64	80.11 ± 2.36
$AlCl_3$	4.98 ± 2.31	4.73 ± 0.27	SDS	29.10 ± 1.08	36.70 ± 6.61
$CaCl_2$	89.85 ± 8.46	87.79 ± 1.60	Triton X-100	89.42 ± 4.03	89.77 ± 0.61
$CoCl_2$	9.31 ± 0.33	16.64 ± 1.30	Tween 20	91.83 ± 1.67	91.18 ± 2.20
$CrCl_3$	2.50 ± 0.04	1.32 ± 4.80	Tween 80	95.41 ± 3.85	92.52 ± 1.60
$CuCl_2$	0.66 ± 0.40	0.84 ± 4.18	Guanidine-HCl	91.14 ± 7.34	84.70 ± 0.68
$FeCl_2$	1.05 ± 0.17	3.29 ± 1.14	Urea	100.14 ± 1.41	93.30 ± 1.06
KCl	93.78 ± 1.00	91.08 ± 1.39	DTT	119.04 ± 1.16	124.61 ± 1.88
$MgCl_2$	92.48 ± 2.42	95.15 ± 1.57	β-Mercaptoethanol	125.03 ± 2.88	124.12 ± 3.42
NaCl	92.41 ± 1.53	86.42 ± 1.17	EDTA	109.32 ± 3.01	95.28 ± 2.20
$NiCl_2$	4.67 ± 0.16	7.06 ± 0.10	Na-Iodoacetate	105.34 ± 1.19	92.45 ± 2.57
RbCl	90.79 ± 1.76	97.47 ± 1.01	Pefabloc	37.79 ± 5.67	76.94 ± 12.14
$SrCl_2$	93.15 ± 1.84	96.20 ± 2.63	CTAB	24.85 ± 1.69	30.68 ± 2.60
$ZnCl_2$	1.85 ± 0.20	3.67 ± 0.29	Na-Azide	102.37 ± 2.14	103.81 ± 6.97

* Laminarinase activity without additives was defined as 100%.

3. Discussion

The two genes *flamA* and *flamB* revealed high sequence similarities of 62% (100% query coverage) with each other. Two highly similar genes within the genome of *Fervidobacterium nodosum* were located in two different operons. The pattern of the operons and adjacent genes as well as the presence of a truncated transposase gene suggest that these structures were generated by a duplication of an around 20 kb genome section. Moreover, it has been shown that *flamA* and *flamB* encode for active laminarinases possessing nearly the same biochemical characteristics, indicating that both genes form a multigene family.

There are other bacterial species which express more than one gene encoding for laminarinases. Nevertheless, these genes contain various modular architectures and the proteins possess different domains and functions like in the marine bacterium *Zobellia galactanivorans* [17]. In *Lysobacter enzymogenes,* two laminarinases are present which have high sequence identities in part of the catalytic domain (86%) but differentiate in an additional carbohydrate binding module (CBM) in one of the proteins [18]. Similar relationships were also observed for related laminarinases from *Streptomyces sioyaensis* [19]. So far, however, only members of the family Fervibacteriaceae are known to harbor two similar genes, such as *flamA* and *flamB*, which encode for laminarinases with similar sequence and function.

It has been shown that gene duplication probably generates no benefit by enhanced gene dosage of laminarinases or adjacent genes, such as those for the ABC transporter system. The increase of gene dosages may cause higher energy consumption leading to significant reduced fitness of the organism [20,21]. Therefore, it is more likely that gene duplication leads to redundancy and, thus, facilitates mutations in one of the copies and possibly results in new benefiting gene functions [22]. A similar operon composition occurs in the closely related *F. pennivorans*, where two genes exist with 100%and 70% sequence identity to *flamA* and *flamB*, respectively.

This demonstrates that *flamA* is highly conserved in both species, whereas differences in *flamB* were probably generated by point mutations. This may lead to beneficial properties, like obviously higher catalytic activities of FLamB towards β-1,3-glucans. If there is no selective advantage, duplicated genes normally become inactivated by mutations and these pseudogenes will disappear from the genome [23].

Phylogenetic analysis of the catalytic residues of FLamA, FLamB, and other characterized laminarinases revealed a solid clade with close a relationship between enzymes from the bacterial order Thermotogales, including FLamA and FLamB, and the Archaeon *Pyrococcus furiosus* (Figure 2). Thermotogales are assumed to have interchanged large numbers of genes with Archaea and Firmicutes by horizontal gene transfer (HGT) [14]. An extraordinarily high number of insertion sequence elements for example in the genome of *F. nodosum* demonstrate the great influence of foreign genes. The high sequence similarity of FLamA, FLamB, and the laminarinases of the bacterium *Thermotoga* to one from the Archaeon *P. furiosus* suggests that an ancestral gene was exchanged by HGT.

Regarding the temperature optima of these enzymes, FLamA and FLamB are both thermoactive enzymes showing highest activity at 90 °C. This optimum is similar to that of the other examined laminarinases from the phylum Thermotogae [24,25] which are to our knowledge the ones with the highest temperature optima among reported β-1,3-glucanases from eubacteria and plants. However, LamA from the Archaeon *P. furiosus* showed among all characterized laminarinases activity at the highest temperatures of 100–105 °C (Table 3) [26].

Both FLamA and FLamB showed high storage stabilities at 4 °C in a broad pH range of 3 to 11 which enables an easy handling of both laminarinases. Their stability at 70 to 90 °C is also comparable to those of the other laminarinases operating at high temperatures (Table 3). All in all, FLamA and FLamB have significantly high temperature stabilities, which are a decisive criterion for industrial application.

The investigation of the substrate specificities of FLamA and FLamB revealed different temperature optima for the tested substrates (Figure 4). Schwarz et al. [27] observed related characteristics of a laminarinase of *Clostridium thermocellum*. The optimal temperature towards curdlan is lower in consideration to the fact that curdlan forms at temperatures above 80 °C irreversible high-set gels [28], which might be less accessible for enzymes. Nevertheless, the reasons for the different temperature optima can only be speculated.

Table 3. Comparison of biochemically characterized laminarinases from thermophilic bacteria and Archaea.

Organism and Enzyme	T_{opt} (°C)	Thermal Stability			pH_{opt}	Activity (U/mg) +				Reference
		T (°C)	t (h)	A * (%)		Lam	Curd	Glu	Lich	
Caldicellulosiruptor sp. F32, Lam16A-GH	75	65	42	72	6.5	172	ND	2961	ND	[29]
Fervidobacterium sp., FLamA	90	80	5	50	6.5	609	270	592	271	Present study
Fervidobacterium sp., FLamB	90	80	1	50	6.5	876	434	648	350	Present study
Laceyella putida, LpGluA	80	75	0.5	45	4.2	48%	100%	ND	ND	[30]
Nocardiopsis sp. F96, BglF	70	ND	ND	ND	9.0	100%	159%	ND	815%	[31]
Pyrococcus furiosus, LamA	100/105	80	80	100	6–6.5	922	ND	99	95	[26]
Rhodothermus marinus 21, BglA	85	80	16	100	7.0	542	ND	1568	1445	[32]
Rhodothermus marinus ITI278, LamR	88	90	0.45	50	5.5	656	ND	2199	3111	[33]
Ruminiclostridium thermocellum, CelC	65	70	10	30	6.5	86	ND	504	245	[27]
Ruminiclostridium thermocellum, Lic16A	70	70	0.17	50	6.0	340	29	268	2404	[34]
Thermotoga maritima, TmβG	80	ND	ND	ND	5.0	Efficient on β-1,3-glucans				[35]
Thermotoga neapolitana, LamA	95	95	0.5	82	6.3	3100	ND	ND	90	[25]
Thermotoga petrophila, TpLam	91	80	16	60	6.2	48	ND	41	21	[24]

* Residual activity detected after incubation at the given temperature and time. + If available, specific activity (U/mg); otherwise relative activities (%) towards laminarin (Lam), Curdlan (Curd), barley β-glucan (Gluc) or lichenin (Lich).

So far, published thermostable β-1,3-linkage hydrolyzing enzymes generally possessed strong preferences either for β-1,3-glucans or mixed-linked glucans (Table 3). Both FLamA and FLamB take an intermediate position between those groups by degrading approximately both types of substrates with activity in the same range. Additionally, FLamA and FLamB exhibit higher substrate affinities towards mixed-linked glucans in comparison to similar enzymes from *Thermotoga neapolitana* [25] and *P. furiosus* [26]. Ilari et al. [36] investigated a deletion mutant and were able to show that the loss of a five amino acid kink at the entrance of the catalytic cleft leads to higher activity towards mixed-linked glucans. This effect was also observed in other laminarinases missing these residues [34,37]. Even in FLamA and FLamB, these amino acid residues are missing (Figure S2). Thus, differences in the kink possibly explain the altered substrate preferences between those enzymes.

The described substrate spectra and hydrolysis products suggest that FLamA and FLamB are able to cleave β-1,3 glycosidic bonds in an endo-acting mode. Nevertheless, the hydrolysis of β-1,4 glycosidic bonds in mixed-linked glucans is also possible. Similar product patterns of laminarin and barley β-glucan were observed with a laminarinase form *Caldicellulosiruptor* sp., when the authors proved that the enzyme was able to degrade β-1,4 glyosidic bonds adjacent to 3-*O*-substituted glucopyranose units [32].

For industrial application, detection of inhibitory effects and comparison with other enzymes the influences of different metal ions and additives were tested. For FLamA and FLamB similar effects were observed. The inhibitory effects of metal ions observed in this study are well known for many glucosidases, probably due to the redox effects on the amino acids [38]. The SDS had a negative influence on the activity of the enzymes by disturbing hydrophobic interactions and generating protein denaturation. Moreover, the detergent CTAB blocked the catalytic cleft of a related laminarinase by hydrophobic interactions [38], which may lead to reduced activities of FLamA and FLamB as well. Pefabloc, a serine protease inhibitor, negatively influenced the enzyme activity probably by binding irreversibley to serine residues nearby the catalytic cleft. In contrary, the reducing properties of DTT and β-mercaptoethanol might positively influence a cysteine residue adjacent to the nucleophile of the active site and therefore increase the laminarinase activity.

Although FLamA and FLamB share many biochemical properties, the two enzymes exhibit differences in substrate specificities, particularly towards β-1,3-glucans. In comparison to the specific activities of FLamA, FLamB exhibits approximately 40% and more than 70% higher activity towards laminarin and amorphous curdlan, respectively (Table 1). These differences were reflected in the kinetic parameters as well. In the analysis of Labourel et al. [39], significant differences in substrate specificity were caused by an additional loop in protein structure which leads to higher affinities towards mixed-linked glucans. Nevertheless, according to structural predictions, those considerable differences were not observed between FLamA and FLamB (Figure S1). Only minor structural modifications particularly in a loop adjacent to the catalytic cleft possibly result in an upwardly more opened cleft of FLamB. Based on the observations of Labourel et al. [39] and Jeng et al. [40] concerning the enzyme-substrate complexes, the enlarged opening might improve the access for β-1,3-glucans that possess helical conformation, whereas the affinity of FLamB towards mixed-linked glucans with linear conformation is not affected. These results and further investigations will help to improve activities of β-1,3-glucananases towards β-1,3-glucans.

4. Materials and Methods

4.1. Cloning of the Endo-β-1,3-Glucanase Encoding Genes *flamA* and *flamB*

Metagenomic DNA was extracted from environmental samples taken from different locations at the hot spring Caldeirão at Furnas Valley (Azores, Portugal) followed by the production and sequencing of a 454 shotgun library as described previously [41]. For the two genes *flamA* and *flamB*, encoding for endo-β-1,3-glucanases from *Fervidobacterium*, no signal peptides were predicted by SignalP [42]. For cloning into the StarGate system (IBA Lifesciences) the two genes were amplified by PCR using the metagenomic DNA as a template and the following primers (primer extending sequences are indicated in boldface):

- *flamA*-for: **AGCGGCTCTTCAATG**AAAGTTAAATATTTCTCAAATATT
- *flamA*-rev: **AGCGGCTCTTCTCCC**CTCATTTTCAAGCTTGTATAC
- *flamB*-for: **AGCGGCTCTTCAATG**AGAGAAAAGTTGCTGT
- *flamB*-rev: **AGCGGCTCTTCTCCC**CTCTTCATCTAATGTATACAC

The PCR products were cloned into the destination vector pASG-IBA33 according to the producer instructions resulting in recombinant fusion genes with C-terminal sequences encoding for hexahistidine tags. After transformation of *Escherichia coli* TOP10 and selection on LB medium containing 100 μg/mL ampicillin and 50 μL/mL X-gal, plasmids of recombinant clones were isolated. The inserts were sequenced for verification. Subsequently, the vectors were used to transform *E. coli* C43(DE3) for protein production.

4.2. Sequence Comparison and Phylogenetic Analysis

From the GenBank database, amino acid sequences of characterized β-1,3-glucanases and β-1,3(4)-glucanases of the family GH 16 were selected. Multiple sequence alignment was performed using ClustalX. Homologous sequence regions were selected and applied in a second multiple sequence alignment. Using MEGA6, a phylogenetic tree was calculated by the neighbor joining method. Bootstrap analysis with resampling of the dataset was performed (n = 100) to test the reliability of the tree. Structure prediction of both proteins were done by SWISS-MODEL using the crystal structure of the laminarinase from *Thermotoga maritima* (PDB ID: 3azx) as a template. Structures of FLamA and FLamB were visualized and compared in the UCFS Chimera program by applying the Needleman–Wunsch algorithm and the scoring matrix Blosum62.

4.3. Heterologous Expression of the *flamA* and *flamB* Genes and Purification of the Endo-β-1,3-Glucanases

Escherichia coli C43(DE3) harboring the plasmids pASG-IBA33::*flamA* or pASG-IBA33::*flamB* were grown in LB media (100 μg/mL ampicillin) at 37 °C and 160 rpm. Gene expression was induced at OD_{600} 0.6 by adding anhydrotetracycline to a final concentration of 200 ng/μL. After four hours of induction cells were harvested by centrifugation at 9000 × *g* at 4 °C for 20 min. The resulting cell pellet was stored at −20 °C.

For purification, 0.2 g cells were resuspended per 1 mL lysis buffer (50 mM NaH2PO4, 300 mM NaCl, 10 mM imidazole, pH 8) and disrupted by three passages through a French pressure cell with constant pressure of 1,000 psi (French Pressure Cell Press, SLM-Aminco). Cell debris was removed by centrifugation (20,000 × *g*, 4 °C, 30 min) and supernatant was loaded onto a 1 ml Ni-NTA Superflow column (Qiagen). Proteins were eluted by an increasing imidazole gradient according to the manufacturer's instructions. Eluted fractions were pooled, washed three times with buffer G (50 mM Na-phosphate buffer, pH 7.2, 150 mM NaCl) by ultrafiltration in an Amicon filter unit (Amicon Ultra-15, 1000 MWCO, Merck Millipore). For final purification via size exclusion chromatography, protein solutions were loaded onto a Superose 12 column (GE Healthcare) previously equilibrated with buffer G. Protein fractions containing the purified β-1,3-glucanases were pooled and stored at 4 °C.

Protein samples were analyzed on a 12.5% SDS-PAGE (12.5%) [43]. Additionally, β-1,3-glucanase activity was determined by zymogram technique. For this, proteins were applied to a denaturing SDS-PAGE and were subsequently renaturated by incubation for 1 h in 1% (*v/v*) Triton X-100 and three successive washings for 5 min in 50 mM sodium phosphate buffer (pH 6.5). Then, the gel slices were incubated for 20 min at 80 °C on an agarose gel containing 0.1% (*w/v*) curdlan. For visualization agarose slides were stained for 1 h with 1% (*w/v*) Congo Red and destained in 1 M NaCl. To increase the contrast, the slides were finally overlaid with 0.1 M acetic acid.

Protein concentrations were determined according to Bradford [44], with bovine serum albumin as the standard.

4.4. β-Glucanase Activity Assay

The standard assay was carried out at 90 °C for 7 min in 500 μL reaction mixture using 0.25% (*w/v*) laminarin from *Laminaria digitata* (Merck) as substrate in 20 mM sodium phosphate buffer (pH 6.5) and 50 μL enzyme sample. In advance, it was ensured that product formation per min was constant in the time interval and with a linear correlation to preclude instability effects of the enzymes. Additionally, blank experiments without enzymes were performed by default for all measurement series. The hydrolytic activities of the purified enzymes FLamA and FLamB were detected by measuring the reducing sugars with 3,5-dinitrosalicylic acid (DNS) according to Miller [45] with glucose as the standard. In brief, after enzyme reaction 500 μl reaction mixture were mixed with 500 μL DNS reagent (1% (*w/v*) DNS, 30% (*w/v*) potassium sodium tartrate, 0.4 M NaOH) and were incubated for 5 min at 100 °C. Samples were subsequently cooled on ice to room temperature and absorption was measured at 546 nm. All measurements were done in triplicates. One unit of enzyme activity was defined as the amount of enzyme required to release 1 μmol of reducing sugars per minute.

The influence of temperature was examined by performing the standard assay at temperatures from 20 to 100 °C. To investigate the temperature stability of FLamA and FLamB, the enzymes were preincubated with a concentration of 0.1 mg/mL in 20 mM sodium phosphate buffer (pH 6.5) at 70, 80, and 90 °C. Samples were taken in time intervals up to 26 h and residual activities were measured by using the standard assay.

To investigate the influence of the pH on enzyme activity, a standard assay was performed using Britton–Robinson buffer (50 mM) in a range of pH 2–11 in the reaction mixture [46]. The pH stabilities of both enzymes were tested by preincubation of the enzymes with a concentration of 0.01 mg/mL in 50 mM Britton–Robinson buffer pH 3–11 for 24 h at 4 °C. Residual activity was determined with the standard assay by dilution the incubation mixtures in Britton–Robinson buffer at pH 6. Enzyme activity previous to incubation was defined as 100%.

Additionally, the influences of metal ions on enzyme activity were analyzed by using a standard assay, but with 20 mM maleate buffer (pH 6.5) and the addition of 5 mM $AgNO_3$, $AlCl_3$, $CaCl_2$, $CoCl_2$, $CrCl_3$, $CuCl_2$, $FeCl_2$, KCl, $MgCl_2$, NaCl, $NiCl_2$, RbCl, $SrCl_2$ or $ZnCl_2$. Furthermore, the influences of 3-((3-cholamidopropyl)dimethylammonio)-1-propanesulfonate (CHAPS), SDS, Triton X-100, Tween 20, Tween 80, guanidine hydrochloride, urea, dithiothreitol (DTT), β-mercaptoethanol, EDTA, iodoacetic acid, Pefabloc, cetyltrimethylammonium bromide (CTAB) and sodium azide were examined by the standard assay procedure. All additives were tested in a concentration of 5 mM under standard conditions.

To measure the specific activities of FLamA and FLamB, substrates were used in a final concentration of 0.25% (*w/v*). The CMC and lichenin were obtained from Merck and β-glucan (barley) and curdlan (*Alcaligenes faecalis*) from Megazyme. In case of curdlan an undissolved and a dissolved (amorphous) form was tested. To achieve an amorphous type of curdlan, 0.2 g were first solubilized in 6 mL alkaline solution (0.6 M NaOH) and subsequently neutralized with HCl to a concentration of 0.5% (*w/v*) and pH 6.5 in 20 mM sodium phosphate buffer.

Kinetic parameters were determined by performing the standard assay with twelve different substrate concentrations varying from 0 to 25 mM. The Michaelis constant K_m and the maximum reaction rate at maximum substrate concentration v_{max} were calculated by non-linear regression applying the Michaelis–Menten equation.

4.5. Determination of the Hydrolysis Products

For determination of the hydrolysis products of laminarin and barley β-glucan, 0.25% (*w/v*) substrate were incubated with FLamA or FLamB in standard reaction mixtures at 70 °C for 18 h. After the inactivation of the enzymes at 100 °C for 10 min, samples were centrifuged (20.000 × *g*, 10 min) and the supernatant was filtered using a 0.22 μm membrane filter unit. Hydrolysis products were analyzed by high-performance liquid chromatography (HPLC) under the following conditions: Hi-Plex Na column (Agilent Technologies), 80 °C, 0.2 mL/min flow rate, water as mobile phase, RI detector (Agilent Technologies). Laminaritetraose, laminaritriose, laminaribiose (all from Megazymes) and glucose (Merck) were used as standards.

5. Conclusions

Laminarinases are enzymes which could be applied in diverse fields, from biomass conversion, over yeast extract production, agents against fungal plant pathogens to the production of antiviral and antitumor therapeutics from β-1,3-glucans. The biochemical characterization of the two laminarinases FLamA and FLamB derived from a *Fervidobacterium* species revealed high specific activities and resistance to elevated temperatures and various additives which make both enzymes suitable candidates for application under harsh conditions. Moreover, the comparative analysis of both enzymes showed differences in their thermal stability and catalytic efficiency towards β-1,3-glucans, like laminarin and curdlan. In conclusion, these results will contribute to our knowledge of sequence-function correlations and will potentially help to improve activity and stability of laminarinases and other related glucanases.

Author Contributions: C.B. designed the study and performed the experimental work. C.S., G.S. and G.A. supervised this study. C.B. prepared the manuscript. C.S., J.C. and G.A. reviewed and edited the manuscript before submission.

References

1. Stone, B.A. Chemistry, biochemistry, and biology of (1-3)-β-glucans and related polysaccharides. In *Chemistry of β-Glucans*, 1st ed.; Bacic, A., Fincher, G.B., Stone, B.A., Eds.; Elsevier Academic Press: Amsterdam, The Netherlands, 2009; pp. 5–46.

2. Ebringerová, A.; Hromádková, Z.; Heinze, T. Hemicellulose. In *Polysaccharides, I. Structure, Characterisation and Use*; Heinze, T.T., Ed.; Springer-Verlag GmbH: Berlin/Heidelberg, Germany, 2005; pp. 1–67.
3. Painter, T.J. Algal polysaccharides. In *The Polysaccharides*; Aspinall, G.O., Ed.; Academic Press (Molecular biology): New York, NY, USA, 1983; pp. 195–285. Volume 2.
4. Schiener, P.; Black, K.D.; Stanley, M.S.; Green, D.H. The seasonal variation in the chemical composition of the kelp species Laminaria digitata, Laminaria hyperborea, Saccharina latissima Alaria esculenta. *J. Appl. Phycol.* **2015**, *27*, 363–373. [CrossRef]
5. Wei, N.; Quarterman, J.; Jin, Y.-S. Marine macroalgae: an untapped resource for producing fuels and chemicals. *Trends Biotechnol.* **2013**, *31*, 70–77. [CrossRef] [PubMed]
6. Jung, K.A.; Lim, S.-R.; Kim, Y.; Park, J.M. Potentials of macroalgae as feedstocks for biorefinery. *Bioresour. Technol.* **2013**, *135*, 182–190. [CrossRef] [PubMed]
7. Hehemann, J.-H.; Boraston, A.B.; Czjzek, M. A sweet new wave: structures and mechanisms of enzymes that digest polysaccharides from marine algae. *Curr. Opin. Struct. Biol.* **2014**, *28*, 77–86. [CrossRef] [PubMed]
8. Shi, P.; Yao, G.; Yang, P.; Li, N.; Luo, H.; Bai, Y.; Wang, Y.; Yao, B. Cloning, characterization, and antifungal activity of an endo-1,3-β-D-glucanase from *Streptomyces* sp. S27. *Appl. Microbiol. Biotechnol.* **2010**, *85*, 1483–1490. [CrossRef] [PubMed]
9. Vetvicka, V. Glucan-immunostimulant, adjuvant, potential drug. *World J. Clin. Oncol.* **2011**, *2*, 115–119. [CrossRef]
10. Cantarel, B.L.; Coutinho, P.M.; Rancurel, C.; Bernard, T.; Lombard, V.; Henrissat, B. The Carbohydrate-Active EnZymes database (CAZy): an expert resource for Glycogenomics. *Nucleic Acids Res.* **2009**, *37*, D233–D238. [CrossRef]
11. Sarmiento, F.; Peralta, R.; Blamey, J.M. Cold and Hot Extremozymes: Industrial Relevance and Current Trends. *Front. Bioeng. Biotechnol.* **2015**, *3*, 348. [CrossRef]
12. Krahe, M.; Markl, H.; Antranikian, G. Fermentation of extremophilic microorganisms. *FEMS Microbiol. Rev.* **1996**, *18*, 271–285. [CrossRef]
13. Schroeder, C.; Elleuche, S.; Blank, S.; Antranikian, G. Characterization of a heat-active archaeal beta-glucosidase from a hydrothermal spring metagenome. *Enzyme Microb. Technol.* **2014**, *57*, 48–54. [CrossRef]
14. Zhaxybayeva, O.; Swithers, K.S.; Lapierre, P.; Fournier, G.P.; Bickhart, D.M.; DeBoy, R.T.; Nelson, K.E.; Nesbø, C.L.; Doolittle, W.F.; Gogarten, J.P.; et al. On the chimeric nature, thermophilic origin, and phylogenetic placement of the Thermotogales. *Proc. Natl. Acad. Sci. USA* **2009**, *106*, 5865–5870. [CrossRef] [PubMed]
15. Patel, B.K.C.; Morgan, H.W.; Daniel, R.M. Fervidobacterium nodosum gen. nov. and spec. nov., a new chemoorganotrophic, caldoactive, anaerobic bacterium. *Arch. Microbiol.* **1985**, *141*, 63–69. [CrossRef]
16. Wang, Y.; Wang, X.; Tang, R.; Yu, S.; Zheng, B.; Feng, Y. A novel thermostable cellulase from Fervidobacterium nodosum. *J. Mol. Catal. B: Enzym.* **2010**, *66*, 294–301. [CrossRef]
17. Labourel, A.; Jam, M.; Legentil, L.; Sylla, B.; Hehemann, J.H.; Ferrières, V.; Czjzek, M.; Michel, G. Structural and biochemical characterization of the laminarinase ZgLamCGH16 from Zobellia galactanivorans suggests preferred recognition of branched laminarin. *Acta Crystallogr. D* **2015**, *71 Pt 2*, 173–184. [CrossRef]
18. Palumbo, J.D.; Sullivan, R.F.; Kobayashi, D.Y. Molecular Characterization and Expression in Escherichia coli of Three β-1,3-Glucanase Genes from Lysobacter enzymogenes Strain N4-7. *J. Bacteriol.* **2003**, *185*, 4362–4370. [CrossRef] [PubMed]
19. Hong, T.-Y.; Huang, J.-W.; Meng, M.; Cheng, C.-W. Isolation and biochemical characterization of an endo-1,3-β-glucanase from Streptomyces sioyaensis containing a C-terminal family 6 carbohydrate-binding module that binds to 1,3-β-glucan. *Microbiology* **2002**, *148 Pt 4*, 1151–1159. [CrossRef]
20. Qian, W.; Zhang, J. Gene dosage and gene duplicability. *Genetics* **2008**, *179*, 2319–2324. [CrossRef]
21. Wagner, A. Energy Constraints on the Evolution of Gene Expression. *Mol. Boil. Evol.* **2005**, *22*, 1365–1374. [CrossRef]
22. Francino, M.P. An adaptive radiation model for the origin of new gene functions. *Nat. Genet.* **2005**, *37*, 573–578. [CrossRef]
23. Bratlie, M.S.; Johansen, J.; Sherman, B.T.; Huang, D.W.; Lempicki, R.A.; Drabløs, F. Gene duplications in prokaryotes can be associated with environmental adaptation. *BMC Genom.* **2010**, *11*, 588. [CrossRef]
24. Cota, J.; Alvarez, T.M.; Citadini, A.P.; Santos, C.R.; Neto, M.D.O.; Oliveira, R.R.; Pastore, G.M.; Ruller, R.; Prade, R.A.; Murakami, M.T.; et al. Mode of operation and low-resolution structure of a multi-domain and

hyperthermophilic endo-β-1,3-glucanase from Thermotoga petrophila. *Biochem. Biophys. Res. Commun.* **2011**, *406*, 590–594. [CrossRef] [PubMed]

25. Zverlov, V.V.; Volkov, I.Y.; Velikodvorskaya, T.V.; Schwarz, W.H. Thermotoga neapolitana bglB gene, upstream of lamA, encodes a highly thermostable beta-glucosidase that is a laminaribiase. *Microbiology* **1997**, *143 Pt 11*, 3537–3542. [CrossRef]
26. Gueguen, Y.; Voorhorst, W.G.B.; van der Oost, J.; de Vos, W.M. Molecular and biochemical characterization of an endo-β -1,3-glucanase of the hyperthermophilic archaeon *Pyrococcus furiosus*. *J. Biol. Chem.* **1997**, *272*, 31258–31264. [CrossRef] [PubMed]
27. Schwarz, W.H.; Schimming, S.; Staudenbauer, W.L. Isolation of a Clostridium thermocellum gene encoding a thermostable β-1,3-glucanase (laminarinase). *Biotechnol. Lett.* **1988**, *10*, 225–230. [CrossRef]
28. Kasai, N.; Harada, T. Ultrastructure of Curdlan. In *Fiber Diffraction Methods*; French, A.D., Ed.; American Chemical Society (ACS symposium series): Washington, DC, USA, 1980; pp. 363–383. Volume 141.
29. Meng, D.D.; Wang, B.; Ma, X.Q.; Ji, S.Q.; Lu, M.; Li, F.L. Characterization of a thermostable endo-1,3(4)-beta-glucanase from Caldicellulosiruptor sp. strain F32 and its application for yeast lysis. *Appl. Microbiol. Biotechnol.* **2016**, *100*, 4923–4934. [CrossRef] [PubMed]
30. Kobayashi, T.; Uchimura, K.; Kubota, T.; Nunoura, T.; Deguchi, S. Biochemical and genetic characterization of beta-1,3 glucanase from a deep subseafloor Laceyella putida. *Appl. Microbiol. Biotechnol.* **2016**, *100*, 203–214. [CrossRef]
31. Masuda, S.; Endo, K.; Koizumi, N.; Hayami, T.; Fukazawa, T.; Yatsunami, R.; Fukui, T.; Nakamura, S. Molecular identification of a novel beta-1,3-glucanase from alkaliphilic Nocardiopsis sp. strain F96. *Extremophiles* **2006**, *10*, 251–255. [CrossRef] [PubMed]
32. Spilliaert, R.; Hreggvidsson, G.O.; Kristjansson, J.K.; Eggertsson, G.; Palsdottir, A. Cloning and Sequencing of a Rhodothermus marinus Gene, bglA, Coding for a Thermostable beta-Glucanase and its Expression in Escherichia coli. *JBIC J. Boil. Inorg. Chem.* **1994**, *224*, 923–930. [CrossRef]
33. Krah, M.; Misselwitz, R.; Politz, O.; Thomsen, K.K.; Welfle, H.; Borriss, R. The laminarinase from thermophilic eubacterium *Rhodothermus marinus*. Conformation, stability, and identification of active site carboxylic residues by site-directed mutagenesis. *Eur. J. Biochem.* **1998**, *257*, 101–111. [CrossRef]
34. Fuchs, K.-P.; Zverlov, V.V.; Velikodvorskaya, G.A.; Lottspeich, F.; Schwarz, W.H. Lic16A of *Clostridium thermocellum*, a non-cellulosomal, highly complex endo-beta-1,3-glucanase bound to the outer cell surface. *Microbiology* **2003**, *149 Pt 4*, 1021–1031. [CrossRef]
35. Woo, C.-B.; Kang, H.-N.; Lee, S.-B. Molecular cloning and anti-fungal effect of endo-β-1,3-glucanase from Thermotoga maritima. *Food Sci. Biotechnol.* **2014**, *23*, 1243–1246. [CrossRef]
36. Ilari, A.; Fiorillo, A.; Angelaccio, S.; Florio, R.; Chiaraluce, R.; Van Der Oost, J.; Consalvi, V. Crystal structure of a family 16 endoglucanase from the hyperthermophile Pyrococcus furiosus- structural basis of substrate recognition. *FEBS J.* **2009**, *276*, 1048–1058. [CrossRef] [PubMed]
37. Gaiser, O.J.; Piotukh, K.; Ponnuswamy, M.N.; Planas, A.; Borriss, R.; Heinemann, U. Structural basis for the substrate specificity of a *Bacillus* 1,3-1,4-beta-glucanase. *J. Mol. Biol.* **2006**, *357*, 1211–1225. [CrossRef] [PubMed]
38. Pereira, J.D.C.; Giese, E.C.; Moretti, M.M.D.S.; Gomes, A.C.D.S.; Perrone, M.B.O.M.; Da Silva, R.; Martins, D.A.B. Effect of Metal Ions, Chemical Agents and Organic Compounds on Lignocellulolytic Enzymes Activities. In *Enzyme Inhibitors and Activators*; IntechOpen: London, UK, 2017; pp. 139–164.
39. Labourel, A.; Jam, M.; Jeudy, A.; Hehemann, J.-H.; Czjzek, M.; Michel, G. The β-glucanase ZgLamA from *Zobellia galactanivorans* evolved a bent active site adapted for efficient degradation of algal laminarin. *J. Biol. Chem.* **2014**, *289*, 2027–2042. [CrossRef] [PubMed]
40. Jeng, W.-Y.; Wang, N.-C.; Lin, C.-T.; Shyur, L.-F.; Wang, A.H.-J. Crystal structures of the laminarinase catalytic domain from *Thermotoga maritima* MSB8 in complex with inhibitors: essential residues for β-1,3- and β-1,4-glucan selection. *J. Biol. Chem.* **2011**, *286*, 45030–45040. [CrossRef] [PubMed]
41. Sahm, K.; John, P.; Nacke, H.; Wemheuer, B.; Grote, R.; Daniel, R.; Antranikian, G. High abundance of heterotrophic prokaryotes in hydrothermal springs of the Azores as revealed by a network of 16S rRNA gene-based methods. *Extremophiles* **2013**, *17*, 649–662. [CrossRef] [PubMed]
42. Petersen, T.N.; Brunak, S.; Von Heijne, G.; Nielsen, H. SignalP 4.0: discriminating signal peptides from transmembrane regions. *Nat. Methods* **2011**, *8*, 785–786. [CrossRef] [PubMed]

43. Laemmli, U.K. Cleavage of Structural Proteins during the Assembly of the Head of Bacteriophage T4. *Nature* **1970**, *227*, 680–685. [CrossRef] [PubMed]
44. Bradford, M.M. A rapid and sensitive method for the quantitation of microgram quantities of protein utilizing the principle of protein-dye binding. *Anal. Biochem.* **1976**, *72*, 248–254. [CrossRef]
45. Miller, G.L. Use of Dinitrosalicylic Acid Reagent for Determination of Reducing Sugar. *Anal. Chem.* **1959**, *31*, 426–428. [CrossRef]
46. Britton, H.T.S.; Robinson, R.A. CXCVIII.—Universal buffer solutions and the dissociation constant of veronal. *J. Chem. Soc.* **1931**, 1456–1462. [CrossRef]

4

Immobilization of β-Galactosidases on the *Lactobacillus* Cell Surface using the Peptidoglycan-Binding Motif LysM

Mai-Lan Pham [1], Anh-Minh Tran [1,2], Suwapat Kittibunchakul [1], Tien-Thanh Nguyen [3], Geir Mathiesen [4] and Thu-Ha Nguyen [1,*]

1 Food Biotechnology Laboratory, Department of Food Science and Technology, BOKU-University of Natural Resources and Life Sciences, A-1190 Vienna, Austria; mailanpham.22@gmail.com (M.-L.P.); tranminh1703@gmail.com (A.-M.T.); suwapatkt@gmail.com (S.K.)

2 Department of Biology, Faculty of Fundamental Sciences, Ho Chi Minh City University of Medicine and Pharmacy, 217 Hong Bang, Ho Chi Minh City, Vietnam

3 School of Biotechnology and Food Technology, Hanoi University of Science and Technology, 1 Dai Co Viet, Hanoi, Vietnam; nguyenthanh.ibft@gmail.com

4 Faculty of Chemistry, Biotechnology and Food Science, Norwegian University of Life Sciences (NMBU), N-1432 Ås, Norway; geir.mathiesen@nmbu.no

* Correspondence: thu-ha.nguyen@boku.ac.at

Abstract: Lysin motif (LysM) domains are found in many bacterial peptidoglycan hydrolases. They can bind non-covalently to peptidoglycan and have been employed to display heterologous proteins on the bacterial cell surface. In this study, we aimed to use a single LysM domain derived from a putative extracellular transglycosylase Lp_3014 of *Lactobacillus plantarum* WCFS1 to display two different lactobacillal β-galactosidases, the heterodimeric LacLM-type from *Lactobacillus reuteri* and the homodimeric LacZ-type from *Lactobacillus delbrueckii* subsp. *bulgaricus*, on the cell surface of different *Lactobacillus* spp. The β-galactosidases were fused with the LysM domain and the fusion proteins, LysM-LacLMLreu and LysM-LacZLbul, were successfully expressed in *Escherichia coli* and subsequently displayed on the cell surface of *L. plantarum* WCFS1. β-Galactosidase activities obtained for *L. plantarum* displaying cells were 179 and 1153 U per g dry cell weight, or the amounts of active surface-anchored β-galactosidase were 0.99 and 4.61 mg per g dry cell weight for LysM-LacLMLreu and LysM-LacZLbul, respectively. LysM-LacZLbul was also displayed on the cell surface of other *Lactobacillus* spp. including *L. delbrueckii* subsp. *bulgaricus*, *L. casei* and *L. helveticus*, however *L. plantarum* is shown to be the best among *Lactobacillus* spp. tested for surface display of fusion LysM-LacZLbul, both with respect to the immobilization yield as well as the amount of active surface-anchored enzyme. The immobilized fusion LysM-β-galactosidases are catalytically efficient and can be reused for several repeated rounds of lactose conversion. This approach, with the β-galactosidases being displayed on the cell surface of non-genetically modified food-grade organisms, shows potential for applications of these immobilized enzymes in the synthesis of prebiotic galacto-oligosaccharides.

Keywords: *Lactobacillus*; β-galactosidase; immobilization; cell surface display; LysM domains

---

1. Introduction

β-Galactosidases catalyze the hydrolysis and transgalactosylation of β-D-galactopyranosides (such as lactose) [1–3] and are found widespread in nature. They catalyze the cleavage of lactose (or related compounds) in their hydrolysis mode and are thus used in the dairy industry to remove lactose from various products. An attractive biocatalytic application is found in the transgalactosylation potential of these enzymes, which is based on their catalytic mechanism [1,4]. β-Galactosidases can be obtained from different sources including microorganisms, plants and animals, however microbial sources of β-galactosidase are of great biotechnological interest because of easier handling, higher multiplication rates, and production yield. Recently, a number of studies have focused on the use of the genus *Lactobacillus* for the production and characterization of β-galactosidases, including the enzymes from *L. reuteri, L. acidophilus, L. helveticus, L. plantarum, L. sakei, L. pentosus, L. bulgaricus, L. fermentum, L. crispatus* [5–15]. β-Galactosidases from *Lactobacillus* species are different at molecular organization [6,8,10,12,16]. The predominant glycoside hydrolase family 2 (GH2) β-galactosidases found in lactobacilli are of the LacLM type, which are heterodimeric proteins encoded by the two overlapping genes, *lacL* and *lacM*, including *lacLM* from *L. reuteri* [16], *L. acidophilus* [6], *L. helveticus* [7], *L. pentosus* [11], *L. plantarum* [8], and *L. sakei* [10]. Di- or oligomeric GH2 β-galactosidases of the LacZ type, encoded by the single *lacZ* gene, are sometimes, but not often found in lactobacilli such as in *L. bulgaricus* [12]. Lactobacilli have been studied intensively with respect to their enzymes for various different reasons, one of which is their 'generally recognized as safe' (GRAS) status and their safe use in food applications. It is anticipated that galacto-oligosaccharides (GOS) produced by these β-galactosidases will have better selectivity for growth and metabolic activity of this bacterial genus in the gut.

An economical, sustainable and intelligent use of biocatalysts can be achieved through immobilization, where the enzyme is bound onto a suitable food-grade carrier. Efforts have been made to immobilize β-galactosidases from *L. reuteri*, a LacLM-type, and *Lactobacillus bulgaricus*, a LacZ-type, on chitin using the chitin binding domain (ChBD) of *Bacillus circulans* WL-12 chitinase A1 [17]. Cell surface display has been shown as a new strategy for enzyme immobilization, which involves the use of food-grade organism *L. plantarum* both as a cell factory for the production of enzymes useful for food applications and as the carrier for the immobilization of the over-expressed enzyme by anchoring the enzyme on the cell surface [18,19]. This enables the direct use of the microbial cells straight after the fermentation step as an immobilized biocatalysts, offering the known advantages of immobilization (reuse of enzyme, stabilization, etc.) together with a significant simplification of the production process since costly downstream processing of the cells producing the enzyme (cell disruption, protein purification, etc.) as well as the use of carrier material will not be necessary. We recently reported cell surface display of mannanolytic and chitinolytic enzymes in *L. plantarum* using two anchors from *L. plantarum*, a lipoprotein-anchor derived from the Lp_1261 protein and a cell wall anchor (cwa2) derived from the Lp_2578 protein [19]. However, this approach works less efficient with dimeric and oligomeric enzymes, such as β-galactosidases from lactobacilli, due to low secretion efficiency of target proteins. Therefore, it is of our interest to find another strategy to display lactobacillal β-galactosidases on *Lactobacillus* cell surface for use as immobilized biocatalysts for applications in lactose conversion and GOS formation processes.

There are two principally different ways of anchoring a secreted protein to the bacterial cell wall: covalently, via the sortase pathway, or non-covalently, via a protein domain that interacts strongly with cell wall components. In sortase-mediated anchoring, the secreted protein carries a C-terminal anchor containing the so-called LP × TG motif followed by a hydrophobic domain and a positively charged tail [20]. The hydrophobic domain and the charged tail keep the protein from being released to the medium, thereby allowing recognition of the LP × TG motif by a membrane-associated transpeptidase called sortase [20–22]. The sortase cleaves the peptide bond between threonine and glycine in the LP × TG motif and links the now C-terminal threonine of the surface protein to a pentaglycine in the cell wall [21–25]. One of the non-covalent cell display systems exploits so-called LysM domains, the peptidoglycan binding motifs, that are known to promote cell wall association of several natural proteins [23,26]. These domains have been used to display proteins in lactic acid bacteria (LAB) by fusing the LysM domain N- or C-terminally to the target protein [27–30]. In *L. plantarum* WCFS1 ten proteins are predicted to be displayed at the cell wall through LysM domains [31].

In this present study, we exploit a single LysM domain derived from the Lp_3014 protein in *L. plantarum* WCFS1 for external attachment of two lactobacillal β-galactosidases, a LacLM-type from *L. reuteri* and a LacZ-type from *L. bulgaricus*, on the cell surface of four *Lactobacillus* species. The immobilization of active β-galactosidases through cell-surface display can be utilized as safe and stable non-GMO food-grade biocatalysts that can be used in the production of prebiotic GOS.

2. Results

2.1. Expression of Recombinant Lactobacillal β-Galactosidases in E. coli

The overlapping *lacLM* genes from *L. reuteri* L103 and the *lacZ* gene from *L. bulgaricus* DSM20081, both encoding β-galactosidases, were fused N-terminally to the LysM motif for expression and later attachment of the hybrid proteins to the peptidoglycan layer of *Lactobacillus* spp. An 88 residue fragment of the LysM motif from the 204-residue-Lp_3014 protein of an extracellular transglycosylase of *L. plantarum* WCFS1 [31,32] was fused to two β-galactosidases for production in *E. coli*. The two hybrid sequences were then cloned into the expression vector pBAD containing an N-terminal 7 × Histidine tag for immunodetection, yielding pBAD3014LacLMLreu and pBAD3014LacZLbul (Figure 1).

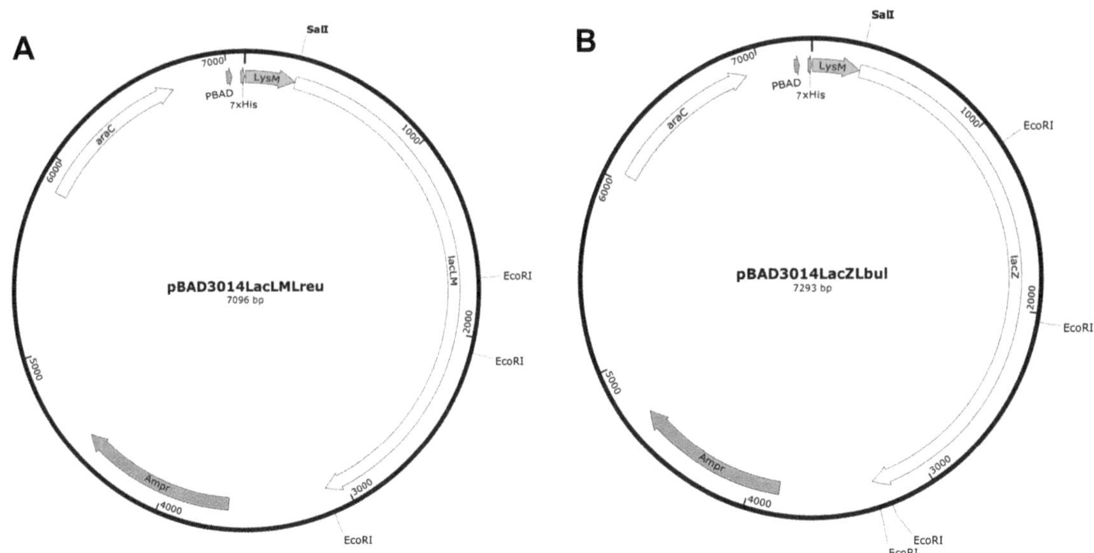

Figure 1. The expression vectors for LysM-LacLMLreu (**A**) and LysM-LacZLbul (**B**) in *E. coli*. The vectors are the derivatives of the pBAD vector (Invitrogen, Carlsbad, CA, USA) containing a 7 × His tag sequence fused to a single LysM domain from Lp_3014, *L. plantarum* WCFS1. LacLMLreu encoded by two overlapping genes *lacLM* and LacZLbul encoded by the *lacZ* gene are the β-galactosidases from *L. reuteri* and *L. delbrueckii* subsp. *bulgaricus* DSM 20081, respectively. See text for more details.

The *E. coli* strains were cultivated in Luria-Bertani (LB) medium, induced for gene expression (as described in Materials and Methods), and the SDS-PAGE and Western blot analyses of cell-free extracts (Figure 2) showed the production of the two recombinant β-galactosidases, LysM-LacLMLreu and LysM-LacZLbul. As judged by SDS-PAGE (Figure 2A), LysM-LacLMLreu shows two bands with apparent molecular masses corresponding to a large subunit (LacL) and a small subunit (LacM) at ~90 kDa and ~35 kDa. These values are in agreement with reported molecular masses of 73 and 35 kDa for these two subunits of β-galactosidase from *L. reutei* [5,16]. The increase in molecular mass of a larger subunit in LysM-LacLMLreu is due to the added His-LysM fragment (~18 kDa). On the other hand, β-galactosidase from *L. bulgaricus* was reported to be a homodimer, consisting of two identical subunits of ~115 kDa [12]. A unique band of ~130 kDa corresponding to the molecular mass of a single subunit of LacZ fused with the 18 kDa-fragment of the histidine-tag and the LysM domain was shown on SDS-PAGE analysis of a cell-free extract of LysM-LacZLbul as expected (Figure 2A). Western blot analysis of the crude, cell-free extracts was performed using anti-His antibody for detection. Figure 2B shows that the recombinant bacteria produced the expected proteins, LysMLacL (lane 2) and LysMLacZ (lane 4). LacM was not detected as it does not contain the histidine-tag.

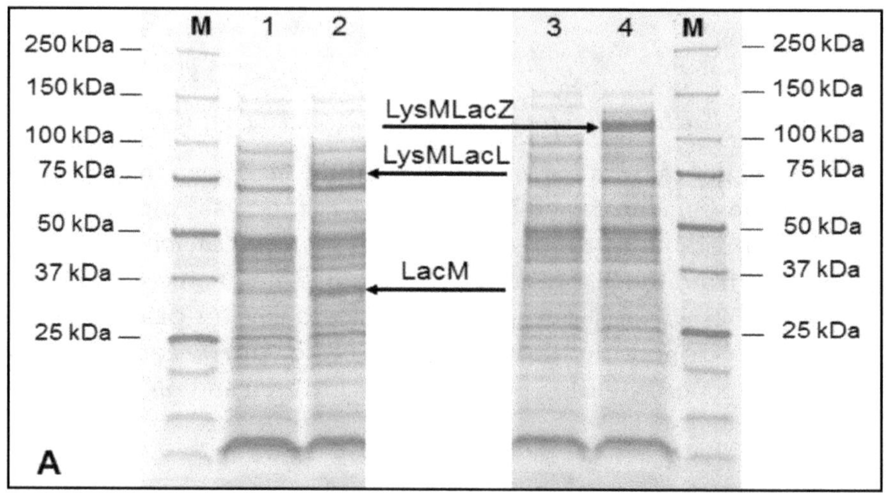

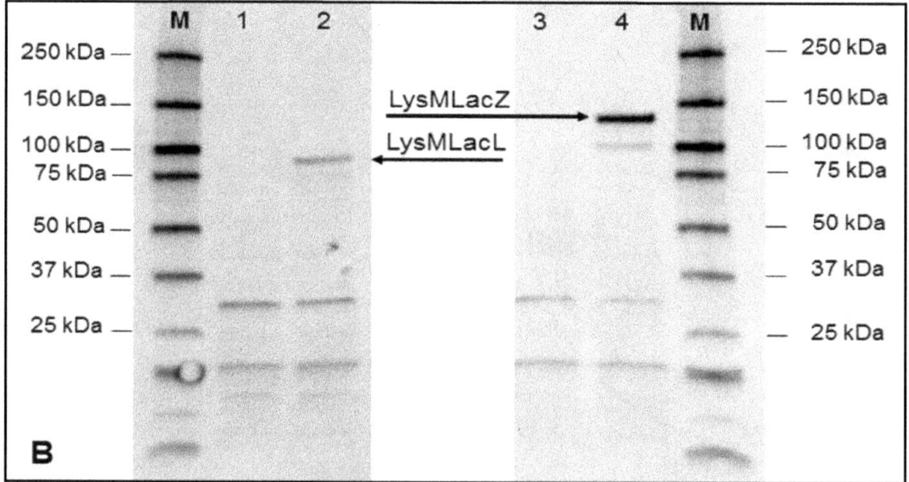

Figure 2. SDS-PAGE analysis (**A**) and Western blot analysis (**B**) of a cell-free extract of crude β-galactosidase fusion proteins, LysM-LacLMLreu (non-induced: lane 1, induced: lane 2) and LysM-LacZLbul (non-induced: lane 3, induced: lane 4), overexpressed in *E. coli* HST08. LacLMLreu encoded by two overlapping genes *lacLM* and LacZLbul encoded by *lacZ* gene are the β-galactosidases from *L. reuteri* and *L. delbrueckii* subsp. *bulgaricus* DSM 20081, respectively. The cultivation and induction conditions are as described in Materials and Methods and samples were taken at different time points after induction during cultivations. The arrows indicate the subunits of the recombinant β-galactosidases. M denotes the Precision protein ladder (Biorad, CA, USA).

To check if the heterologously produced enzymes were functionally active, β-galactosidase activities of cell-free lysates of *E. coli* cells carrying different expression vectors were measured. The highest yields obtained for the two recombinant enzymes were 11.1 ± 1.6 k·U_{oNPG} per L of medium with a specific activity of 6.04 ± 0.03 U·mg^{-1} for LysM-LacLMLreu and 46.9 ± 2.7 kU_{oNPG} per L of medium with a specific activity of 41.1 ± 0.9 U·mg^{-1} for LysM-LacZLbul, respectively (Table 1). The β-galactosidase activities in non-induced *E. coli* cells were negligible for both LysM-LacLMLreu and LysM-LacZLbul showing that the activity is from the overproduced β-galactosidases (Table 1).

Table 1. β-Galactosidase activities in cell-free lysates of *E. coli* cells carrying different expression vectors.

Expression Vector	Volumetric Activity (k·U/L Culture Medium)		Specific Activity (U/mg Protein)	
	Non-Induced	Induced	Non-Induced	Induced
pBAD3014LacLMLreu	n.d.	11.1 ± 1.5	n.d.	6.04 ± 0.03
pBAD3014LacZLbul	n.d.	46.9 ± 2.7	n.d.	41.1 ± 0.9

n.d.: not detected.

2.2. *Display of Lactobacillal β-Galactosidases on Lactobacillus Cell Surface*

To investigate the attachment of the two hybrid proteins, LysM-LacLMLreu and LysM-LacZLbul, to the cell wall of *L. plantarum,* cell-free crude extracts from *E. coli* harboring β-galactosidases corresponding to 50 U_{oNPG} (~5–6 mg protein) were incubated with *L. plantarum* cells collected from one mL cultures at OD_{600} ~4.0. The enzymes and *L. plantarum* were incubated at 37 °C with gentle agitation, and after 24 h of incubation, the residual activities in the supernatant as well as on the cell surface were determined for both enzymes (Table 2A). The immobilization yield (IY) is a measure of how much of the applied protein bound to the surface of *Lactobacillus* cells. Immobilizations yields for LysM-LacLMLreu and LysM-LacZLbul were 6.5% and 31.9%, respectively. SDS-PAGE analysis of the samples after the immobilization procedure showed strong bands of LysM-LacL and LacM or LysM-LacZ in the residual supernatants (Figure 3A, lane 2; Figure 3B, lane 2), indicating relatively high amounts of non-anchored proteins in the supernatants. Two successive washing steps with 50 mM sodium phosphate buffer (NaPB, pH 6.5) did not release the enzymes showing that the immobilization is both effective and stable (Figure 3A, lanes 4, 5; Figure 3B, lanes 3, 4). The low immobilization yield for LysM-LacLMLreu was confirmed by the SDS-PAGE analysis (Figure 3A, lane 3). Western blot analysis of the crude, cell-free extracts of *L. plantarum* LacZLbul-displaying cells was performed using an anti-His antibody for detection showing the presence of LacZLbul (Figure 3C; lane 3). Flow cytometry confirmed the surface localization of both enzymes LysM-LacLMLreu and LysM-LacZLbul as clear shifts in the fluorescence signals for *L. plantarum* LacLMLreu- and LacZLbul-displaying cells in comparison to the control strain were observed (Figure 4A,B). The surface-displayed enzymes were shown to be functionally active. β-Galactosidase activities obtained for *L. plantarum* displaying cells were 179 and 1153 U per g dry cell weight, corresponding to approximately 0.99 and 4.61 mg of active, surface-anchored β-galactosidase per g dry cell mass for LysM-LacLMLreu and LysM-LacZLbul (Table 2A), respectivel.

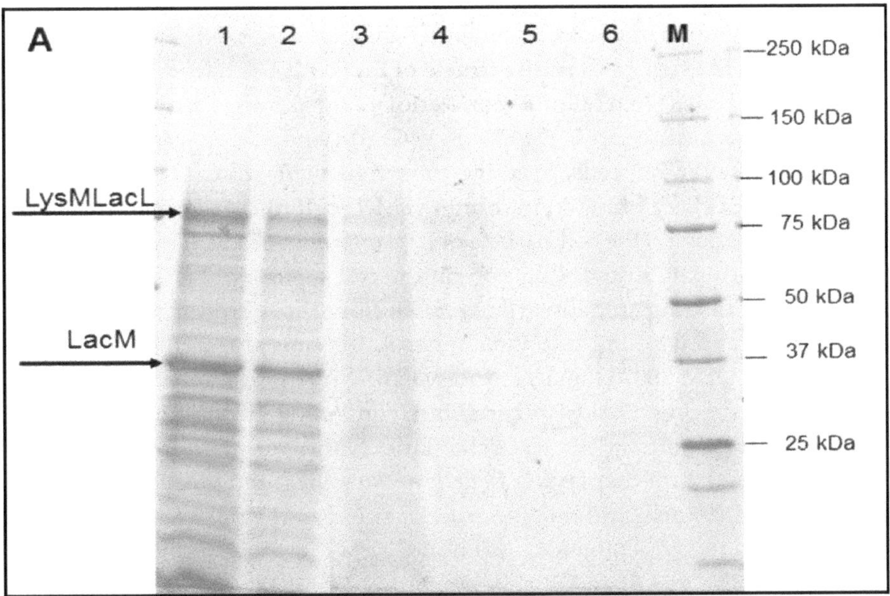

Figure 3. *Cont.*

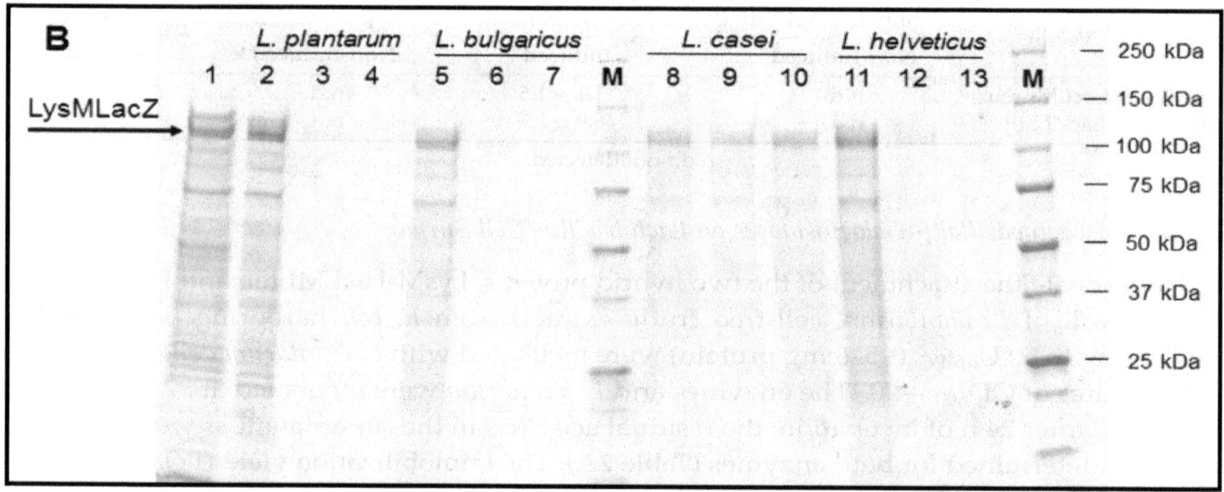

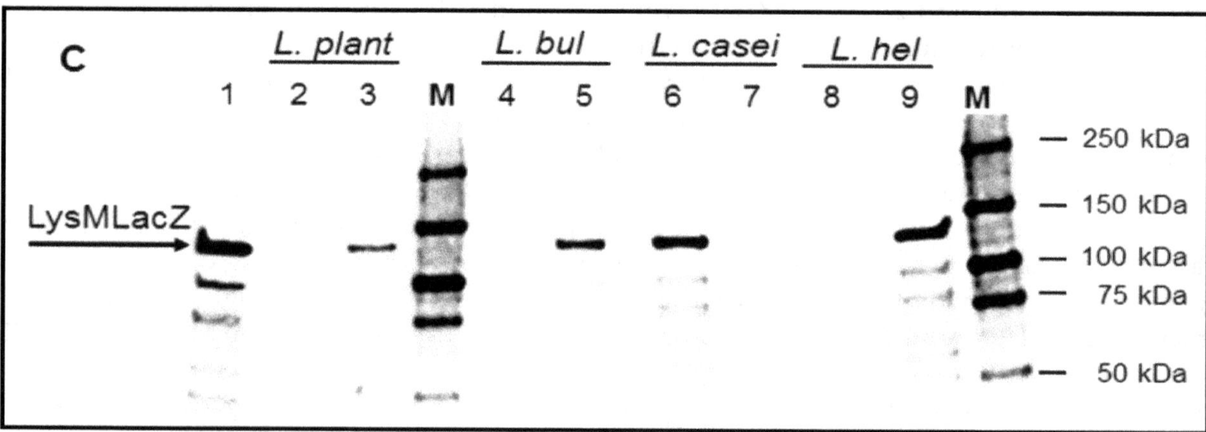

Figure 3. SDS-PAGE analysis (**A**,**B**) and Western blot analysis (**C**) of immobilization of recombinant enzymes. LacLMLreu encoded by two overlapping genes *lacLM* and LacZLbul encoded by *lacZ* gene are the β-galactosidases from *L. reuteri* and *L. delbrueckii* subsp. *bulgaricus* DSM 20081, respectively. The arrows indicate the subunits of the recombinant β-galactosidases. M denotes the Precision protein ladder (Biorad, CA, USA). (**A**) Cell-free crude extracts of *E. coli* HST08 harboring pBAD3014LacLMLreu (containing LysM-LacLMLreu) at 18 h of induction (lane1); flow through during immobilization (lane 2); surface anchored-LysM-LacLMLreu in *L. plantarum* WCFS1 (lane 3) and washing fractions (lanes 4, 5); non-displaying *L. plantarum* WCFS1 cells, negative control (lane 6). (**B**) Cell-free crude extracts of *E. coli* HST08 harboring pBAD3014LacZLbul (containing LysM-LacZLbul) at 18 h of induction (lane1); flow through during immobilization on the cell surface of *L. plantarum* WCFS1 (lane 2) and washing fractions (lanes 3, 4); flow through during immobilization on the cell surface of *L. delbrueckii* subsp. *bulgaricus* DSM 20081 (lane 5) and washing fractions (lanes 6, 7); flow through during immobilization on cell surface of *L. casei* (lane 8) and washing fractions (lanes 9, 10); flow through during immobilization on cell surface of *L. helveticus* DSM 20075 (lane 11) and washing fractions (lanes 12, 13). (**C**) Cell-free crude extracts of *E. coli* HST08 harboring pBAD3014LacZLbul (containing LysM-LacZLbul) at 18 h of induction (lane 1); non-displaying *L. plantarum* WCFS1 cells (lane 2) and surface anchored-LysM-LacZLbul in *L. plantarum* WCFS1 (lane 3); non-displaying *L. delbrueckii* subsp. *bulgaricus* DSM 20081 cells (lane 4) and surface anchored-LysM-LacZLbul in *L. delbrueckii* subsp. *bulgaricus* DSM 20081 (lane 5); surface anchored-LysM-LacZLbul in *L. casei* (lane 6) and non-displaying *L. casei* cells (lane 7); non-displaying *L. helveticus* DSM 20075 cells (lane 8) and surface anchored-LysM-LacZLbul in *L. helveticus* DSM 20075 (lane 9).

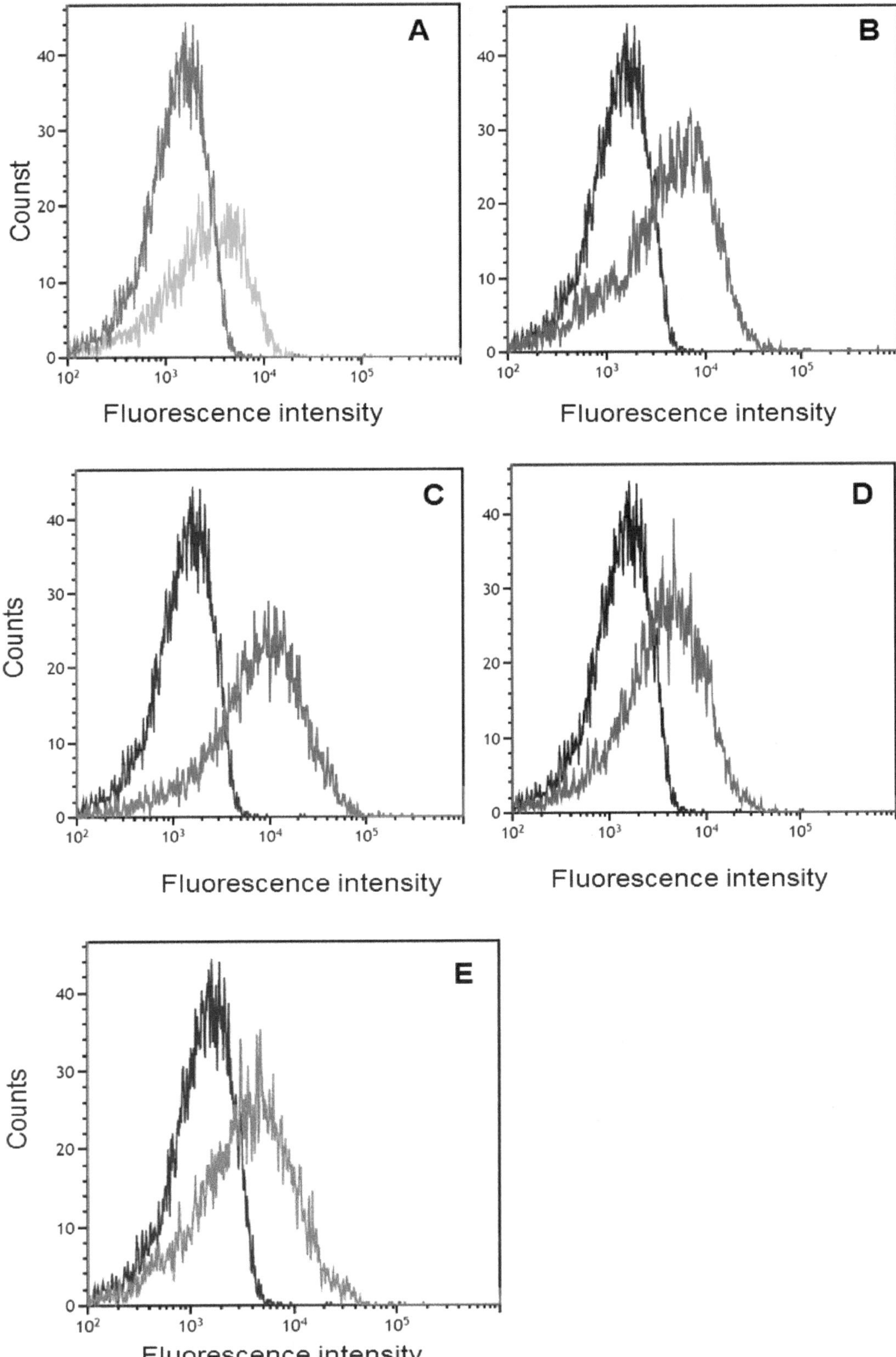

Figure 4. Analysis of surface localization of LysM-LacLMLreu and LysM-LacZLbul in *Lactobacillus* cells by using flow cytometry: surface anchored-LysM-LacLMLreu in *L. plantarum* WCFS1 (**A**, green line); surface anchored-LysM-LacZLbul in *L. plantarum* WCFS1 (**B**, blue line), in *L. delbrueckii* subsp. *bulgaricus* DSM 20081 (**C**, red line), in *L. casei* (**D**, purple line) and in *L. helveticus* DSM 20075 (**E**, olive line). Non-displaying *Lactobacillus* cells were used as negative controls (A–E, black line).

Table 2. Immobilization of (**A**) recombinant lactobacillal β-galactosidases on *L. plantarum* WCFS1 cell surface and (**B**) recombinant β-galactosidase from *L. bulgaricus* DSM 20081 (LysM-LacZLbul) on the cell surface of different *Lactobacillus* spp.

	Residual Activities in Supernatant	Immobilization Yield [a] (IY)	Activity on Cell Surface [b]		Activity Retention [c] (AR)	Amount of Active Surface Anchored β-gal [d]
	(%)	(%)	(%)	U/g DCW	(%)	mg/g DCW
(A) Enzyme (on *L. plantarum* WCFS1 cell surface)						
LysM-LacLMLreu	93.5 ± 1.2	6.53	3.06 ± 0.08	179 ± 5	46.9	0.99 ± 0.02
LysM-LacZLbul	68.1 ± 0.1	31.9	20.3 ± 0.2	1153 ± 12	63.5	4.61 ± 0.05
(B) *Lactobacillus* spp. (with enzyme LysM-LacZLbul)						
L. plantarum WCFS1	68.1 ± 0.1	31.9	20.3 ± 0.2	1153 ± 12	63.5	4.61 ± 0.05
L. bulgaricus DSM 20081	71.3 ± 0.9	28.7	14.0 ± 0.9	795 ± 53	48.5	3.18 ± 0.11
L. casei	76.1 ± 0.9	23.9	15.1 ± 0.8	861 ± 48	63.2	3.44 ± 0.09
L. helveticus DSM20075	75.3 ± 0.9	24.7	14.3 ± 0.5	812 ± 27	57.7	3.25 ± 0.11

[a] IY (%) was calculated by subtraction of the residual enzyme activity (%) in the supernatant after immobilization from the total activity applied (100%). [b] Activity on the cell surface (%) is the percentage of enzyme activity measured on the cell surface to the total applied activity. Activity on the cell surface (U/g DCW) is calculated as the amount of enzyme (Units) per g dry cell weight. [c] Activity retention, AR (%), is the ratio of activity on the cell surface (%) to IY (%). [d] It was calculated based on specific activities of purified LacLMLreu of 180 U/mg protein [16] and of purified LacZLbul (His Tagged) of 250 U/mg protein [12]. Values given are the average value from at least two independent experiments, and the standard deviation was always less than 5%.

Due to higher immobilization yields and increased amounts of active surface-anchored protein in *L. plantarum*, LysM-LacZLbul was chosen for further analysis of its display on the cell surface of other food-relevant *Lactobacillus* spp. including *L. bulgaricus*, *L. casei* and *L. helveticus*. The parameters of residual activities in the supernatant after the anchoring experiment, activity on the cell surface, immobilization yields, activity retention and amounts of active surface-anchored LysM-LacZLbul were determined and are presented in Table 2B.

It was shown that surface-anchored LysM-LacZLbul was released from the cell surface of *L. casei* during the subsequent washing steps (Figure 3B, lanes 9, 10). Western blot analysis of the crude, cell-free extracts of *Lactobacillus* LysM-LacZLbul-displaying cells indicated the binding of LysM-LacZLbul to all four *Lactobacillus* spp. tested (Figure 3C; lanes 3, 5, 6, 9) as was also confirmed by flow cytometry (Figure 4B–E). *L. plantarum* bound most efficiently among the tested *Lactobacillus* species shown by the highest immobilization yield and the highest amount of active, surface-anchored LysM-LacZLbul (Table 2B).

2.3. *Enzymatic Stability of β-Galactosidase-Displaying Cells*

Both temperature stability and reusability of β-galactosidase displaying cells were determined. For temperature stability, *L. plantarum* galactosidase-displaying cells were incubated in 50 mM sodium phosphate buffer (NaPB), pH 6.5 at different temperatures, and at certain time intervals, the residual β-galactosidase activities on *L. plantarum* cell surface were measured. Both LysM-LacLMLreu and LysM-LacZLbul-displaying cells are very stable at −20 °C with a half-life time of activity ($\tau_{\frac{1}{2}}$) of approximately 6 months (Table 3). The half-life time of activity of LysM-LacLMLreu-displaying cells at 30 °C is 55 h, whereas half-life times of activity of LysM-LacZLbul-displaying cells at 30 °C and 50 °C are 120 h and 30 h, respectively (Table 3).

Table 3. Stability of *L. plantarum* β-galactosidase-displaying cells at various temperatures [a].

LysM-LacLMLreu		LysM-LacLZLbul	
Temperature	$\tau_{\frac{1}{2}}$	**Temperature**	$\tau_{\frac{1}{2}}$
−20 °C	6 months	−20 °C	6 months
4 °C	3 months	4 °C	Nd [b]
30 °C	55 h	30 °C	120 h
50 °C	nd [b]	50 °C	30 h

[a] *L. plantarum* galactosidase-displaying cells were incubated in 50mM sodium phosphate buffer (NaPB), pH 6.5 at different temperatures. Experiments were performed at least in duplicates. [b] not determined.

To test the reusability of LysM-LacLMLreu- and LysM-LacZLbul-displaying cells, the enzyme activity was measured during several repeated rounds of lactose conversion with two washing steps between each cycle. The enzymatic activities of *L. plantarum* LysM-LacZLbul-displaying cells decreased by ~23% and 27% at 30 °C and 50 °C (Figure 5), respectively, after three conversion/washing cycles, indicating that these displaying cells can be reused for several rounds of biocatalysis at tested temperatures. LysM-LacLMLreu-displaying cells are less stable than LysM-LacZLbul-displaying cells as only 56% of the initial β-galactosidase activity are retained at 30 °C after the third cycle (Figure 5). LysM-LacZLbul-displaying cells retained 35% of β-galactosidase activity after the fourth cycle at 50 °C, 57% and 51% after the fourth and fifth cycle, respectively, at 30 °C (Figure 5). These observations indicate that immobilized fusion LysM-β-galactosidases can be reused for at least four to five repeated rounds of lactose conversion.

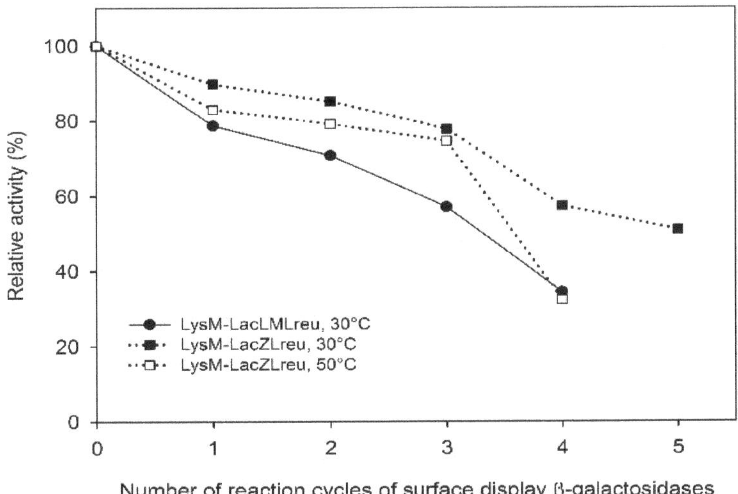

Figure 5. Enzymatic activity of surface display β-galactosidases, LysM-LacLMLreu- and LysM-LacZLbul, during several repeated rounds of lactose conversion using *L. plantarum* WCFS1 displaying cells. Experiments were performed in duplicates, and the standard deviation was always less than 5%.

2.4. Formation of Galacto-Oligosaccharides (GOS)

Figure 6 shows the formation of GOS using *L. plantarum* cells displaying β-galactosidase LacZ from *L. bulgaricus* (LysM-LacZLbul) with 1.0 U_{Lac} β-galactosidase activity per mL of the reaction mixture and 205 g/L initial lactose in 50 mM sodium phosphate buffer (pH 6.5) at 30°C. The maximal GOS yield was around 32% of total sugars obtained at 72% lactose conversion after 7 h of conversion. This observation shows that surface-displayed LacZ is able to convert lactose to form galacto-oligosaccharides. We could identify the main GOS products of transgalactosylation, which are β-D-Gal*p*-(1→6)-D-Glc, β-D-Gal*p*-(1→3)-D-Lac, β-D-Gal*p*-(1→3)-D-Glc, β-D-Gal*p*-(1→3)-D-Gal, β-D-Gal*p*-(1→6)-D-Gal, and β-D-Gal*p*-(1→6)-D-Lac. This is similar to the product profile when performing the conversion reaction with the free enzyme as previously reported [12].

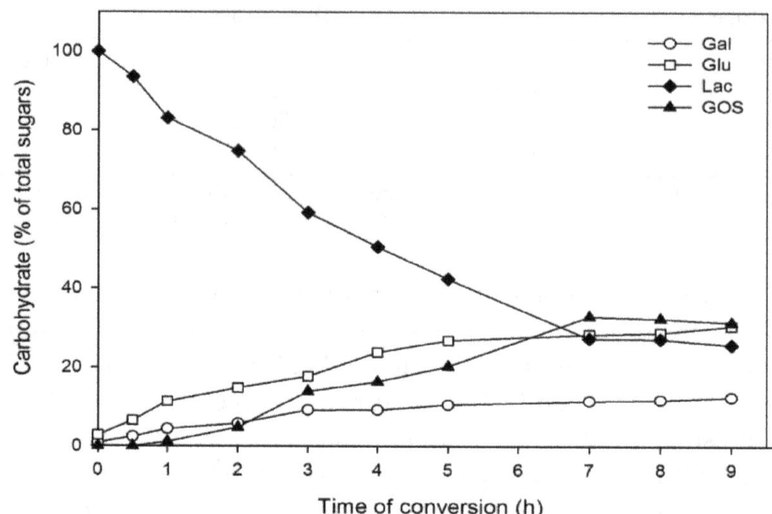

Figure 6. Course of reaction for lactose conversion by surface display β-galactosidase from *L. bulgaricus* (LysM-LacZLbul) in *L. plantarum* WCFS1 as determined by HPLC. The batch conversion was carried out at 30 °C using 205 g/L initial lactose concentration in 50 mM NaPB (pH 6.5) and constant agitation (500 rpm). *L. plantarum* LysM-LacZLbul displaying cells were added to equivalent concentrations of 1.0 U_{Lac}/mL of the reaction mixture. Experiments were performed in duplicates, and the standard deviation was always less than 5%.

3. Discussion

Surface display of proteins on cells of lactic acid bacteria (LAB) generally requires genetic modifications, which might have limitations in food and medical applications due to the sensitive issue of the use of genetically modified organisms (GMO). Anchoring heterologous proteins on the cell surface of non-genetically modified LAB (non-GMO) via mediated cell wall binding domains including surface layer domain (SLPs) [33,34], LysM domain [26,30,35–37], W × L domains [38] attracts increasing interest.

Lysin motif (LysM) domains are found in many bacterial peptidoglycan hydrolases [26,38,39]. Peptidoglycan contains sugar (glycan) chains, which consist of *N*-acetylglucosamine (NAG) and *N*-acetylmuramic acid (NAM) units joined by glycosidic linkages. Proteins harboring LysM motifs have been shown to bind non-covalently to the peptidoglycan layer and have been employed to display heterologous proteins on the bacterial cell-surface [26,40,41]. These domains can contain single or multiple LysM motifs [41], and they have been used to display proteins in LAB by fusion either to the N- or C-terminus of a target protein [27–30]. Interestingly, the LysM motif derived from the *L. plantarum* Lp_3014 transglycosylase has been used successfully for surface display of invasin [36] and a chemokine fused to an HIV antigen [37] previously.

In this work, we used the single LysM domain derived from Lp_3014 to anchor two different lactobacillal β-galactosidases, a heterodimeric type from *L. reuteri* and a homodimeric type from *L. bulgaricus*, on the cell surface of four species of lactobacilli. Functional active fusion proteins, LysM-LacLMLreu and LysM-LacZLbul, were successfully expressed in *E. coli*. However, the expression yield of LysM-LacLMLreu was ten-fold lower than that of the β-galactosidase from *L. reuteri* (LacLMLreu) without LysM expressed previously in *E. coli*, which was reported to be 110 kU of β-galactosidase activity per liter of cultivation medium [16]. This may indicate that the fusion of the LysM domain has a negative effect on the expression level. Interestingly, the expression yields of LysM-LacZLbul were 4-fold and 7-fold higher in terms of volumetric and specific activities, respectively, than that of LysM-LacLMLreu using the same host, expression system and induction conditions.

β-Galactosidase from *L. bulgaricus* (LysM-LacZLbul) is a homodimer whereas β-galactosidase from *L. reuteri* (LysM-LacLMLreu) is a heterodimer, and hence the fusion of the LysM domain only to the LacL subunits might lead to the discrepancy between the yields of these two fusion proteins due to different folding mechanisms.

Not surprisingly, the affinity for peptidoglycan of homodimeric LysM-LacZLbul is significantly higher than LysM-LacLMLreu as shown by the immobilization yield (Table 2A). As aforementioned LacLMLreu from *L. reuteri* is a heterodimer and the LysM domain is fused N-terminally to only LacL, while LacZLbul from *L. bulgaricus* is a homodimer, hence each of the identical subunits will carry its own LysM domain leading to stronger attachment of LacZ on the *L. plantarum* cell wall. This could be a likely explanation for the higher immobilization yields observed for LysM-LacZLbul. Even though the immobilization yields obtained in this study were significantly lower than the immobilization yields for these same enzymes when a chitin binding domain (ChBD) together with chitin was used [17], the activity retention (AR) on the *L. plantarum* cell surface (46.9% and 63.5% for LysM-LacLMLreu and LysM-LacZLbul, respectively) were significantly higher. The AR values for ChBD-LacLM, LacLM-ChBD and LacZ-ChBD using chitin beads were 19%, 26% and 13%, respectively [17]. Notably, the amount of active surface anchored LysM-LacLMLreu (0.99 ± 0.02 mg per g dry cell weight) on the cell surface of *L. plantarum* WCFS1 is significantly lower than LysM-LacZLbul (4.61 ± 0.05 mg per g dry cell weight). This is mainly due to the low immobilization yield of LysM-LacLMLreu. *L. plantarum* collected from one mL cultures at OD_{600} ~4.0 was used in immobilization reactions, hence the amount of *L. plantarum* cells was estimated to be ~3.0×10^9 cfu/mL. Therefore, we calculated that 8.22 μg LysM-LacLMLreu and 38.3 μg LysM-LacZLbul anchored on 3.0×10^9 *L. plantarum* cells or 0.002 pg LysM-LacLMLreu and 0.012 pg LysM-LacZLbul per *L. plantarum* cell. Xu et al. (2011) reported the use of the putative muropeptidase MurO (Lp_2162) from *L. plantarum* containing two putative LysM repeat regions for displaying a green fluorescent protein (GFP) and a β-galactosidase from *Bifidobacterium bifidum* on the surface of *L. plantarum* cells [42]. They reported that 0.008 pg of GFP was displayed per cell on non-treated *L. plantarum* cells, while the amount of active surface anchored β-galactosidase from *B. bifidum* on the surface of *L. plantarum* cells was not reported in that study.

Further, we tested the capability of binding the fusion protein LysM-LacZLbul to the cell wall of three other *Lactobacillus* species. *L. plantarum* showed the best capacity among the tested *Lactobacillus* for surface anchoring of LysM-LacZLbul (Table 2B), whereas *L. bulgaricus*, *L. casei* and *L. helveticus* are comparable in term of the amount of active surface-anchored enzyme.

The highest GOS yield of 32% obtained with the surface-immobilized enzyme is lower than the yield obtained with the free enzyme LacZ from *L. bulgaricus* (Figure 6), which was previously reported to be approximately 50% [12]. This could be due to the binding of LysM-LacZLbul to the peptidoglycan and the attachment of the enzyme on *Lactobacillus* cell surface, which might hinder the access of the substrate lactose to the active site of the enzyme. Interestingly, the GOS yield obtained from lactose conversion using *L. plantarum* cells displaying β-galactosidase (LysM-LacZLbul) from *L. bulgaricus* is significantly higher than the yield obtained with immobilized β-galactosidase (LacZ-ChBD) on chitin, which was previously reported around 23%–24% [12]. It indicates that β-galactosidase from *L. bulgaricus* anchored on *L. plantarum* cell surface is more catalytically efficient than its immobilized form on chitin.

4. Materials and Methods

4.1. Bacterial Strains and Culture Conditions

The bacterial strains and plasmids used in this study are listed in Table 4. *Lactobacillus plantarum* WCFS1, isolated from human saliva as described by Kleerebezem et al. [32], was originally obtained from NIZO Food Research (Ede, The Netherlands) and maintained in the culture collection of the Norwegian University of Life Sciences, Ås, Norway. *L. helveticus* DSM 20075 (ATCC 15009) and *L. delbrueckii* subsp. *bulgaricus* DSM 20081 (ATCC 11842) were obtained from the German Collection of Microorganisms and Cell Cultures (DSMZ; Braunschweig, Germany). *L. casei* was obtained from the culture collection of the Food Biotechnology Laboratory, BOKU-University of Natural Resources and Life Sciences Vienna. *Lactobacillus* strains were cultivated on MRS medium (*Lactobacillus* broth according to De Man, Rogosa and Shape [43]) (Carl Roth, Karlsruhe, Germany) at 37 °C without agitation. *E. coli* NEB5α (New England Biolabs, Frankfurt am Main, Germany) was used as cloning hosts in the transformation of DNA fragments; whereas *E. coli* HST08 (Clontech, Mountain View, CA, USA) was used as the expression host strain. *E. coli* strains were cultivated in Luria-Bertani (LB) medium (10g/L tryptone, 10 g/L NaCl, and 5 g/L yeast extract) at 37 °C with shaking at 140 rpm. Agar media were prepared by adding 1.5% agar to the respective media. When needed, ampicillin was supplemented to media to a final concentration of 100 μg/mL for *E. coli* cultivations.

Table 4. Strains and plasmids used in the study.

Strains or Plasmids	Relevant Characteristics	Reference Source
Strains		
L. plantarum WCFS1		[32]
L. delbrueckii subsp. *bulgaricus* DSM 20081		DSMZ
L. casei		BOKU
L. helveticus DSM 20075		DSMZ
E. coli HST08	Host strain	Clontech
Plasmids		
pBAD_3014_AgESAT_DC	Amp^r; pBAD derivate with the LysM domain sequence from Lp3014 fused to the hybrid antigen AgESAT_DC	[44]
pBAD3014LacLMLreu	Amp^r; pBAD_3014_AgESAT_DC derivative with a fragment of *lacLM* genes instead of the gene fragment encoding AgESAT_DC	This study
pBAD3014LacZLbul	Amp^r; pBAD_3014_AgESAT_DC derivate with *lacZ* fragment instead of the gene fragment encoding AgESAT_DC	This study
pHA1032	Amp^r; pET21d derivative for expression of *lacLM* from *L. reuteri* in *E. coli*	[16]
pTH103	Erm^r; *spp*-based expression vector pSIP409 for expression of *lacZ* from *L. bulgaricus* DSM 20081 in *L. plantarum* WCFS1	[12]

4.2. Chemicals, Enzymes and Plasmids

All chemicals and enzymes were purchased from Sigma (St. Louis, MO, USA) unless stated otherwise and were of the highest quality available. All restriction enzymes, Phusion high-fidelity DNA polymerase, T4 DNA ligase, and corresponding buffers were from New England Biolabs (Frankfurt am Main, Germany). Staining dyes, DNA and protein standard ladders were from Bio-Rad (Hercules, CA, USA). All plasmids used in this study are listed in Table 4.

4.3. DNA Manipulation

Plasmids were isolated from *E. coli* strains using Monarch Plasmid Miniprep Kit (New England Biolabs, Frankfurt am Main, Germany) according to the manufacturer's instructions. PCR amplifications of DNA were done using Q5 High-Fidelity 2X Master Mix (New England Biolabs). The primers used in this study, which were supplied by VBC-Biotech Service (Vienna, Austria), are listed in Table 5. PCR products and DNA fragments obtained by digestion with restriction enzymes were purified using Monarch DNA Gel Extraction Kit (New England Biolabs); and the DNA amounts were estimated using Nanodrop 2000 (Thermo Fisher Scientific, Waltham, MA, USA). The sequences of PCR-generated fragments were verified by DNA sequencing performed by a commercial provider (Microsynth, Vienna, Austria). The ligation of DNA fragments was performed using NEBuilder HiFi Assembly Cloning Kit (New England Biolabs). All plasmids were transformed into *E. coli* NEB5α chemical competent cells following the manufacturer's protocol for obtaining the plasmids in sufficient amounts. The constructed plasmids (Table 4) were chemically transformed into expression host strain *E. coli* HST08.

Table 5. Primers used in the study.

Primer	Sequence* 5′→3′	Restriction Site Underlined
Fwd1LreuSalI	GAGTTCAACTGTCGACCAAGCAAATATAAA	SalI
Rev1LreuEcoRI	AGCCAAGCTTCGAATTCTTATTTTGCATTC	EcoRI
Fwd2LbulSalI	GTTCAACTGTCGACAGCAATAAGTTAGTAAAAGAAAAAAGAG	SalI
Rev2LbulEcoRI	CAGCCAAGCTTCGAATTCTTATTTTAGTAAAAGGGGCTGAATC	EcoRI

* The nucleotides in italics are the positions that anneal to the DNA of the target genes (*lacLM* or *lacZ*).

4.4. Plasmid Construction

Two recombinant fusion proteins were constructed. The first fusion protein was based on LacLM from *L. reuteri* and the LysM domain attached upstream of LacLM (termed LysM-LacLMLreu). The second fusion protein was based on LacZ from *L. delbrueckii* subsp. *bulgaricus* DSM 20081 and the LysM domain attached upstream of LacZ (termed LysM-LacZLbul). Plasmid pBAD_3014AgESAT_DC (Table 4) [44] (was used for the construction of the expression plasmids. This plasmid is a derivate of pBAD vector (Invitrogen, Carlsbad, CA, USA) containing a 7 × His tag sequence and a single LysM domain from Lp_3014, which is a putative extracellular transglycosylase with LysM peptidoglycan binding domain from *L. plantarum* WCFS1 (NCBI reference sequence no. NC_004567.2) [31,32], fused to the hybrid tuberculosis antigen AgESAT-DC [44]. The fragment of *lacLM* genes from *L. reuteri* was amplified from the plasmid pHA1032 (Table 4) [16] with the primer pair Fwd1LreuSalI and Rev1LreuEcoRI (Table 5), whereas the *lacZ* gene from *L. bulgaricus* was amplified from the plasmid pTH103 (Table 4) [12] with the primer pair Fwd2LbulSalI and Rev2LbulEcoRI (Table 5). The PCR-generated products were then cloned into *Sal*I and *Eco*RI cloning sites of the pBAD_3014AgESAT_DC vector using and NEBuilder HiFi DNA Assembly Cloning Kit (New England Biolabs) following the manufacturer's instructions, resulting in two expression plasmids pBAD3014LacLMLreu and pBAD3014LacZLbul (Figure 1).

4.5. Gene Expression in E. coli

The constructed plasmids pBAD3014LacLMLreu and pBAD3014LacZLbul were chemically transformed into expression host *E. coli* HST08. For gene expression, overnight cultures of *E. coli* HST08 were diluted in 300 mL of fresh LB broth containing 100 μg/mL ampicillin to an OD_{600} of ~0.1 and incubated at 37 °C with shaking at 140 rpm to an OD_{600} ~0.6. Gene expression was then induced by L-arabinose to a final concentration of 0.7 mg/mL and the cultures were incubated further at 25 °C for 18 h with shaking at 140 rpm. Cells were harvested at an OD_{600} of ~3.0 by centrifugation at 4000× *g* for 30 min at 4 °C, washed twice, and resuspended in 50 mM sodium phosphate buffer (NaPB), pH 6.5. Cells were disrupted by using a French press (AMINCO, Maryland, USA). Debris was removed by centrifugation (10,000× *g* for 15 min at 4 °C) to obtain the crude extract.

4.6. Immobilization of β-Galactosidases on Lactobacillus Cell Surface

One mL of *Lactobacillus* cultures were collected at OD_{600} ~4.0 by centrifugation (4000× *g* for 5 min at 4 °C) and the cells were washed with 50 mM sodium phosphate buffer (NaPB), pH 6.5. The cell pellets were then mixed with one mL of diluted cell-free crude extracts of 50 U_{oNPG}/mL (~5–6 mg protein/mL) of fused LysM-β-galactosidases (LysM-LacLMLreu or LysM-LacZLbul) and incubated at 37 °C for 24 h with gentle agitation. *Lactobacillus* β-galactosidase displaying cells were separated from the supernatants by centrifugation (4000× *g* for 5 min at 4 °C). Cells were then washed with NaPB (pH 6.5) two times; the supernatants and wash solutions were collected for SDS-PAGE analysis and activity and protein measurements. *Lactobacillus* β-galactosidase displaying cells were resuspended in NaPB (pH 6.5) for further studies.

4.7. Protein Determination

Protein concentrations were determined using the method of Bradford [45] with bovine serum albumin (BSA) as standard.

4.8. β-Galactosidase Assays

β-Galactosidase activity was determined using *o*-nitrophenyl-β-D-galactopyranoside (*o*NPG) or lactose as the substrates as previously described [5] with modifications. When chromogenic substrate *o*NPG was used, the reaction was started by adding 20 μL of *Lactobacillus* β-galactosidase displaying cell suspension to 480 μL of 22 mM *o*NPG in 50 mM NaBP (pH 6.5) and stopped by adding 750 μL of 0.4 M Na_2CO_3 after 10 min of incubation at 30 °C. The release of *o*-nitrophenol (*o*NP) was measured by determining the absorbance at 420 nm. One unit of *o*NPG activity was defined as the amount of β-galactosidase releasing 1 μmol of *o*NP per minute under the defined conditions.

When lactose was used as the substrate, 20 μL of *Lactobacillus* β-galactosidase displaying cell suspension was added to 480 μL of a 600 mM lactose solution in 50 mM sodium phosphate buffer, pH 6.5. After 10 min of incubation at 30 °C, the reaction was stopped by heating the reaction mixture at 99 °C for 5 min. The reaction mixture was cooled to room temperature, and the release of D-glucose was determined using the test kit from Megazyme. One unit of lactase activity was defined as the amount of enzyme releasing 1 μmol of D-glucose per minute under the given conditions.

4.9. Gel Electrophoresis Analysis

For visual observation of the expression level of the two recombinant β-galactosidases (LysM-LacLMLreu and LysM-LacZLbul) in *E. coli* and the effectiveness of the immobilization, cell-free extracts, supernatants, and wash solutions were analyzed by Sodium Dodecyl Sulfate Polyacrylamide Gel Electrophoresis (SDS-PAGE). Protein bands were visualized by staining with Bio-safe Coomassie (Bio-Rad). The determination of protein mass was carried out using Unstained Precision plus Protein Standard (Bio-Rad).

4.10. Western Blotting

Proteins in the cell-free extracts were separated by SDS-PAGE. Protein bands were then transferred to a nitrocellulose membrane using the Trans-Blot Turbo™ Transfer System (Biorad) following the manufacturer's instructions. Monoclonal mouse anti-His antibody (Penta His Antibody, BSA-free) was obtained from Qiagen (Hilden, Germany), diluted 1:5000 and used as recommended by the manufacturer. The protein bands were visualized by using polyclonal rabbit anti-mouse antibody conjugated with horseradish peroxidase (HRP) (Dako, Denmark) and the Clarity™ Western ECL Blotting Substrate from Bio-Rad (Hercules, CA, USA).

4.11. Flow Cytometry

Lactobacillus β-galactosidase displaying cells were resuspended in 50 μL of phosphate buffered saline (PBS) (137 mM NaCl, 2.7 mM KCl, 2 mM KH_2PO_4, and 10 mM Na_2HPO_4, pH 7.4) containing 2% of BSA (PBS-B) and 0.1 μL of Penta His Antibody, BSA-free (Qiagen; diluted 1:500 in PBS-B). After incubation at RT for 40 min, the cells were centrifuged at 4000× *g* for 5 min at 4 °C and washed three times with 500 μL PBS-B. The cells were subsequently incubated with 50 μL PBS-B and 0.1 μL anti-mouse IgG H&L/Alexa Flour 488 conjugate (Cell Signaling Technology, Frankfurt am Main, Germany, diluted 1:750 in PBS-B) for 40 min in the dark at room temperature. After washing five times with 500 μL PBS-B, the stained cells were analyzed by flow cytometry using a CytoFLEX Flow Cytometer (Beckman Coulter, Brea, CA, USA) following the manufacturer's instructions.

4.12. Temperature Stability and Reusability of Immobilized Enzymes

The temperature stability of immobilized enzymes was studied by incubating *L. plantarum* LysM-LacLMLreu- and LysM-LacZLbul-displaying cells in 50 mM NaPB (pH 6.5) at various temperatures (−20, 4, 30, 50 °C). At certain time intervals, samples were withdrawn, the residual activity was measured using *o*NPG as the substrate under standard assay conditions and the $\tau_{1/2}$ value was determined.

To test the reusability of immobilized enzymes, several repeated rounds of lactose conversion at 30 °C using LysM-LacLMLreu- and LysM-LacZLbul-displaying cells and at 50 °C using LysM-LacZLbul-displaying cells were carried out with 600 mM initial lactose in 50mM NaBP (pH 6.5) and constant agitation (500 rpm). The enzyme activity during these repeated cycles with intermediate two washing steps was measured using *o*NPG as the substrate under standard assay conditions.

4.13. Lactose Conversion and Formation of Galacto-Oligosaccharides (GOS)

The conversion of lactose was carried out in discontinuous mode using *L. plantarum* cells displaying β-galactosidase LacZ from *L. bulgaricus* (LysM-LacZLbul). The conversion was performed at 30 °C using 205 g/L initial lactose concentration in 50 mM NaPB (pH 6.5) and constant agitation (500 rpm). *L. plantarum* LysM-LacZLbul displaying cells were added to equivalent concentrations of 1.0 U_{Lac}/mL of reaction mixture. Samples were withdrawn at intervals, heated at 99 °C for 5 min and further analyzed for lactose, galactose, glucose and GOS present in the samples.

4.14. Analysis of Carbohydrate Composition

The carbohydrate composition in the reaction mixture was analyzed by high-performance liquid chromatography (HPLC) equipped with a Dionex ICS-5000+ system (Thermo Fisher Scientific) consisting of an ICS-5000+ dual pump (DP) and an electrochemical detector (ED). Separations were performed at room temperature on CarboPac PA-1 column (4 × 250 mm) connected to a CarboPac PA-1 guard column (4 × 50 mm) (Thermo Fisher Scientific) with flow rate 1 mL/min. All eluents A (150 mM NaOH), B (150 mM NaOH and 500 mM sodium acetate) and C (deionized water) were degassed by flushing with helium for 30 min. Separation of D-glucose, D-galactose, lactose and allolactose was carried out with a run with the following gradient: 90% C with 10% A for 45 min at 1.0 mL/min, followed by 5 min with 100% B. The concentration of saccharides was calculated by interpolation from external standards. Total GOS concentration was calculated by subtraction of the quantified saccharides (lactose, glucose, galactose) from the initial lactose concentration. The GOS yield (%) was defined as the percentage of GOS produced in the samples compared to initial lactose.

4.15. Statistical Analysis

All experiments and measurements were conducted at least in duplicate, and the standard deviation (SD) was always less than 5%. The data are expressed as the mean ± SD when appropriate.

5. Conclusions

This work describes the immobilization of two lactobacillal β-galactosidases, a β-galactosidase from *L. reuteri* of the heterodimeric LacLM-type and one from *L. bulgaricus* of the homodimeric LacZ-type, on the *Lactobacillus* cell surface using a peptidoglycan-binding motif as an anchor, in this case, the single LysM domain Lp_3014 from *L. plantarum* WCFS1. The immobilized fusion LysM-β-galactosidases are catalytically efficient and can be reused for several repeated rounds of lactose conversion. Surface anchoring of β-galactosidases in *Lactobacillus* results in safe, non-GMO and stable biocatalysts that can be used in the applications for lactose conversion and production of prebiotic galacto-oligosaccharides.

Author Contributions: Conceptualization, G.M. and T.-H.N.; Data Curation, M.-L.P., G.M. and T.-H.N.; Investigation, M.-L.P., A.-M.T., S.K. and T.-T.N.; Methodology, M.-L.P. and T.-H.N.; Supervision, T.-H.N.; Writing-Original Draft Preparation, M.-L.P.; Writing-Review & Editing, T.-H.N.

References

1. Nakayama, T.; Amachi, T. β-Galactosidase, enzymology. In *Encyclopedia of Bioprocess Technology, Fermentation, Biocatalysis, and Bioseparation*; Flickinger, M.C., Drew, S.W., Eds.; John Willey and Sons: New York, NY, USA, 1999; Volume 3, pp. 1291–1305.
2. Pivarnik, L.F.; Senegal, A.G.; Rand, A.G. Hydrolytic and transgalactosylic activities of commercial β-galactosidase (lactase) in food processing. *Adv. Food Nutr. Res.* **1995**, *38*, 1–102.
3. Prenosil, J.E.; Stuker, E.; Bourne, J.R. Formation of oligosaccharises during enzymatic lactose hydrolysis: Part I: State of art. *Biotechnol. Bioeng.* **1987**, *30*, 1019–1025. [CrossRef] [PubMed]
4. Petzelbauer, I.; Zeleny, R.; Reiter, A.; Kulbe, K.D.; Nidetzky, B. Development of an ultra-high-temperature process for the enzymatic hydrolysis of lactose: II. Oligosaccharide formation by two thermostable β-glycosidases. *Biotechnol. Bioeng.* **2000**, *69*, 140–149. [CrossRef]
5. Nguyen, T.H.; Splechtna, B.; Steinböck, M.; Kneifel, W.; Lettner, H.P.; Kulbe, K.D.; Haltrich, D. Purification and characterization of two novel β-galactosidases from *Lactobacillus reuteri*. *J. Agric. Food Chem.* **2006**, *54*, 4989–4998. [CrossRef]
6. Nguyen, T.H.; Splechtna, B.; Krasteva, S.; Kneifel, W.; Kulbe, K.D.; Divne, C.; Haltrich, D. Characterization and molecular cloning of a heterodimeric β-galactosidase from the probiotic strain *Lactobacillus acidophilus* R22. *FEMS Microbiol. Lett.* **2007**, *269*, 136–144. [CrossRef] [PubMed]
7. Kittibunchakul, S.; Pham, M.-L.; Tran, A.-M.; Nguyen, T.-H. β-Galactosidase from *Lactobacillus helveticus* DSM 20075: Biochemical characterization and recombinant expression for applications in dairy industry. *Int. J. Mol. Sci.* **2019**, *20*, 947. [CrossRef] [PubMed]
8. Iqbal, S.; Nguyen, T.H.; Nguyen, T.T.; Maischberger, T.; Haltrich, D. β-galactosidase from *Lactobacillus plantarum* WCFS1: Biochemical characterization and formation of prebiotic galacto-oligosaccharides. *Carbohydr. Res.* **2010**, *345*, 1408–1416. [CrossRef]
9. Gobinath, D.; Prapulla, S.G. Permeabilized probiotic *Lactobacillus plantarum* as a source of β-galactosidase for the synthesis of prebiotic galactooligosaccharides. *Biotechnol. Lett.* **2013**, *36*, 153–157. [CrossRef]
10. Iqbal, S.; Nguyen, T.H.; Nguyen, H.A.; Nguyen, T.T.; Maischberger, T.; Kittl, R.; Haltrich, D. Characterization of a heterodimeric GH2 β-galactosidase from *Lactobacillus sakei* Lb790 and formation of prebiotic galacto-oligosaccharides. *J. Agric. Food Chem.* **2011**, *59*, 3803–3811. [CrossRef] [PubMed]
11. Maischberger, T.; Leitner, E.; Nitisinprasert, S.; Juajun, O.; Yamabhai, M.; Nguyen, T.H.; Haltrich, D. β-galactosidase from *Lactobacillus pentosus*: Purification, characterization and formation of galacto-oligosaccharides. *Biotechnol. J.* **2010**, *5*, 838–847. [CrossRef]
12. Nguyen, T.T.; Nguyen, H.A.; Arreola, S.L.; Mlynek, G.; Djinović-Carugo, K.; Mathiesen, G.; Nguyen, T.H.; Haltrich, D. Homodimeric β-galactosidase from *Lactobacillus delbrueckii* subsp. *bulgaricus* DSM 20081: Expression in *Lactobacillus plantarum* and biochemical characterization. *J. Agric. Food Chem.* **2012**, *60*, 1713–1721. [CrossRef]
13. Black, B.A.; Lee, V.S.Y.; Zhao, Y.Y.; Hu, Y.; Curtis, J.M.; Ganzle, M.G. Structural identification of novel oligosaccharides produced by *Lactobacillus bulgaricus* and *Lactobacillus plantarum*. *J. Agric. Food Chem.* **2012**, *60*, 4886–4894. [CrossRef]
14. Liu, G.X.; Kong, J.; Lu, W.W.; Kong, W.T.; Tian, H.; Tian, X.Y.; Huo, G.C. Beta-galactosidase with transgalactosylation activity from *Lactobacillus fermentum* K4. *J. Dairy Sci.* **2011**, *94*, 5811–5820. [CrossRef]
15. Nie, C.; Liu, B.; Zhang, Y.; Zhao, G.; Fan, X.; Ning, X.; Zhang, W. Production and secretion of *Lactobacillus crispatus* β-galactosidase in *Pichia pastoris*. *Protein Expr. Purif.* **2013**, *92*, 88–93. [CrossRef] [PubMed]
16. Nguyen, T.H.; Splechtna, B.; Yamabhai, M.; Haltrich, D.; Peterbauer, C. Cloning and expression of the β-galactosidase genes from *Lactobacillus reuteri* in *Escherichia coli*. *J. Biotechnol.* **2007**, *129*, 581–591. [CrossRef]
17. Pham, M.L.; Leister, T.; Nguyen, H.A.; Do, B.C.; Pham, A.T.; Haltrich, D.; Yamabhai, M.; Nguyen, T.H.; Nguyen, T.T. Immobilization of β-galactosidases from *Lactobacillus* on chitin using a chitin-binding domain. *J. Agric. Food Chem.* **2017**, *65*, 2965–2976. [CrossRef]
18. Michon, C.; Langella, P.; Eijsink, V.G.; Mathiesen, G.; Chatel, J.M. Display of recombinant proteins at the surface of lactic acid bacteria: Strategies and applications. *Microb. Cell Fact.* **2016**, *15*, 70. [CrossRef]
19. Nguyen, H.M.; Mathiesen, G.; Stelzer, E.M.; Pham, M.L.; Kuczkowska, K.; Mackenzie, A.; Agger, J.W.; Eijsink, V.G.; Yamabhai, M.; Peterbauer, C.K.; et al. Display of a β-mannanase and a chitosanase on the cell

surface of *Lactobacillus plantarum* towards the development of whole-cell biocatalysts. *Microb. Cell Fact.* **2016**, *15*, 169. [CrossRef] [PubMed]

20. Schneewind, O.; Missiakas, D.M. Protein secretion and surface display in Gram-positive bacteria. *Philos. Trans. R. Soc. B Biol. Sci.* **2012**, *367*, 1123–1139. [CrossRef] [PubMed]
21. Leenhouts, K.; Buist, G.; Kok, J. Anchoring of proteins to lactic acid bacteria. *Antonie van Leeuwenhoek* **1999**, *76*, 367–376. [PubMed]
22. Proft, T. Sortase-mediated protein ligation: An emerging biotechnology tool for protein modification and immobilisation. *Biotechnol. Lett.* **2009**, *32*, 1–10.
23. Diep, D.B.; Mathiesen, G.; Eijsink, V.G.H.; Nes, I.F. Use of lactobacilli and their pheromone-based regulatory mechanism in gene expression and drug delivery. *Curr. Pharm. Biotechnol.* **2009**, *10*, 62–73. [CrossRef]
24. Boekhorst, J.; De Been, M.W.H.J.; Kleerebezem, M.; Siezen, R.J. Genome-wide detection and analysis of cell wall-bound proteins with LPxTG-like sorting motifs. *J. Bacteriol.* **2005**, *187*, 4928–4934. [CrossRef] [PubMed]
25. Marraffini, L.A.; Dedent, A.C.; Schneewind, O. Sortases and the art of anchoring proteins to the envelopes of gram-positive bacteria. *Microbiol. Mol. Biol. Rev.* **2006**, *70*, 192–221. [CrossRef] [PubMed]
26. Visweswaran, G.R.; Leenhouts, K.; van Roosmalen, M.; Kok, J.; Buist, G. Exploiting the peptidoglycan-binding motif, LysM, for medical and industrial applications. *Appl. Microbiol. Biotechnol.* **2014**, *98*, 4331–4345. [CrossRef]
27. Turner, M.S.; Hafner, L.M.; Walsh, T.; Giffard, P.M. Identification and characterization of the novel LysM domain-containing surface protein Sep from *Lactobacillus fermentum* BR11 and its use as a peptide fusion partner in *Lactobacillus* and *Lactococcus*. *Appl. Environ. Microbiol.* **2004**, *70*, 3673–3680. [CrossRef] [PubMed]
28. Raha, A.R.; Varma, N.R.S.; Yusoff, K.; Ross, E.; Foo, H.L. Cell surface display system for *Lactococcus lactis*: A novel development for oral vaccine. *Appl. Microbiol. Biotechnol.* **2005**, *68*, 75–81. [CrossRef]
29. Steen, A.; Buist, G.; Leenhouts, K.J.; El Khattabi, M.; Grijpstra, F.; Zomer, A.L.; Venema, G.; Kuipers, O.P.; Kok, J. Cell wall attachment of a widely distributed peptidoglycan binding domain is hindered by cell wall constituents. *J. Biol. Chem.* **2003**, *278*, 23874–23881. [CrossRef]
30. Okano, K.; Zhang, Q.; Kimura, S.; Narita, J.; Tanaka, T.; Fukuda, H.; Kondo, A. System using tandem repeats of the cA peptidoglycan-binding domain from *Lactococcus lactis* for display of both N-and C-terminal fusions on cell surfaces of lactic acid bacteria. *Appl. Environ. Microbiol.* **2008**, *74*, 1117–1123. [CrossRef] [PubMed]
31. Boekhorst, J.; Wels, M.; Kleeberezem, M.; Siezen, R.J. The predicted secretome of *Lactobacillus plantarum* WCFS1 sheds light on interactions with its environment. *Microbiology* **2006**, *152*, 3175–3183. [CrossRef] [PubMed]
32. Kleerebezem, M.; Boekhorst, J.; van Kranenburg, R.; Molenaar, D.; Kuipers, O.P.; Leer, R.; Tarchini, R.; Peters, S.A.; Sandbrink, H.M.; Fiers, M.W.; et al. Complete genome sequence of *Lactobacillus plantarum* WCFS1. *Proc. Natl. Acad. Sci. USA* **2003**, *100*, 1990–1995. [CrossRef] [PubMed]
33. Mesnage, S.; Tosi-Couture, E.; Fouet, A. Production and cell surface anchoring of functional fusions between the SLH motifs of the *Bacillus anthrasis* S-layer proteins and the *Bacillus subtilis* levansucrase. *Mol. Microbiol.* **1999**, *31*, 927–936. [CrossRef]
34. Mesnage, S.; Weber-Levy, M.; Haustant, M.; Mock, M.; Fouet, A. Cell surface-exposed tetanus toxin fragment C produced by recombinant *Bacillus anthracis* protects against tetanus toxin. *Infect. Immun.* **1999**, *67*, 4847–4850.
35. Bosma, T.; Kanninga, R.; Neef, J.; Audouy, S.A.; van Roosmalen, M.L.; Steen, A.; Buist, G.; Kok, J.; Kuipers, O.P.; Robillard, G.; et al. Novel surface display system for proteins on non-genetically modified gram-positive bacteria. *Appl. Environ. Microbiol.* **2006**, *72*, 880–889. [CrossRef] [PubMed]
36. Fredriksen, L.; Kleiveland, C.R.; Olsen Hult, L.T.; Lea, T.; Nygaard, C.S.; Eijsink, V.G.H.; Mathiesen, G. Surface display of N-terminally anchored invasin by *Lactobacillus plantarum* activates NF-κB in monocytes. *Appl. Environ. Microbiol.* **2012**, *78*, 5864–5871. [CrossRef] [PubMed]
37. Kuczkowska, K.; Mathiesen, G.; Eijsink, V.G.H.; Øynebråten, I. *Lactobacillus plantarum* displaying CCL3 chemokine in fusion with HIV-1 Gag derived antigen causes increased recruitment of T cells. *Microb. Cell Fact.* **2015**, *14*, 1. [CrossRef] [PubMed]
38. Desvaux, M.; Dumas, E.; Chafsey, I.; Hebraud, M. Protein cell surface display in Gram-positive bacteria: From single protein to macromolecular protein structure. *FEMS Microbiol. Lett.* **2006**, *256*, 1–15. [CrossRef]
39. Joris, B.; Englebert, S.; Chu, C.-P.; Kariyama, R.; Daneo-Moore, L.; Shockman, G.D.; Ghuysen, J.-M. Modular design of the *Enterococcus hirae* muramidase-2 and *Streptococcus faecalis* autolysin. *FEMS Microbiol. Lett.* **1992**, *91*, 257–264. [CrossRef]

40. Buist, G.; Steen, A.; Kok, J.; Kuipers, O.P. LysM, a widely distributed protein motif for binding to (peptido) glycans. *Mol. Microbiol.* **2008**, *68*, 838–847. [CrossRef] [PubMed]
41. Mesnage, S.; Dellarole, M.; Baxter, N.J.; Rouget, J.B.; Dimitrov, J.D.; Wang, N.; Fujimoto, Y.; Hounslow, A.M.; Lacroix-Desmazes, S.; Fukase, K.; et al. Molecular basis for bacterial peptidoglycan recognition by LysM domains. *Nat. Commun.* **2014**, *5*, 4269. [CrossRef] [PubMed]
42. Xu, W.; Huang, M.; Zhang, Y.; Yi, X.; Dong, W.; Gao, X.; Jia, C. Novel surface display system for heterogonous proteins on *Lactobacillus plantarum*. *Lett. Appl. Microbiol.* **2011**, *53*, 641–648. [CrossRef] [PubMed]
43. De Man, J.C.; Rogosa, M.; Sharpe, M.E. A medium for the cultivation of lactobacilli. *J. Appl. Bacteriol.* **1960**, *23*, 130–135. [CrossRef]
44. Målbakken, N. Development of a Non-GMO Tuberculosis Vaccine, Using *Lactobacillus* as a Delivery Vehicle. Master's Thesis, Norwegian University of Life Sciences, Ås, Norway, 2014.
45. Bradford, M.M. A rapid and sensitive method for the quantitation of microgram quantities of protein utilizing the principle of protein-dye binding. *Anal. Biochem.* **1976**, *72*, 248–254. [CrossRef]

Deciphering the Role of V88L Substitution in NDM-24 Metallo-β-Lactamase

Zhihai Liu [1,2], Alessandra Piccirilli [3], Dejun Liu [2], Wan Li [2], Yang Wang [2] and Jianzhong Shen [2,*]

1. Agricultural Bio-pharmaceutical Laboratory, College of Chemistry and Pharmaceutical Sciences, Qingdao Agricultural University, Qingdao 266109, China
2. Beijing Advanced Innovation Center for Food Nutrition and Human Health, College of Veterinary Medicine, China Agricultural University, Beijing 100193, China
3. Dipartimento di Scienze Cliniche Applicate e Biotecnologiche, Università degli Studi dell'Aquila, 67100 L'Aquila, Italy

* Correspondence: sjz@cau.edu.cn

Abstract: The New Delhi metallo-β-lactamase-1 (NDM-1) is a typical carbapenemase and plays a crucial role in antibiotic-resistance bacterial infection. Phylogenetic analysis, performed on known NDM-variants, classified NDM enzymes in seven clusters. Three of them include a major number of NDM-variants. In this study, we evaluated the role of the V88L substitution in NDM-24 by kinetical and structural analysis. Functional results showed that V88L did not significantly increase the resistance level in the NDM-24 transformant toward penicillins, cephalosporins, meropenem, and imipenem. Concerning ertapenem, *E. coli* DH5α/NDM-24 showed a MIC value 4-fold higher than that of *E. coli* DH5α/NDM-1. The determination of the k_{cat}, K_m, and k_{cat}/K_m values for NDM-24, compared with NDM-1 and NDM-5, demonstrated an increase of the substrate hydrolysis compared to all the β-lactams tested, except penicillins. The thermostability testing revealed that V88L generated a destabilized effect on NDM-24. The V88L substitution occurred in the β-strand and low β-sheet content in the secondary structure, as evidenced by the CD analysis data. In conclusion, the V88L substitution increases the enzyme activity and decreases the protein stability. This study characterizes the role of the V88L substitution in NDM-24 and provides insight about the NDM variants evolution.

Keywords: New Delhi metallo-β-lactamase; NDM-24; kinetic profile; secondary structure

---

1. Introduction

Metallo-β-lactamases (MBLs) are a group of enzymes that confer high resistance to most β-lactams. The active site of these enzymes contains one or two zinc ions, that are crucial for catalytic mechanism [1]. Based on their amino acid sequences, MBLs have been divided into subclasses B1, B2, and B3 [2]. Among subclass B1, the New Delhi metallo-β-lactamase (NDM-1) is one of the most widespread carbapenemase. NDM-1 was first identified in 2008 in a clinical strain of *Klebsiella pneumoniae* [3]. NDM-1 producing bacteria can hydrolyse all β-lactams (except monobactams), including carbapenems, the "last resort" antibiotics used in clinical therapy. NDM-1 genes are located on plasmids that mediate their dissemination across different bacterial strains [4,5]. However, the clinical success of NDM is also due to the fact that it is a lipoprotein anchored to the outer membrane, resulting in an unusual stability of NDM-1 and enabling secretion, in Gram-negative bacteria [6–8].

To date, more than 26 variants differing by a limited number of substitutions have been identified [9]. Previous studies revealed that these substitutions have contributed to NDM-1 to increase the hydrolytic activity toward several β-lactams resulting in an increment of resistance in the host bacteria [10]. Crystal structures showed that NDM-1 presents the typical αβ/βα fold of MBLs [11,12]. In this enzyme, the zinc ions are coordinated by six conserved residues: H120, H122, and H189 for Zn1

(BBL numbering) and D124, C208, and H250 for Zn2 (BBL numbering). The active site is surrounded by Loop 3 (residues 67-73) and Loop 10 (residues 210-230), involved in the substrate accommodation [12]. The most frequent substitution in NDM-1 is M154L, found in 11 NDM variants (NDM-4, -5, -7, -8, -12, -13, -15, -17, -19, -20 and -21) [9,13–16]. Indeed, V88L has been reported in five NDM variants (NDM-5, -17, -20, -21 and -24). Other frequent substitutions are A233V (NDM-6, -15, -19 and -27), D130G (NDM-8 and -14), D130N (NDM-7 and -19), and D95N (NDM-3 and -27) [10]. The single substitutions of M154L and D130G seem to increase the carbapenemase activity in NDM-4 and NDM-14, respectively [17,18]. Moreover, the combination D130G/M154L (NDM-8), reduces the hydrolysis toward carbapenems [19]. The main goal of the study was to evaluate the role of the V88L substitution in the NDM-24 enzyme. The NDM-24 was generated in the laboratory by a site-directed mutagenesis of NDM-1 and the enzyme properties, protein structure, and thermal stability were studied compared with NDM-1 and NDM-5.

2. Results and Discussion

2.1. Phylogenetic Analysis

A phylogenetic analysis of NDM-1 variants was performed in order to classify these enzymes based on their amino acid similarities. Overall, the NDM variants were classified into three major clusters (NDM-1, NDM-4, and NDM-24), two minor clusters (NDM-3 and NDM-6), and two divergent proteins (NDM-14 and NDM-10). As shown in Figure 1, the NDM variants are well categorized. The NDM-1 cluster includes eight variants that showed only one amino acid replacement, except for NDM-18 where an insertion of five amino acids have been found (position 48-52). In the NDM-4 group, all variants possess the replacement at position 154. In particular, except for NDM-11 containing the M154V substitution, all variants shared M154L. In the NDM-24 group, Valine at position 88 has been replaced by a Leucine (V88L). Concerning the two minor groups, similar characteristics were observed with the D95N and A233V substitution for the NDM-3 and NDM-6 clusters, respectively.

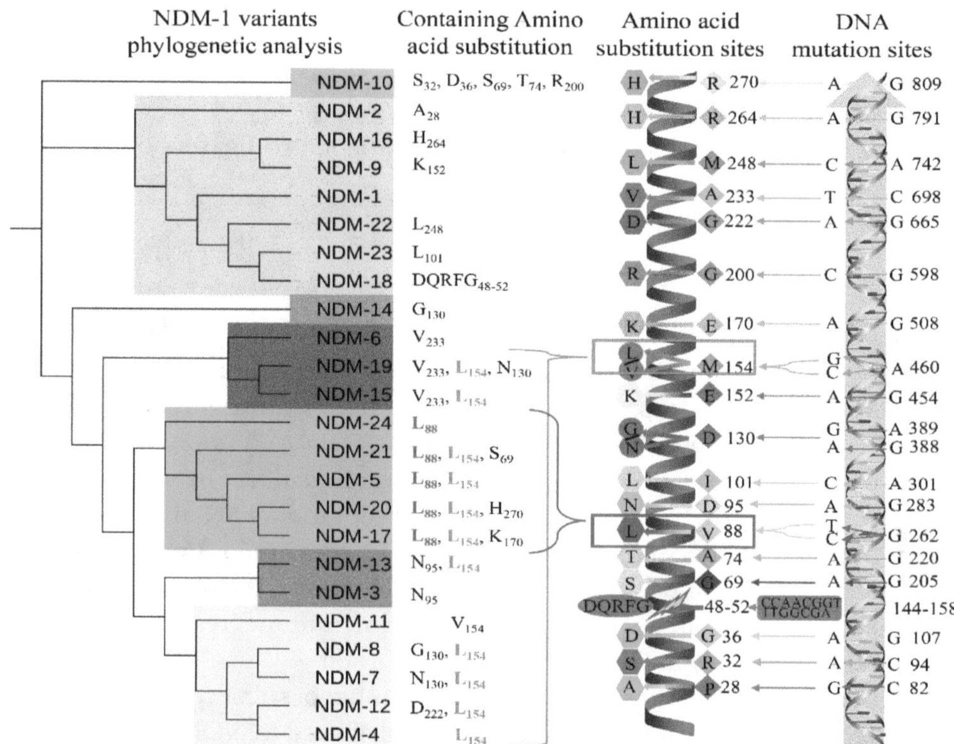

Figure 1. New Delhi metallo-β-lactamase-1 (NDM-1) variants phylogenetic analysis. Phylogenetic groups were differently coloured: For example, the NDM-24 cluster is coloured in green.

2.2. Functional Study

The NDM-24 variant was obtained by a site-directed mutagenesis by using the NDM-1 as template. All genes (*bla*$_{NDM-1}$, *bla*$_{NDM-5}$, and *bla*$_{NDM-24}$) were cloned into pHSG398, which were controlled by the same promoter and thus the same expression. The *E. coli* DH5α recombinant strains were used to test susceptibility to a wide panel of β-lactams. As shown in Table 1, the results of the susceptibility test revealed that NDM-1, NDM-5, and NDM-24 confer resistance to most β-lactams with similar MIC values, suggesting that the NDM enzymes were successfully expressed in the host cells. A different behaviour was observed for carbapenems, for which the MIC values were markedly lower than those of penicillins and cephalosporins with the exception of cefepime, as previously reported [15–20]. Concerning ertapenem, NDM-24 and NDM-5 showed an increase of the MIC values of the 4- and 8-fold with respect to NDM-1. Based on the data obtained, the V88L substitution enhances resistance toward ertapenem.

Table 1. Antimicrobial susceptibility of *E. coli* BL21(DE3)-DH5α carrying *bla*$_{NDM-1}$, *bla*$_{NDM-5}$, and *bla*$_{NDM-24}$.

Antibiotic	MIC (μg/mL)			
	E. coli DH5α/pHSG398	*E. coli* DH5α/pHSG398 -NDM-24	*E. coli* DH5α/pHSG398 -NDM-1	*E. coli* DH5α/pHSG398 -NDM-5
Ampicillin	2	>256	>256	>256
Penicillin G	16	>256	>256	>256
Aztreonam	0.031	0.031	0.031	0.031
Cefepime	0.031	2	1	2
Cefotaxime	0.062	32	64	64
Cefoxitin	2	128	128	128
Ceftazidime	0.125	256	256	256
Cefazolin	2	128	128	256
Ertapenem	0.015	1	0.25	2
Imipenem	0.062	2	2	2
Meropenem	0.031	1	1	2

2.3. Characteristics of Enzyme Activity

In order to obtain soluble and active enzymes, the recombinant plasmids were successfully expressed in the *E. coli* BL21 (DE3) cells as described in the methods section. After purification, the enzymes were checked on SDS-PAGE to confirm the solubility and purity (>90%) (Figure S1). The MALDI-TOF mass spectrometry was used to confirm the molecular mass of the three enzymes, which corresponds to 24884,024 Da (Figure S2). To investigate the enzyme activity, the kinetic parameters for NDM-1, NDM-5, and NDM-24 were determined (Table 2).

All the NDM variants of this study were able to hydrolyse all the β-lactams tested. Compared with NDM-1, NDM-24 showed lower K_m values for penicillins and ceftazidime whereas for carbapenems they are quite similar. Comparing the k_{cat} values, NDM-24 hydrolyses all β-lactams, except penicillins, better than NDM-1 and NDM-5. In particular, the k_{cat} values of NDM-24 are 2.26-, 1.61-, 2.73-, 2.02-, 2.17-, and 1.75-fold higher than NDM-1 towards penicillin G, ceftazidime, cefepime, imipenem, meropenem, and ertapenem, respectively. This was also confirmed by a slight increase of catalytic efficiency. This result stated the important role of V88L in the substrate hydrolysis. The contribution of V88L is likely that of M154L as demonstrated by the calculation of the k_{cat}/K_m rates (Table 2). This was possibly due to differences in the intrinsic properties, such as the enzyme stability, protein expression, and adaptability [21–24], and nutritional conditions of bacteria in vivo/vitro. Comparing the k_{cat}/K_m values of carbapenems, the carbapenemase activity of NDM-5 was similar to NDM-24, but higher than NDM-1. A recent study showed that an increase of the catalytic efficiency (k_{cat}/K_m) for meropenem has been ascertained in NDM-5 (V88L and M154L). In NDM-4, which contains only M154L, no significant change has been observed, suggesting that V88L might play a role in enhancing the NDM enzymes activity rather than M154L. Moreover, an increase of the carbapenemase activity was also observed in the evolutionary NDM variants, such as NDM-17 (V88L, M154L, and E170K) and NDM-20 (V88L, M154L, and R270H) [10,15,16].

Table 2. Kinetic parameters of NDM-1, NDM-5, and NDM-24 toward β-lactams *a*.

Kinetic Parameters	Enzyme	β-Lactams [b]						
		AMP	PEN	TAG	FEP	MEM	IPM	ETP
K_m(μM)	NDM-24	638.79 ± 23.86	331.30 ± 29.43	173.85 ± 9.73	318.93 ± 10.86	266.24 ± 27.03	338.20 ± 24.23	125.23 ± 19.08
	NDM-1	1249.98 ± 210.94	224.57 ± 13.57	213.90 ± 11.01	173.55 ± 19.46	284.24 ± 7.87	234.83 ± 7.44	105.54 ± 3.09
	NDM-5	825.00 ± 0.29	315.21 ± 46.68	76.45 ± 4.76	179.64 ± 12.19	275.16 ± 36.87	292.97 ± 13.76	82.18 ± 3.86
k_{cat} (s^{-1})	NDM-24	259.94 ± 23.52	179.10 ± 8.17	43.13 ± 1.06	22.98 ± 0.34	151.75 ± 6.69	173.16 ± 8.83	110.31 ± 7.62
	NDM-1	254.34 ± 28.96	79.28 ± 1.96	26.73 ± 0.71	8.42 ± 0.63	75.18 ± 3.44	79.81 ± 5.15	62.89 ± 1.15
	NDM-5	346.13 ± 31.30	214.13 ± 12.11	26.96 ± 0.75	13.05 ± 0.24	142.48 ± 17.91	149.63 ± 2.02	83.18 ± 1.67
k_{cat}/K_m ($\mu M^{-1}\ s^{-1}$)	NDM-24	0.41	0.54	0.25	0.072	0.57	0.51	0.88
	NDM-1	0.20	0.35	0.13	0.046	0.26	0.34	0.60
	NDM-5	0.40	0.68	0.35	0.073	0.52	0.51	1.01
k_{cat}/K_m ($\mu M^{-1}\ s^{-1}$) ratio for:	NDM-24/NDM-1	2.00	1.53	1.98	1.49	2.16	1.51	1.46
	NDM-5/NDM-24	1.03	1.26	1.42	1.01	0.91	1.00	1.15
	NDM-5/NDM-1	2.07	1.92	2.82	1.50	1.96	1.50	1.68

[a] The proteins were initially purified with a His-tag, which was removed after purification. Each kinetic value is the mean of three different measurements; the error was below 5%. [b] β-lactams: AMP, ampicillin; TAG, ceftazidime; PEN, penicillin G; FEP, cefepime; IPM, imipenem; MEM, meropenem; ETP, ertapenem.

2.4. Thermal Stability

As previously reported, mutations in the NDM variants can affect the enzymes stability, resulting in changing the persistence lifetime in the bacterial host, and consequent antibiotics resistance [25]. For determining whether the V88L substitution influences the NDM-24 stability property, circular dichroism CD was used to assay the protein stability by recording signal changes. NDM-5 was used as reference to analyze the effect of M154L. Compared with NDM-1 and NDM-5, NDM-24 (59.41 ± 0.06 °C) possessed the lowest melting temperature (Figure 2). Notably, the V88L destabilized effect was compensated by M154L in NDM-5 with a remarkable higher thermal temperature than NDM-24 (69.13 ± 3.6 °C compared to 59.41 ± 0.06 °C). Moreover, NDM-5 showed a higher stability than NDM-1 suggesting the destabilized role of M154L. This was in agreement with a previous document that the M154L mutation would be a turning point for the NDM variants, in which combing M154L with additional substitutions benefit for the NDM enzymes exhibiting increased thermostability [10]. In the NDM-24 group there are four variants (NDM-5, -17, -20, and -21) in which the combination of the V88L and M154L substitutions takes favorable results in terms of the stability and environmental selection.

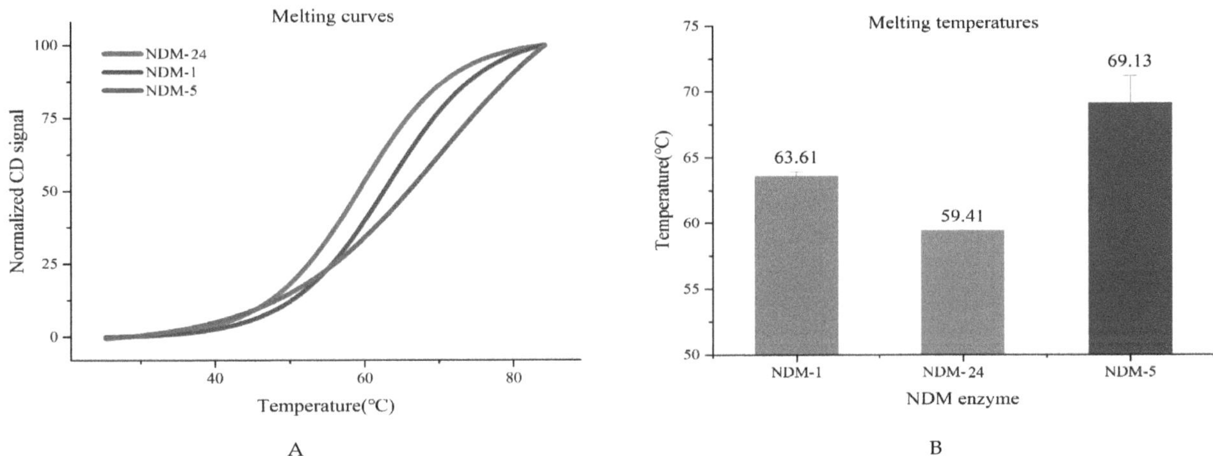

Figure 2. (**A**) Thermal stability melting curves. (**B**) Melting temperatures of the NDM enzymes as determined by the circular dichroism analysis: NDM-1, 63.61 ± 0.57 °C; NDM-V88L, 59.41 ± 0.06 °C; and NDM-5, 69.13 ± 3.6 °C. Data are the means of triplicate experiments, with error bars showing the standard deviation (±SD).

2.5. Structure Analysis

Previous reports indicated that mutations in the NDM influence the α-helical, β-sheets content, and loop flexibility [26]. For example, the Q123A substitution in NDM-1 leads to a decrease of the α-helical content [27]. To know if the V88L substitution could modify the NDM-24 structure, a secondary structure was determined by the Far-UV CD spectrum. All NDM variants CD spectrum data were fitted and shown in Figure 3. The spectrum signals were superimposable at most wavelengths, and showed characteristics of αβ/βα fold, a typical and conservative protein structure in MBLs [28]. The presence of positive bands at 192 nm and two negative peaks at 208 nm, a minimum peak, and 220 nm, suggesting the dominance of the β-sheets and α-helical content. The major differences were observed in the nearby 192 nm, symbolizing α-helical peak, and 208–220 nm, a α-helical and β-sheets bonds. Overall, the α-helical content was found ranging between 13%–20% in NDM-1, NDM-5, and NDM-24 (Table 3), in agreement with previous reports and the content of the β-sheet was high around 30% [27]. Compared to NDM-1, NDM-24 possesses a higher α-helical content and lower β-sheet content, suggesting that V88L was responsible for the secondary structure content changes of NDM-24. Furthermore, the secondary predicted result (Figure S3) confirmed that the V88L substitution occurred in the β-strand terminal, which may be prematurely terminated, leading to a decrease in the β-sheet content. Kumar et al. claimed that 152A, located in the β-strand, drastically influenced the NDM-5 activity and protein thermolability, by reducing the β-sheet content [26]. Our analysis demonstrates that the emergence of M154L (in NDM-5) caused the α-helical to decrease and the β-sheet to increase relative to NDM-24, while the α-helical and β-sheet content of NDM-5 were between NDM-1 and NDM-24. In addition, the 3D model of NDM-24 (Figure 4) was generated by using NDM-1 (PDB accession: 5N0H, 4RBS, 4HKY, and 4EYL) and NDM-5 (PDB accession: 6MGY, and 4TZE) as a template. Although the residue 88L is away from the active site groove and far to the active loops (Loop 3 and Loop 10), differences in the structure content, stability, and enzyme activity were ascertained. Several studies confirmed that non-activity sites substitution can influence the NDM catalytic efficiency [29], and our results about the V88L substitution support this theory.

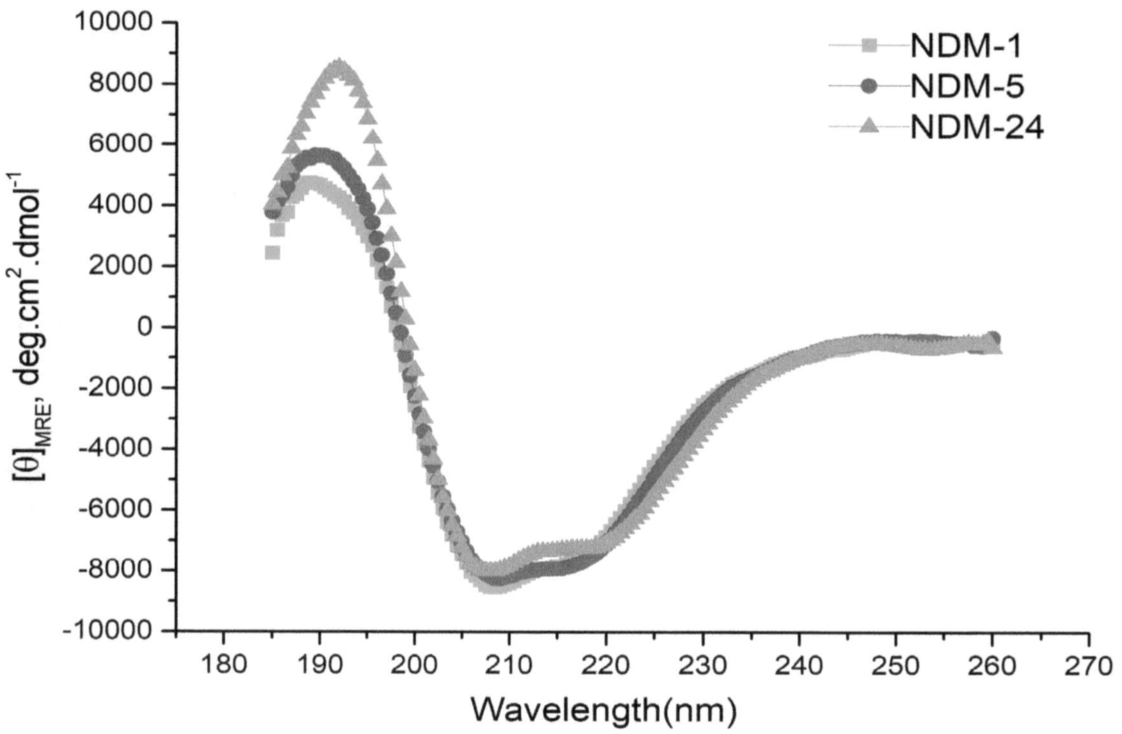

Figure 3. Normalized circular dichroism (CD) spectra of the NDM enzymes tested. MRE: Mean residue ellipticity.

Table 3. Proportions of various secondary structural elements in the NDM-1, NDM-5, and NDM-24 enzymes.

Program Algorithms [a]	Structural Elements [b]	SMP50(9) [c]			SP37(3) [c]			SP29(1) [c]		
		NDM-1	NDM-5	NDM-24	NDM-1	NDM-5	NDM-24	NDM-1	NDM-5	NDM-24
SELCON3	H(r)	0.070	0.078	0.092	0.062	0.074	0.092	0.059	0.079	0.087
	H(d)	0.085	0.088	0.089	0.081	0.088	0.089	0.078	0.087	0.086
	S(r)	0.215	0.199	0.195	0.228	0.214	0.195	0.231	0.191	0.196
	S(d)	0.115	0.109	0.108	0.117	0.113	0.108	0.118	0.107	0.108
	Trn	0.214	0.211	0.194	0.218	0.214	0.194	0.226	0.214	0.215
	Unrd	0.284	0.287	0.261	0.282	0.279	0.261	0.287	0.292	0.285
	H(r)+H(d)	0.155	0.166	0.181	0.143	0.162	0.181	0.137	0.166	0.173
	S(r)+S(d)	0.33	0.308	0.303	0.345	0.327	0.303	0.349	0.298	0.304
CONTINLL	H(r)	0.054	0.075	0.091	0.046	0.079	0.097	0.071	0.078	0.093
	H(d)	0.079	0.092	0.101	0.089	0.095	0.103	0.092	0.096	0.100
	S(r)	0.217	0.208	0.187	0.202	0.205	0.182	0.197	0.197	0.183
	S(d)	0.114	0.113	0.108	0.112	0.111	0.107	0.113	0.111	0.107
	Trn	0.233	0.220	0.220	0.248	0.216	0.216	0.231	0.222	0.225
	Unrd	0.303	0.292	0.293	0.304	0.293	0.294	0.297	0.297	0.292
	H(r)+H(d)	0.133	0.167	0.192	0.135	0.174	0.200	0.163	0.174	0.193
	S(r)+S(d)	0.331	0.321	0.295	0.314	0.316	0.289	0.310	0.308	0.290

[a] The CDPro program package was used to analyse the data using two algorithms: CONTINLL and SELCON3.
[b] H(r), regular α-helix; H(d), distorted α-helix; S(r), regular β-strand; S(d), distorted β-strand; Trn, turns; Unrd, unordered. [c] The reference protein sets (IBasis sets) were adopted.

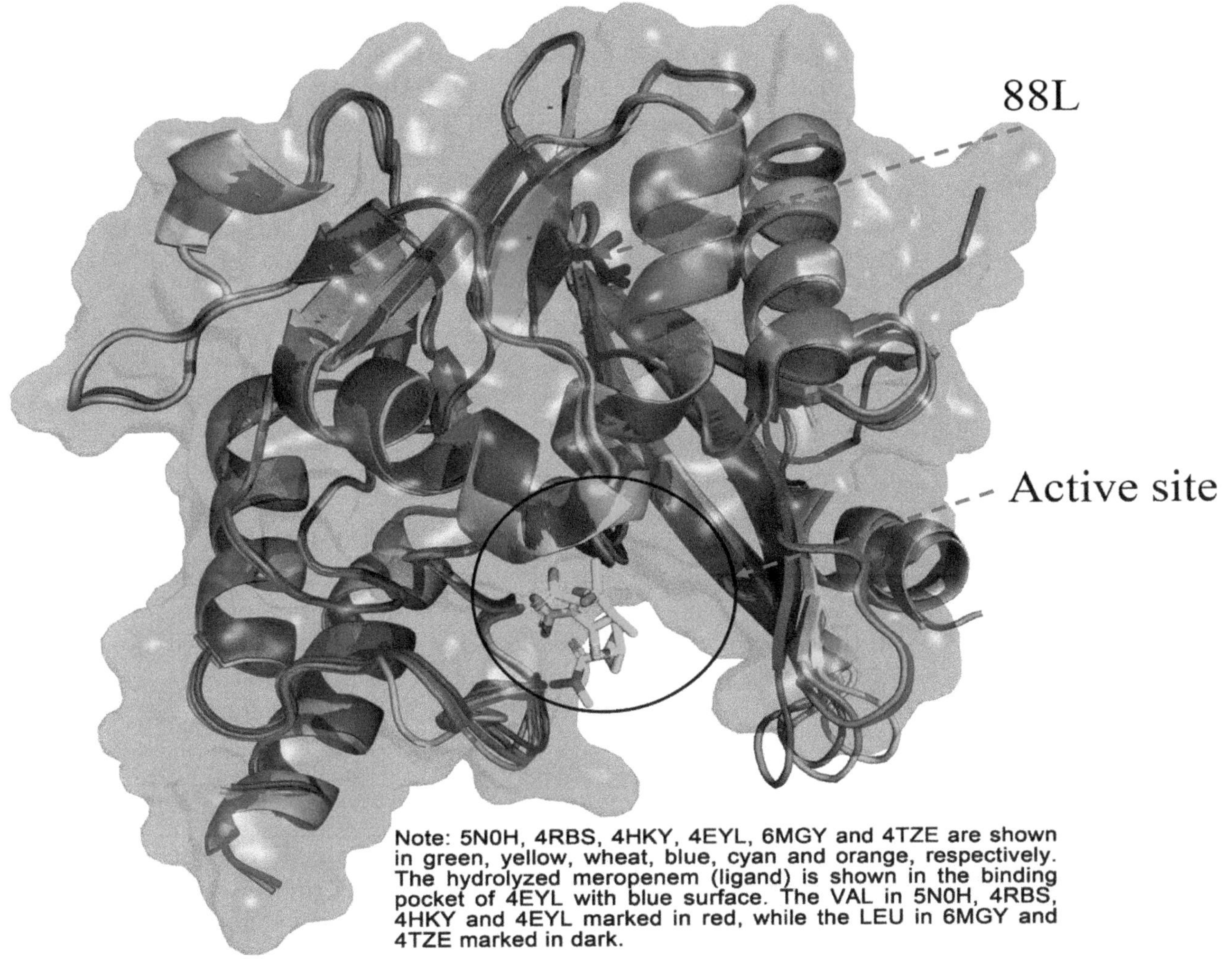

88L
Active site
Note: 5N0H, 4RBS, 4HKY, 4EYL, 6MGY and 4TZE are shown in green, yellow, wheat, blue, cyan and orange, respectively. The hydrolyzed meropenem (ligand) is shown in the binding pocket of 4EYL with blue surface. The VAL in 5N0H, 4RBS, 4HKY and 4EYL marked in red, while the LEU in 6MGY and 4TZE marked in dark.

Figure 4. Cartoon model of NDM-24. To acquire a credible model, the 6 NDM crystal structure (NDM-1(5N0H, 4RBS, 4HKY, and 4EYL) and NDM-5(6MGY, and 4TZE)) were adopted. The residue 88L and active pocket were labelled.

3. Material and Methods

3.1. Site-Directed Mutagenesis, Cloning and Expression of NDM Variants

The $bla_{\text{NDM-1}}$ and $bla_{\text{NDM-5}}$ encoding genes were obtained from clinical *Escherichia coli* strains as previously described [15,16]. Site-directed mutagenesis was performed to generate $bla_{\text{NDM-24}}$ using the pHSG398/NDM-1 plasmid as template and primers listed in Table S1, as previously described [30].

First, the bla_{NDM}-genes were cloned into a pHSG398 vector (TaKaRa Bio, Dalian, China) using the *BamHI* and *XhoI* restriction sites. In a second PCR experiment, the bla_{NDM} variants were amplified without a signal peptide introducing the Tobacco Etch Virus (TEV) at the N-terminal sequence. In order to obtain enzymes overexpression, the amplicons were inserted into a pET-28a(+) plasmid. The *E. coli* DH5α competent cells were used as a non-expression host. *E. coli* BL21(DE3) was used for enzymes overexpression. The authenticity of recombinant plasmids was verified by PCR and sequencing was with Sanger.

3.2. Antimicrobial Susceptibility Tests

The phenotypic profile has been characterized by a microdilution method using a bacterial inoculum of 5 x 10^5 CFU/ml according to the Clinical Laboratory and Standards Institute [31,32]. *E. coli* ATCC25922 was used as a negative control.

3.3. *Production and Purification of NDM-1, NDM-5, and NDM-24*

NDM-1, NDM-5, and NDM-24 were extracted from 0.5 L of culture of *E. coli* BL21(DE3)/pETNDM-1, *E. coli* BL21 (DE3) /pETNDM-5, and *E. coli* BL21 (DE3)/pETNDM-24, respectively. The cultures were grown at 37 °C to achieve an A_{600} of 0.5 L, and 0.4 mM of IPTG was added. After addition of the IPTG, the cultures were incubated at 20 °C for 16 h, under aerobic conditions. Thereafter, a cell supernatant containing the soluble NDM protein was obtained from the lytic bacteria by centrifuging at 10^4 rpm. Proteins purification followed the manufacturer's instructions (Qiagen, Hilden, Germany) by using the Ni-nitrilotriacetic acid (NTA) agarose. The turbo tobacco etch virus (TEV) protease (Accelagen, San Diego, CA, USA) was used to gain the untagged protein without the His tags. The SDS-PAGE was performed to estimate the NDM purity enzymes. The protein concentration was determined using a BCA Protein Quantification Kit (Vazyme Biotech Co., Ltd, Nanjing, China). The β-lactamase activity was monitored at each purification step using the colour change of nitrocefin 1 mg/mL, a chromogenic cephalosporin, according to the previous report [20].

3.4. *Determination of Kinetic Parameters*

Steady-state kinetic experiments were performed following the hydrolysis of each substrate at 25 °C in a 50 mM phosphate buffer, pH 7.0 in the presence of 20 μM Zn SO_4. The data were collected with a SpectraMax M5 multi-detection microplate reader (Molecular Devices, Sunnyvale, CA, USA) as previously described [33]. Kinetic parameters were determined under initial-rate conditions using the GraphPad Prism® version 5.01 (San Diego, CA, USA) to generate the Michaelis-Menten curves, or by analysing the complete hydrolysis time courses [34,35]. Each kinetic value is the mean of the results of three different measurements. The error was below 5%. NDM-5 was used as a reference to normalize.

3.5. *Circular Dichroism and Structure Analysis*

The circular dichroism (CD) spectra (185 to 260 nm) were determined with a Chirascan Plus CD spectrophotometer (Applied Photophysics Ltd, Leatherhead, UK) equipped with a Peltier temperature-controlled cell holder, at 25 °C using a 1-mm pathlength cuvette and the Savitzky-Golay filter was explored to the baseline-correct spectra data. Protein concentrations were diluted to 0.05-0.2 mg/mL with a 5 mM sodium phosphate buffer pH 7.0 [36]. 207 nm spectrum data was used as the baseline to normalize and calibrate data for eliminating minor errors due to different concentrations [37]. The analysis was performed using the CONTINLL and SELCON3 algorithms with reference data sets three and nine, respectively [38]. The super-secondary (tertiary) structures of the proteins were analysed by the CDPro software package, which is available at the CDPro website: https://sites.bmb.colostate.edu/sreeram/CDPro/ [38,39]. To assay the location of the V88L substitution and analyse its effect on the structure, the pharmacophore modeling and screening software program Discovery Studio (version 2018) was employed to generate a three-dimensional (3D) interconnected model of NDM-24 using NDM-1 and NDM-5 as a template, in which reliable data of the crystal structure were collected from the PDB database.

3.6. *Thermal Stability Testing*

The melting temperature (T_m) was used to show the protein stability. The determination of T_m was performed by recording the CD signal change at 222 nm. Data were collected at a ramp of 1 °C /min with a temperature range from 20-90 °C. The two-state model using nonlinear regression (Boltzmann) in the OriginPro 9.1.64 (OriginLab, Northampton, MA, USA) was applied to analyse the data. When 50% of the protein melts, the temperature is defined as the Tm, representing thermal stability.

4. Conclusions

Our study explored the NDM-24 biological function and probed the V88L substitution role on the structure, enzyme activity, and stability. In brief, this non-active site change enhances the enzyme activity by increasing the turnover rate, related with an indirect effect on conformation. However, the loss cost caused by V88L significantly decreased the protein stability, and would shorten the persistence lifetime in the cell, so that the resistance to antibiotics hardly exhibits an outstanding elevation for the NDM-24-producing transformants. Meanwhile, alterations in the secondary content, such as lowering the β-sheet, have an interesting role in the NDM instability, being relevant to the V88L substitution occurring in the β-strand. According to previous data, the V88L/M154L combination appears to be favorable in the NDM evolution under an environmental pressure selection [14]. Further analysis about the significance of non-active-site residues will help in the comprehension of the resistance mechanism and broaden insight in the development of inhibitors, such as potential antibiotics candidate by reducing the protein stability lifetime.

Author Contributions: J.S. designed the study. Z.L., D.L., and W.L. collected the data. Z.L., Y.W., and D.L. analyzed and interpreted the data. Z.L., A.P., D.L., Y.W., and J.S. wrote the report. All authors revised, reviewed and approved the final report.

References

1. Palzkill, T. Metallo-beta-lactamase structure and function. *Ann. N. Y. Acad. Sci.* **2013**, *1277*, 91–104. [CrossRef] [PubMed]
2. Garau, G.; García-Sáez, I.; Bebrone, C.; Anne, C.; Mercuri, P.; Galleni, M.; Frère, J.M.; Dideberg, O. Update of the standard numbering scheme for class B beta-lactamases. *Antimicrob. Agents Chemother.* **2004**, *48*, 2347–2349. [CrossRef] [PubMed]
3. Yong, D.; Toleman, M.A.; Giske, C.G.; Cho, H.S.; Sundman, K.; Lee, K.; Walsh, T.R. Characterization of a New Metallo-β-Lactamase Gene, blaNDM-1, and a Novel Erythromycin Esterase Gene Carried on a Unique Genetic Structure in Klebsiella pneumoniae Sequence Type 14 from India. *Antimicrob. Agents Chemother.* **2009**, *53*, 5046–5054. [CrossRef] [PubMed]
4. Bonnin, R.A.; Poirel, L.; Carattoli, A.; Nordmann, P. Characterization of an IncFII Plasmid Encoding NDM-1 from Escherichia coli ST131. *PLoS ONE* **2012**, *7*, 34752. [CrossRef] [PubMed]
5. Dolejska, M.; Villa, L.; Poirel, L.; Nordmann, P.; Carattoli, A. Complete sequencing of an IncHI1 plasmid encoding the carbapenemase NDM-1, the ArmA 16S RNA methylase and a resistance-nodulation-cell division/multidrug efflux pump. *J. Antimicrob. Chemother.* **2013**, *68*, 34–39. [CrossRef] [PubMed]
6. King, D.; Strynadka, N. Crystal structure of New Delhi metallo-betalactamase reveals molecular basis for antibiotic resistance. *Protein Sci.* **2011**, *20*, 1484–1491. [CrossRef] [PubMed]
7. Gonzalez, L.J.; Bahr, G.; Nakashige, T.G.; Nolan, E.M.; Bonomo, R.A.; Vila, A.J. Membrane anchoring stabilizes and favors secretion of New Delhi metallo-beta-lactamase. *Nat. Chem. Biol.* **2016**, *12*, 516–522. [CrossRef] [PubMed]
8. Gonzalez, L.J.; Bahr, G.; Vila, A.J. Lipidated beta-lactamases: From bench to bedside. *Future Microbiol.* **2016**, *11*, 1495–1498. [CrossRef]
9. Khan, A.U.; Maryam, L.; Zarrilli, R. Structure, genetics and worldwide spread of New Delhi metallo-beta-lactamase (NDM): A threat to public health. *BMC Microbiol.* **2017**, *17*, 101. [CrossRef]
10. Cheng, Z.; Thomas, P.W.; Ju, L.; Bergstrom, A.; Mason, K.; Clayton, D.; Miller, C.; Bethel, C.R.; Vanpelt, J.; Tierney, D.L.; et al. Evolution of New Delhi metallo-β-lactamase (NDM) in the clinic: Effects of NDM mutations on stability, zinc affinity, and mono-zinc activity. *J. Biol. Chem.* **2018**, *293*, 12606–12618. [CrossRef]
11. Zhang, H.O.; Hau, Q. Crystal structure of NDM-1 reveals a common β-lactam hydrolysis mechanism. *FASEB J.* **2011**, *25*, 2574–2582. [CrossRef] [PubMed]
12. Kim, Y.; Cunningham, M.A.; Mire, J.; Tesar, C.; Sacchettini, J.; Joachimiak, A. NDM-1, the ultimate promiscuous enzyme: Substrate recognition and catalytic mechanism. *FASEB J.* **2013**, *27*, 1917–1927. [CrossRef] [PubMed]
13. Liu, L.; Feng, Y.; McNally, A.; Zong, Z. bla NDM-21, a new variant of blaNDM in an Escherichia coli clinical isolate carrying blaCTX-M-55 and rmtB. *J. Antimicrob. Chemother.* **2018**, *73*, 2336–2339. [CrossRef] [PubMed]
14. Bahr, G.; Vitor-Horen, L.; Bethel, C.R.; Bonomo, R.A.; González, L.J.; Vila, A.J. Clinical evolution of New Delhi Metallo-β-lactamase (NDM) optimizes resistance under Zn(II) deprivation. *Antimicrob. Agents Chemother.* **2018**, *62*, 1817–1849. [CrossRef] [PubMed]
15. Liu, Z.; Li, J.; Wang, X.; Liu, D.; Ke, Y.; Wang, Y.; Shen, J. Novel Variant of New Delhi Metallo-β-lactamase, NDM-20, in Escherichia coli. *Front. Microbiol.* **2018**, *9*, 248. [CrossRef] [PubMed]

16. Liu, Z.; Wang, Y.; Walsh, T.R.; Liu, D.; Shen, Z.; Zhang, R.; Yin, W.; Yao, H.; Li, J.; Shen, J. Plasmid-mediated novel *bla*$_{NDM-17}$ gene encoding a carbapenemase with enhanced activity in a ST48 *Escherichia coli* strain. *Antimicrob. Agents Chemother.* **2017**, *61*, 2216–2233. [CrossRef] [PubMed]
17. Nordmann, P.; Boulanger, A.E.; Poirel, L. NDM-4 Metallo-β-Lactamase with Increased Carbapenemase Activity from Escherichia coli. *Antimicrob. Agents Chemother.* **2012**, *56*, 2184–2186. [CrossRef]
18. Zou, D.; Huang, Y.; Zhao, X.; Liu, W.; Dong, D.; Li, H.; Wang, X.; Huang, S.; Wei, X.; Yan, X.; et al. A Novel New Delhi Metallo-β-Lactamase Variant, NDM-14, Isolated in a Chinese Hospital Possesses Increased Enzymatic Activity against Carbapenems. *Antimicrob. Agents Chemother.* **2015**, *59*, 2450–2453. [CrossRef]
19. Tada, T.; Miyoshi-Akiyama, T.; Dahal, R.K.; Sah, M.K.; Ohara, H.; Kirikae, T.; Pokhrel, B.M. NDM-8 Metallo-β-Lactamase in a Multidrug-Resistant Escherichia coli Strain Isolated in Nepal. *Antimicrob. Agents Chemother.* **2013**, *57*, 2394–2396. [CrossRef]
20. Tada, T.; Shrestha, B.; Miyoshi-Akiyama, T.; Shimada, K.; Ohara, H.; Kirikae, T.; Pokhrel, B.M. NDM-12, a Novel New Delhi Metallo-β-Lactamase Variant from a Carbapenem-Resistant *Escherichia coli* Clinical Isolate in Nepal. *Antimicrob. Agents Chemother.* **2014**, *58*, 6302–6305. [CrossRef]
21. Ines, S.; Emma, K.; Rudolf, R.; Rumyana, M.; Anne Marie, Q.; Adolf, B. VIM-15 and VIM-16, two new VIM-2-like metallo-beta-lactamases in *Pseudomonas aeruginosa* isolates from Bulgaria and Germany. *Antimicrob. Agents Chemother.* **2008**, *52*, 2977.
22. Patricia, M.; Tomatis, P.E.; Mussi, M.A.; Fernando, P.; Viale, A.M.; Limansky, A.S.; Vila, A.J. Biochemical characterization of metallo-beta-lactamase VIM-11 from a *Pseudomonas aeruginosa* clinical strain. *Antimicrob. Agents Chemother.* **2008**, *52*, 2250.
23. Jose-Manuel, R.M.; Patrice, N.; Nicolas, F.; Laurent, P. VIM-19, a metallo-beta-lactamase with increased carbapenemase activity from *Escherichia coli* and *Klebsiella pneumoniae*. *Antimicrob. Agents Chemother.* **2010**, *54*, 471–476.
24. Pierre, B.; Carine, B.; Te-Din, H.; Warda, B.; Yves, D.; Ariane, D.; Kurt, H.; Youri, G. Detection and characterization of VIM-31, a new variant of VIM-2 with Tyr224His and His252Arg mutations, in a clinical isolate of *Enterobacter cloacae*. *Antimicrob. Agents Chemother.* **2012**, *56*, 3283.
25. Corbin, B.D.; Seeley, E.H.; Raab, A.; Feldmann, J.; Miller, M.R.; Torres, V.J.; Anderson, K.L.; Dattilo, B.M.; Dunman, P.M.; Gerads, R.; et al. Metal Chelation and Inhibition of Bacterial Growth in Tissue Abscesses. *Science* **2008**, *319*, 962–965. [CrossRef] [PubMed]
26. Kumar, G.; Issa, B.; Kar, D.; Biswal, S.; Ghosh, A.S. E152A substitution drastically affects NDM-5 activity. *FEMS Microbiol. Lett.* **2017**, *364*. [CrossRef] [PubMed]
27. Ali, A.; Azam, M.W.; Khan, A.U. Non-active site mutation (Q123A) in New Delhi metallo-β-lactamase (NDM-1) enhanced its enzyme activity. *Int. J. Biol. Macromol.* **2018**, *112*, 1272–1277. [CrossRef] [PubMed]
28. Carfi, A.; Pares, S.; Duée, E.; Galleni, M.; Duez, C.; Frère, J.M.; Dideberg, O. The 3-D structure of a zinc metallo-beta-lactamase from Bacillus cereus reveals a new type of protein fold. *EMBO J.* **1995**, *14*, 4914–4921. [CrossRef] [PubMed]
29. Piccirilli, A.; Brisdelli, F.; Aschi, M.; Celenza, G.; Amicosante, G.; Perilli, M. Kinetic Profile and Molecular Dynamic Studies Show that Y229W Substitution in an NDM-1/L209F Variant Restores the Hydrolytic Activity of the Enzyme toward Penicillins, Cephalosporins, and Carbapenems. *Antimicrob. Agents Chemother.* **2019**, *63*, e02270-18. [CrossRef]
30. Meini, M.-R.; Tomatis, P.E.; Weinreich, D.M.; Vila, A.J. Quantitative Description of a Protein Fitness Landscape Based on Molecular Features. *Mol. Boil. Evol.* **2015**, *32*, 1774–1787. [CrossRef]
31. Clinical and Laboratory Standards Institute. *Methods for Dilution Antimicrobial Susceptibility Tests for Bacteria that Grow Aerobically: Approved Standard*, 11th ed.; CLSI document M07; Clinical and Laboratory Standards Institute: Wayne, PA, USA, 2018.
32. Clinical and Laboratory Standards Institute. *Performance Standards for Antimicrobial Susceptibility Testing*, 28th ed.; CLSI document M100; Clinical and Laboratory Standards Institute: Wayne, PA, USA, 2018.
33. Crowder, M.W.; Walsh, T.R.; Banovic, L.; Pettit, M.; Spencer, J. Overexpression, purification, and characterization of the cloned metallo-beta-lactamase L1 from Stenotrophomonas maltophilia. *Antimicrob. Agents Chemother.* **1998**, *42*, 921. [CrossRef] [PubMed]
34. De Meester, F.; Joris, B.; Reckinger, G. Automated analysis of enzyme inactivation phenomena. Application to β-lactamases and DD-peptidases. *Biochem. Pharmacol.* **1987** *36*, 2393–2403. [CrossRef]

35. Segel, I.H. *Biochemical Calculations*, 2nd ed.; John Wiley & Sons: New York, NY, USA, 1976; pp. 236–241.
36. Liu, Z.; Zhang, R.; Li, W.; Yang, L.; Liu, D.; Wang, S.; Shen, J.; Wang, Y. Amino acid changes at the VIM-48 C-terminus result in increased carbapenem resistance, enzyme activity and protein stability. *J. Antimicrob. Chemother.* **2018**, *74*, 885–893. [CrossRef] [PubMed]
37. Raussens, V.; Ruysschaert, J.-M.; Goormaghtigh, E. Protein concentration is not an absolute prerequisite for the determination of secondary structure from circular dichroism spectra: A new scaling method. *Anal. Biochem.* **2003**, *319*, 114–121. [CrossRef]
38. Sreerama, N.; Venyaminov, S.Y.; Woody, R.W. Estimation of the number of α-helical and β-strand segments in proteins using circular dichroism spectroscopy. *Protein Sci.* **1999**, *8*, 370–380. [CrossRef] [PubMed]
39. Sreerama, N.; Venyaminov, S.Y.; Woody, R.W. Analysis of Protein Circular Dichroism Spectra Based on the Tertiary Structure Classification. *Anal. Biochem.* **2001**, *299*, 271–274. [CrossRef] [PubMed]

6

An Innovative Biocatalyst for Continuous 2G Ethanol Production from Xylo-Oligomers by *Saccharomyces cerevisiae* through Simultaneous Hydrolysis, Isomerization and Fermentation (SHIF)

Thais S. Milessi-Esteves [1,†], Felipe A.S. Corradini [2,†], Willian Kopp [1], Teresa C. Zangirolami [1,2], Paulo W. Tardioli [1,2], Roberto C. Giordano [1,2] and Raquel L.C. Giordano [1,2,*]

1 Department of Chemical Engineering, Federal University of São Carlos, Rodovia Washington Luiz, km 235, 13565-905, São Carlos, SP, Brazil; thais.milessi@gmail.com (T.S.M.-E.); willkopp@gmail.com (W.K.); teresacz@ufscar.br (T.C.Z.); pwtardioli@ufscar.br (P.W.T.); roberto@ufscar.br (R.C.G.)

2 Graduate Program of Chemical Engineering, Federal University of São Carlos (PPGEQ-UFSCar), Rodovia Washington Luiz, km 235, 13565-905, São Carlos, SP, Brazil; eq.felipe.silva@gmail.com

* Correspondence: raquel@ufscar.br

† Both authors contributed equally to this work

Abstract: Many approaches have been considered aimed at ethanol production from the hemicellulosic fraction of biomass. However, the industrial implementation of this process has been hindered by some bottlenecks, one of the most important being the ease of contamination of the bioreactor by bacteria that metabolize xylose. This work focuses on overcoming this problem through the fermentation of xylulose (the xylose isomer) by native *Saccharomyces cerevisiae* using xylo-oligomers as substrate. A new concept of biocatalyst is proposed, containing xylanases and xylose isomerase (XI) covalently immobilized on chitosan, and co-encapsulated with industrial baker's yeast in Ca-alginate gel spherical particles. Xylo-oligomers are hydrolyzed, xylose is isomerized, and finally xylulose is fermented to ethanol, all taking place simultaneously, in a process called simultaneous hydrolysis, isomerization, and fermentation (SHIF). Among several tested xylanases, Multifect CX XL A03139 was selected to compose the biocatalyst bead. Influences of pH, Ca^{2+}, and Mg^{2+} concentrations on the isomerization step were assessed. Experiments of SHIF using birchwood xylan resulted in an ethanol yield of 0.39 g/g, (76% of the theoretical), selectivity of 3.12 $g_{ethanol}/g_{xylitol}$, and ethanol productivity of 0.26 g/L/h.

Keywords: 2G ethanol; hemicellulose usage; *S. cerevisiae*; enzyme immobilization; cell immobilization; SHIF

---

1. Introduction

Biofuels will have a significant role in the energetic matrix of the low-carbon economy, helping to meet the goals established at Conference of the Parties (COP 21) [1,2]. Among biofuels, bioethanol production from lignocellulosic materials has been intensively studied once it was shown that these byproducts had high availability, had a low cost, and did not compete with the production of food [3]. Lignocellulosic raw materials are mainly composed of cellulose and hemicellulose (up to 70%), which are polysaccharides that, after a hydrolysis step, generate fermentable sugars, mostly xylose from hemicellulose and glucose from cellulose [4]. The use of these two polysaccharides is important for the economic feasibility of the biofuel production process.

Some microorganisms that naturally ferment pentoses to ethanol have been tested for industrial use, such as *Scheffersomyces stipitis* and *Pachysolen tannophilus* [5,6]. However, these microorganisms have a low tolerance to ethanol and slow fermentation rates and are inhibited by compounds generated during the biomass pretreatment step, such as furfural [7].

Saccharomyces cerevisiae is the most common microorganism used for ethanol production from hexoses, due to its high rate of fermentation and superior ethanol yield. In addition, this yeast exhibits unbeatable tolerance to ethanol, to inhibitors, and to high concentrations of sugar [8,9]. However, in its wild form *S. cerevisiae* is unable to efficiently metabolize D-xylose.

The genetic modification of *S. cerevisiae* aimed at xylose fermentation has been extensively studied [8,10–12]. However, the low specific growth rate, high xylitol production, reduced yeast tolerance, and possible genetic instability are still hindrances for the application of recombinant strains on an industrial scale [7].

In spite of the inability of *S. cerevisiae* to metabolize xylose, it is capable of fermenting its isomer, xylulose, to ethanol. Hence, an alternative for the utilization of the hemicellulose fraction for bioethanol production would be to isomerize xylose to xylulose ex vivo, followed by fermentation by *S. cerevisiae* [7]. The enzyme xylose isomerase (XI) (EC 5.3.1.5) is widely used in the industry for the production of fructose syrup from corn starch and also catalyzes the reversible isomerization of xylose to xylulose [13]. Although the xylose/xylulose chemical equilibrium is unfavorable (3.5:1 at 60 °C) [14], the reaction can be displaced by the simultaneous isomerization and fermentation (SIF) process, where the continuous conversion of xylulose to ethanol might allow the complete depletion of the available xylose [15].

The use of catalysts with immobilized enzymes may be crucial for the application of multi-enzymatic processes on an industrial scale. This approach allows the continuous operation of the reactor and facilitates the product recovery as well as the use of high loads of cells and enzymes [16]. The production of an active and stable enzyme derivative using a non-expensive support is also an important issue in enzyme immobilization [17]. The literature reports successful applications of immobilized enzymes on an industrial scale [18,19] and immobilized XI is one of the most successful and established examples [13]. Silva et al. [7] developed a biocatalyst containing chitosan-immobilized XI, co-immobilized with *S. cerevisiae* in calcium alginate gel. Calcium alginate gel was chosen for being a natural polymer widely studied as a support for the immobilization of viable cells [20]. However, this system showed to be susceptible to contamination by xylose-consuming bacteria. High concentrations of xylose in the medium disfavored the *S. cerevisiae* population, due to its low uptake rates of xylulose.

An alternative to tackling the contamination problem is to use a cultivation medium containing non-readily fermentable substrates, such as xylo-oligomers obtained by the solubilization of hemicellulose under mild conditions [21]. The hetero-polysaccharides that compose hemicellulose are polymers with about 100 units of monomers, mainly xylose, and their solubility depends on the number of monomeric units in the chain [22]. Thus, the extraction of xylan in the form of large oligomers must be carried out under conditions that allow a sufficient number of glycosidic bonds to be broken, so that soluble polymers with lower molecular weight (xylo-oligomers) are released.

Xylanases (β-1,4-D-xylanase) are enzymes that catalyze the hydrolysis of the glycosidic bonds between xylose units. The enzymatic complex is commonly composed of endoxylanases, exoxylanases, β-D-xylosidases as well as accessory enzymes such as glucuronidase and arabinofuranosidase that act on the ramifications of the xylan chain [23]. The addition of these enzymes to the biocatalyst proposed by Silva et al. [7] would allow the feeding of xylo-oligomers to the bioreactor. This substrate might decrease the probability of contamination during the operation of the bioreactor for long periods, which are typical in industry. Preliminary results showed the technical viability of this process [24].

The development of viable processes to increase ethanol yields from lignocellulosic materials is crucial, despite the challenges that still remain for the production of 2G ethanol from xylose. Considering the higher production cost of 2G ethanol (compared to 1G ethanol), the use of pentoses as raw material could make its production more profitable and might overcome the costs of 2G ethanol extra steps [25]. The integration of several biocatalytic transformations in a multi-enzymatic cascade system is particularly appealing to the development of cleaner and more efficient biochemical processes.

Multi-enzymatic cascade reactions offer advantages such as lower demand of time, reduced costs, easier recovery of products, completion of reversible reactions as well as concentrations of inhibitory compounds restrained to a minimum [26].

In this context, the simultaneous hydrolysis, isomerization, and fermentation (SHIF) process stands out for 2G ethanol production since unmodified *S. cerevisiae* remains the preferred microorganism in industry, due to its robustness, high ethanol tolerance, and production rates. The use of wild strains to produce ethanol from xylose is an important issue in countries like Brazil, where biosafety regulations are strict [15]. In addition, one advantage of this approach is that XI, along with amylases and proteases, is among the most widely and cheaply available commercial enzymes [27]. The present work reports the results of using this new biocatalyst for the simultaneous hydrolysis, isomerization, and fermentation of xylan derived from the hemicellulose fraction of biomass, aimed at the production of ethanol (Figure 1).

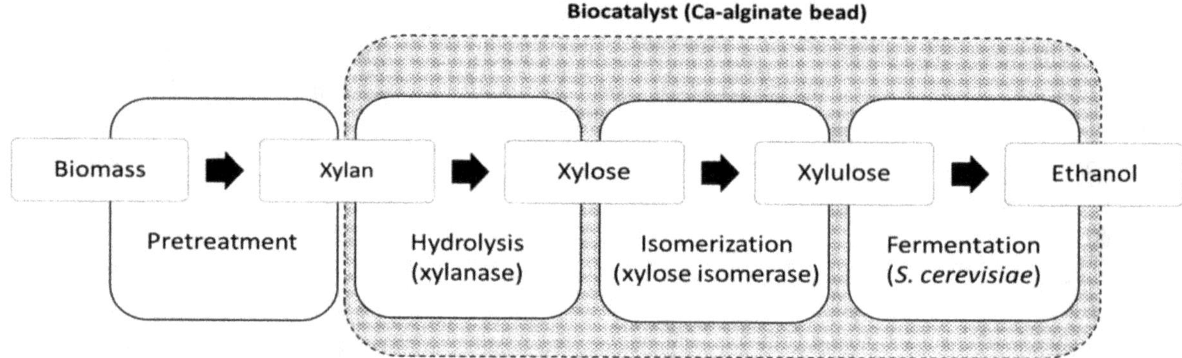

Figure 1. Xylan biomass simultaneous hydrolysis, isomerization, and fermentation (SHIF). Biocatalyst composed of xylanases, xylose isomerase, and co-immobilized *S. cerevisiae*.

2. Results

2.1. Application of the New Biocatalyst in the SHIF Process

The biocatalyst is designed for the industrial production of second-generation ethanol in continuous, fixed-bed reactors through long-term operation, by applying simultaneous hydrolysis, isomerization, and fermentation (SHIF) of the hemicellulosic fraction of biomass. First, xylanase and xylose isomerase were covalently immobilized on chitosan. The obtained XI derivative presented an activity of 252.5 ± 1.6 IU/g (immobilization yield of 93% and recovered activity of 91%), whereas the Accellerase derivative exhibited 346.3 ± 9.2 IU/g (immobilization yield of 54% and recovered activity of 12%). Both derivatives were co-encapsulated with *S. cerevisiae* in Ca-alginate gel and this biocatalyst was used to produce ethanol from commercial birchwood xylan.

Birchwood xylan, which is a heteropolymer composed of long chains, was first hydrolyzed to smaller xylo-oligomers by the action of recombinant endoxylanase of *Bacillus subtilis* (XynA) in order to increase the concentration of xylo-oligomers with smaller chains that may diffuse into the catalyst beads [28]. This step was carried out to make xylan more similar to lignocellulosic hydrolysates obtained from the pretreatment of biomass (data not shown). The composition of the substrate after xylan pre-hydrolysis is shown in Table 1.

As expected, there was no xylose production since XynA is a strict endoxylanase [28]. The solubilized fraction corresponded to 67% (w/w) of the offered xylan. According to Gray et al. [22], the solubility of the xylan oligomers depends on the degree of polymerization of each compound. Under the used conditions, 33% (w/w) of birchwood xylan is insoluble. Therefore, the substrate obtained for the SHIF process had 73 g/L of xylo-oligomers.

Table 1. Characterization of SHIF substrate: birchwood xylan, 108 g/L after 24 h hydrolysis by endoxylanase XynA (150 IU/g_{xylan}) at 50 °C and pH 5.6.

Component	Hydrolyzed Birchwood Xylan	
	(g/L)	(%)
Xylohexaose (X6) or bigger	62.3	57.4
Xylopentaose (X5)	0.0	0.0
Xylotetraose (X4)	1.2	1.1
Xylotriose (X3)	2.5	2.2
Xylobiose (X2)	7.1	6.5
Xylose	0.0	0.0
Total Soluble	73.0	67.2
Total Insoluble	35.6	32.8
Total	108.6	100.0

For the SHIF assays, the offered enzyme activity in the reactor was 1.7×10^4 $IU/L_{reactor}$ for xylanase (Accellerase XY) and 3.7×10^4 $IU/L_{reactor}$ of xylose isomerase. Accellerase XY was used due to the presence of β-xylosidase, which is necessary to xylose formation. Commercial baker's yeast (Itaiquara®) concentration was 50 $g_{dry\ mass}/L_{reactor}$ at the beginning of the SHIF assays. Results in Figure 2 show the production of ethanol through SHIF using the developed biocatalyst. Ethanol production was higher compared to xylitol, presenting a selectivity of 2.61 (2.2 g/L ethanol and 0.84 g/L xylitol). Ethanol productivity of 0.092 g/L/h and yield ($Y_{P/S}$) of 0.160 $g_{ethanol}/g_{potentialxylose}$ (32% of theoretical, calculated on the basis of potential xylose in the xylan) were achieved at the end of the SHIF run.

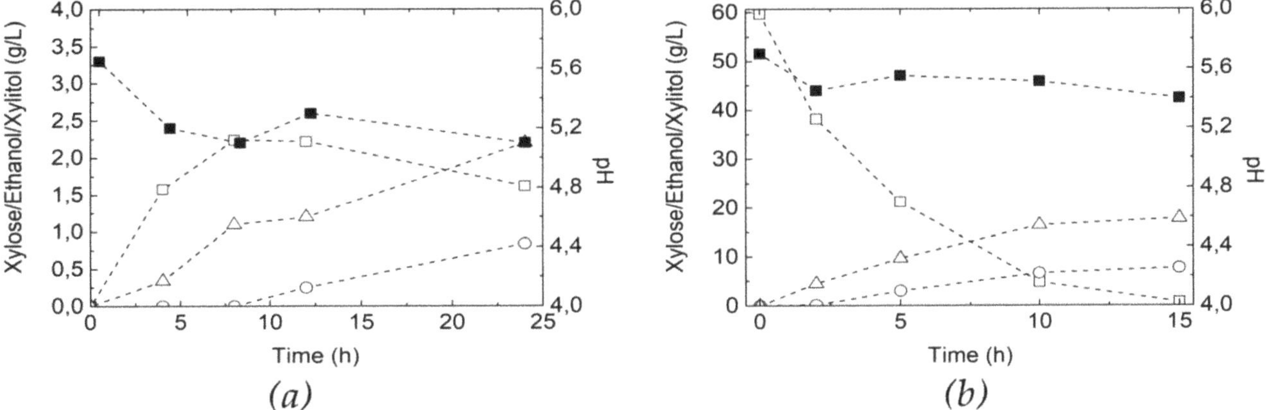

Figure 2. (**a**) SHIF of pre-hydrolyzed birchwood xylan using a biocatalyst containing 5 w % of Accellerase XY, 15 w % of XI, and 10 w % dry mass yeast (1.7×10^4 IU/L, 3.7×10^4 IU/L, and 50 g/L, respectively), pH 5.6, 150 rpm stirring, and 32 °C; (**b**) control experiment using xylose as substrate (SIF): pH (■), xylose (□), ethanol (Δ), and xylitol (○).

Figure 2 shows that ethanol is produced from xylo-oligomers. The slower velocity of SHIF compared to the control experiment (where the substrate is xylose, resulting in a simultaneous isomerization, and fermentation (SIF) process) was expected since xylose concentration in SHIF depends on the velocity of hydrolysis of the xylo-oligomers. However, a decrease in the xylose consumption was observed after eight hours of SHIF (Figure 2a), which indicates that the xylose isomerization was impaired. There are two possible reasons for this occurrence: XI is not catalyzing isomerization of xylose in the required velocity; or the yeast is not consuming the generated xylulose, which would be accumulating and consequently stopping the isomerization due to the xylose:xylulose equilibrium ratio. Since the yeast remained viable (initial and final cell viability unchanged: 96%) and there was no accumulation of xylulose in the medium, the accumulation of xylose seemed to be related to the isomerization step.

In the control experiment, using only xylose as substrate, XI catalyzed the isomerization reaction effectively, almost until depletion of the pentose (60 g/L of xylose were consumed in 12 h, a productivity of 1.2 g/L/h, with $Y_{P/S}$ of 0.303 g/g and selectivity of 2.3 with respect to xylitol). Thus, an increase of the isomerization velocity must be sought. Some of the possible causes for the hindrance of the isomerization step were then investigated. High calcium concentrations are known to inhibit XI action [29], and the hydrolysis reaction could be demanding a higher release of the calcium ion (increased hydrolysis of $CaCO_3$ to control the pH). However, even with a higher release of calcium, the results shown in Figure 2 indicate that the hydrolysis step led to a drop of pH to approximately 5.1. XI shows maximum activity at pH 8, being highly sensitive to a drop of pH to this range [15]. On the other hand, it is known that the magnesium ion is an important cofactor for XI, as an activator of this enzyme [30]. The influence of pH and Ca^{2+} and Mg^{2+} ions in the isomerization step was then investigated.

2.2. *Influence of pH, Ca^{2+}, and Mg^{2+} on XI Activity*

In order to investigate the influence of calcium and magnesium ions on XI, the activity of this enzyme to catalyze fructose–glucose isomerization was measured at different pHs (5.0 to 8.0) in the presence of different concentrations of Ca^{2+} and Mg^{2+} ions. The standard medium for assessing activity was 2 M fructose, pH 8.0 (50 mM tris-maleate buffer) supplemented with 50 mM of $MgSO_4$ and 2.5 mM of $CoCl_2$ at 60 °C. Both Co^{2+} and Mg^{2+} are essential for the activity of XI, however they play differentiated roles. Mg^{2+} is superior to Co^{2+} as an activator, while the latter is responsible for the stabilization of the enzyme and maintenance of its conformation, especially the quaternary structure [31]. Table 2 shows the measured activities of XI in each condition studied, referred to the test performed with the standard medium at pH 8.0 as 100%.

Table 2. Influence of Mg^{2+} and Ca^{2+} on XI activity at different pHs (isomerization of fructose 2M, 60 °C). The activity measured in standard medium (pH 8.0, 50 mM $MgSO_4$, 2.5 mM $CoCl_2$) was taken as 100%.

	Medium	pH 8.0 IU/mL	%	pH 7.0 IU/mL	%	pH 6.0 IU/mL	%	pH 5.0 IU/mL	%
1	Standard	1896 ± 68	100.0	1590 ± 23	83.9	1466 ± 12	77.4	442 ± 21	23.4
2	2× Mg^{2+}	2041 ± 31	107.7	1890 ± 7	99.7	1820 ± 3	96.9	940 ± 24	49.6
3	5× $Mg^{2+\prime}$	2202 ± 67	116.2	2287 ± 43	120.6	2333 ± 12	123.1	1166 ± 35	61.5
4	Standard + Ca^{2+*}	1454 ± 97	76.7	1312 ± 20	69.2	572 ± 85	30.2	380 ± 45	20.1
5	2× Ca^{2+}	1366 ± 72	72.1	944 ± 69	49.8	503 ± 94	26.6	352 ± 45	18.6
6	5× Ca^{2+}	1063 ± 56	56.1	797 ± 56	42.1	356 ± 31	18.8	255 ± 44	13.5
7	Ca^{2+*} + 2× Mg^{2+}	1542 ± 81	81.3	1673 ± 27	88.3	1073 ± 56	56.6	803 ± 20	42.4
8	Ca^{2+*} + 5× Mg^{2+}	1672 ± 88	88.2	2236 ± 39	117.9	1250 ± 16	65.9	966 ± 17	51.0

* [$CaCl_2$] = 4 g/L (same as used during SHIF supplementation).

The Ca^{2+} ion proved to be an inhibitor of this enzyme, since a significant decrease in the XI activity occurred when the Ca^{2+} concentration in the medium increased. The Mg^{2+} ion, in turn, was able to activate the enzyme, increasing its activity in 16.2% at pH 8.0 and 163% at pH 5.0, both in calcium-free medium. This cofactor was still able to bypass the inhibition caused by calcium, since it reactivated the enzyme in the presence of this ion, increasing its catalytic activity in all studied pHs.

Xylose isomerization catalyzed by XI is initiated by opening the sugar ring, followed by isomerization through the exchange of hydride and finally stabilization of the product by ring closure [32]. Although there is no relationship between the presence of magnesium and the ring opening step, this cation is essential for the isomerization [33]. According to Kasumi et al. [34], the reaction mechanism demands the formation of a binary divalent enzyme–cation complex, since the substrate will bind only to the active site of this complex. Xylose isomerase has two active sites, each containing two divalent cations [33]. Thus, the presence of higher concentrations of magnesium in the reaction medium, improving the probability of the presence of this ion in the active sites of the enzyme, would increase the rates of the isomerization reaction.

Although calcium is a divalent cation and belongs to the same family as magnesium (same configuration in the valence layer) in the periodic table, Ca^{2+} has a larger ionic radius than Mg^{2+}. This fact could be the reason why Mg^{2+} is an activator of the enzyme while Ca^{2+} inhibits XI, that is, the difference in their atomic radii would cause a different interaction with the active site of the enzyme.

In addition to the significant influence of Ca^{2+} and Mg^{2+}, XI showed great sensitivity to pH, losing activity significantly at pH 5.0. The sensitivity of XI to pH was previously observed by Milessi et al. [15], who emphasized the importance of pH control during the simultaneous isomerization and fermentation (SIF) of xylose. However, the activation provided by Mg^{2+} is potentiated at lower pHs. When Ca^{2+} (4 g/L) was added, the magnesium ion was able to recover XI activity more effectively at pH 5.0 than at pH 8.0 (Table 2, media 4 and 7).

Data presented in Table 2 also prove that XI activity was greatly reduced at the SHIF pH range (5.0 to 6.0). The quaternary structure of this enzyme is composed of four subunits that are delicately folded and associated with noncovalent links and without interchain disulfide bonds [31,32]. At pH 8.0, the enzyme structure is composed of all four subunits combined, resulting in the maximum catalytic activity. However, as the pH of the reaction medium lowers, the enzyme is more likely to suffer structure distortions, unfolding and dissociating its tetrameric structure.

The presence of Ca^{2+} ions at the low pH of the reaction medium results in a combined effect, acting both on the 3D structure and on the active site. Therefore, the performance of the SHIF process will certainly benefit from a pH control system (pH 5.6). Unfortunately, the substitution of calcium chloride by magnesium chloride in the coagulation solution during sodium alginate gelation was not possible since the resulting beads were not stable. For this reason, SHIF experiments with pH control and excess of magnesium were run, in order to minimize the inhibition of XI caused by the hydrolysis reactions, which release acids from the structures of the xylo-oligomers [35].

2.3. Xylanase Selection

Xylan hydrolysis has to occur efficiently to enable the cascade SHIF process. Hydrolysis cannot be the rate-determining step of these reactions in series: due to the unfavorable equilibrium of xylose to xylulose isomerization, the supply of xylose must not control the reaction [7]. Considering that the composition of each xylanase complex influences the hydrolysis efficiency, different xylanases were evaluated with the purpose of selecting the most efficient for the SHIF process.

The xylanase family is strongly related to the profile of products generated in the process [36]. A xylanase capable of depolymerizing xylan into xylose efficiently is required to ensure that the SHIF process will proceed as expected. Thus, besides Accellerase XY A03304, used in previous SHIF tests, two additional xylanases were evaluated: recombinant *B. subtilis* endoxylanase (XynA) and Multifect CX XL A03139.Hydrolysis profiles and xylooligosaccharide (XOS) composition are reported in Table 3 and Figure 3, respectively.

Table 3. Composition of xylooligosaccharides (XOS) after enzymatic hydrolysis ofbirchwood xylan (25.4 g/L, 3.8 IU/mL, 24 h of reaction at 50 °C, pH 5.6).

Xylanase	X6 (g/L)	X5 (g/L)	X4 (g/L)	X3 (g/L)	X2 (g/L)	Xylose (g/L)	Conversion (%)
XynA	0	0.20	1.22	4.43	3.49	0.00	44.8
Multifect	0	0.18	0.31	1.85	6.48	13.24	78.7
Accellerase	0	0.00	0.34	1.51	6.45	8.08	58.8

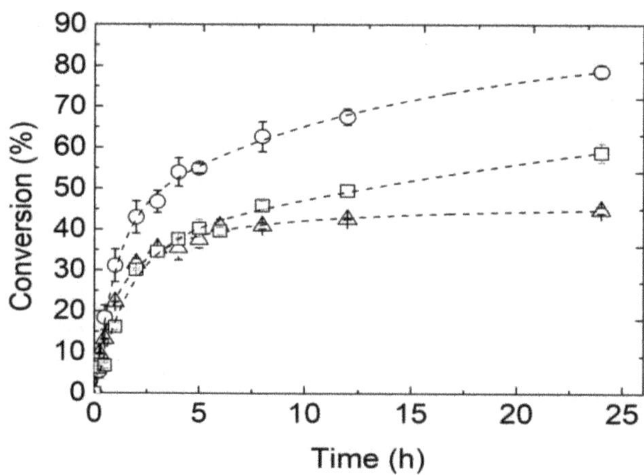

Figure 3. Hydrolysis of birchwood xylan soluble fraction (25,4 g/L) by the studied xylanases (150 IU/g_{xylan} = 3810 IU/L) at 50 °C, 24 h, pH 5.6. XynA (Δ); Multifect (○), and Accellerase (□). Bars are standard errors of triplicates.

Figure 3 shows that Multifect stands out, with a xylan conversion of 78.7%. Moreover, the higher xylose concentration achieved with this enzyme at the end of the experiments indicates that this xylanase complex has a more stable β-xylosidase enzyme, responsible for catalyzing the hydrolysis of xylobiose, the essential final step for the complete xylan hydrolysis. Hence, this xylanase seems to be the most suitable for the production of xylose in the SHIF process, among the studied enzymes.

Accellerase has the highest enzymatic activity under standard conditions. However, in long-term reaction it was able to convert only 58.8% of the available xylan. Several factors may have contributed to Accellerase's inferior performance, such as the amount of each enzyme in the complex, thermal inactivation, substrate affinity, and inhibitory effects.

XynA was already known to have a strictly endoxylanase action, lacking β-xylosidase activity and consequently not producing xylose when hydrolyzing xylan [28]. Accordingly, it presented the lowest conversion (44.8%), probably due to the absence of debranching enzymes.

None of the tested xylanases reached 100% of xylan conversion. Indeed, the incapacity of xylanases to completely hydrolyze xylan has been previously reported. Akpinar et al. [37] observed a yield of 13.8% for tobacco xylan using *Aspergillus niger* xylanase (200 IU/g) at 50 °C after 24 h. Aragon et al. [38] achieved 13% of conversion in the hydrolysis of birch xylan (18 g/L) using *Aspergillus versicolor* endoxylanase immobilized on agarose-glyoxyl at 25 °C and pH 5.0. In fact, since xylan is not a linear polymer of pure xylose, its complete depolymerization requires the use of a varied pool of enzymes [21,23,39]. In this context, the xylanase Multifect CX XL A03139 was selected to be co-immobilized with XI and *S. cerevisiae* in the SHIF process.

2.4. SHIF Assay with pH Control and Excess of Mg^{2+}

In order to overcome the possible inhibition of Ca^{2+} in XI activity, a SHIF assay was performed with pH control and excess of magnesium. Beads without $CaCO_3$ in its composition were prepared, since this salt is only necessary to sustain the pH at the desired range. Moreover, an isomerization free of $CaCO_3$ would contribute to reduce the undesired Ca^{2+} effects. After all these modifications, the obtained derivative of xylanase Multifect presented 330.2 ± 8.1 IU/g (immobilization yield of 96% and recovered activity of 77%). For the SHIF experiment, the medium supplemented with 100 mM $MgSO_4$ (24.6 g/L) and 4 g/L of $CaCl_2$ (to maintain the integrity of the beads) was added together with the beads to the pH-stat stirred reactor. It is important to note that the moderate agitation used during the process did not affect the integrity of the beads. According to Rahim et al. [40], damages to Ca-alginate immobilized biocatalysts due to stirring are usually observed above 200 rpm. In fact, Carvalho et al. [41], in experiments carried out at 300 rpm, noticed a 30% reduction in the size of Ca-alginate beads during experiments with immobilized *Candida guilliermondii*. Accordingly, in the present work, an agitation of 150 rpm was employed and the structural characteristics of the biocatalyst beads were preserved.

The obtained results, showed in Figure 4, indicated a higher ethanol production in SHIF using pH control and excess of Mg^{2+} (3.1 g/L of ethanol) compared to the value of 2.2 g/L, which was achieved under the original SHIF conditions (Figure 2). Ethanol yield (0.39 g/g, 76% of the theoretical), selectivity (3.12), and productivity (0.26 g/L/h) were also improved.

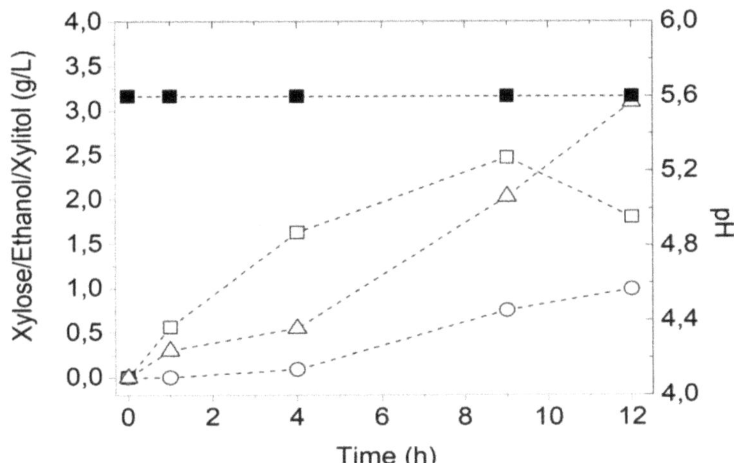

Figure 4. SHIF of previously hydrolyzed birchwood xylan with pH control and excess of Mg^{2+} using immobilized biocatalyst containing 5 w % Multifect xylanase derivative, 15 w % XI derivative, and 10 w % yeast (1.7 x 10^4 IU/L, 3.7 x 10^4 IU/L, 50 g/L), initial pH 5.6, 150 rpm stirring and 32 °C; pH (■), xylose (□), ethanol (Δ), and xylitol (○).

To the best of our knowledge, these are the highest yield and productivity reported in the literature for ethanol production from xylan through the simultaneous hydrolysis, isomerization, and fermentation (SHIF) process. Only a few works have studied ethanol production from pentoses using ex vivo isomerization and native *S. cerevisiae*, due to the differences in optimal pH and temperature ranges for each step (and none of them with co-immobilized enzymes/cells). The inclusion of the xylan hydrolysis step should not be a problem in relation to the temperature and pH of the process. Xylanases have the highest catalytic pH and temperature at approximately 5.5 and 50 °C, respectively, whereas XI optimal conditions are pH 7.0–8.0 and 70 °C [7]. Alcoholic fermentation, on the other hand, operates at pH 5.0 and 30 °C. Due to these facts, the process integration for ethanol production is still a challenge. Nakata et al. [42] studied ethanol production of hot-compressed water pretreated Japanese beech using soluble β-xylosidase, XI, and *S. cerevisiae*. It should be stressed that in this work the absence of exo- and endoxylanases would be a restraint to the saccharification step. The best results reported (0.62 g/L of ethanol, corresponding to 13% of theoretical yield) were achieved at pH 5.0, 30 °C after 100 h. Hence, the immobilized biocatalyst containing enzymes and yeast co-encapsulated reported in the present work was significantly more efficient than using enzymes and microorganisms in their soluble form, leading to a better yield and productivity.

Although there are only a few studies addressing the SHIF of xylan, the simultaneous isomerization and fermentation (SIF) of xylose has been more frequently reported. Rao et al. [27] studied the xylose SIF in the presence of 0.05 M borax to shift equilibrium concentration of xylulose/xylose and improve the isomerization step. However, although the isomerization was enhanced, only half of the available xylose was consumed. Lastick et al. [43] observed an ethanol titer of 2.1% (w/v) from the SIF of 6% xylose using XI and *Schizosaccharomyces pombe* (Y-164). Silva et al. [7] studied the SIF of 65 g/L of xylose at 30 °C, using a biocatalyst containing 32.5 x 10^3 IU/L of xylose isomerase and 20 g/L of yeast co-immobilized in Ca-alginate gel, and reported an ethanol productivity of 0.25 g/L/h. However, the isomerization step became a limiting factor, due to the decrease of the pH from 5.3 (initial) to 4.8 (final). Milessi et al. [15] incorporated $CaCO_3$ into SIF beads to control the pH of the process. The biocatalysts were prepared with 20% chitosan-immobilized XI and 10% fresh yeast. An ethanol yield of 0.35 g/g (70% of the theoretical yield) and 2 g/L/h productivity was observed. However, the long time needed by *S. cerevisiae* to ferment xylulose makes the SIF process susceptible to contamination by bacteria capable of metabolizing the xylose.

In this context, the proposed SHIF process appears as a promising approach for 2G ethanol production from hemicellulose. Process conditions and enzyme loads in the biocatalyst can be optimized to achieve higher yields and productivity as well as to overcome the difference between optimal pH ranges for each step of the process. Despite the improvement achieved after pH control and supplementation with excess of Mg^{2+}, a small accumulation of xylose was still observed, which suggests that the isomerization step may be still limiting the process. There are other factors that might be affecting the isomerization and/or the fermentation steps, such as the presence of xylooligosaccharides (2–8 xylose units) or other intermediate products released during the xylan hydrolysis, which are not present in the SIF process. The understanding of the influence of these compounds as well as the optimization of the biocatalyst composition regarding the balance of the enzyme pool are important issues to be addressed in order to improve ethanol production rates.

In general, the SHIF process using co-immobilized enzymes and cells stands out for 2G ethanol production. Besides presenting the advantages of multi-enzymatic cascade reactions, it also enables an easy recovery of the biocatalyst, which could be applied in continuous or repeated batch ethanol production runs using a medium that inhibits contamination. In addition, it builds on the advantage of using the same native yeast, already employed in 1G ethanol industry, which simplifies the operation of the fermentation unit. The low genetic stability of recombinant microorganisms together with the strict Brazilian biosafety regulations for genetically modified organisms (GMO) application in the industrial environment make the 2G ethanol production process based on a native yeast an attractive alternative.

3. Materials and Methods

3.1. Materials

GENSWEET® SGI (3400 IU/mL, 127 $mg_{protein}$/mL, DuPont™ Genencor®, Palo Alto, CA, USA), an enzymatic extract of commercial xylose isomerase (EC 5.3.1.5) from *Streptomyces rubiginosus*, and the commercial enzyme preparations Accellerase XY (3670 IU/mL, 9.8 $mg_{protein}$/mL) and Multifect CX XL A03139 (785 IU/mL, 35 $mg_{protein}$/mL) were kindly donated by DuPont™ Genencor® (Palo Alto, CA, USA). The *Bacillus subtilis* recombinant endoxylanase (502 IU/mL, 9.2 $mg_{protein}$/mL) was donated by Verdartis (Ribeirão Preto, SP, Brazil). Powdered chitosan (85.2% deacetylation degree) was supplied by Polymar Ind. Ltda (Fortaleza, CE, Brazil) and 25% glutaraldehyde solution was purchased from Vetec (Duque de Caxias, RJ, Brazil). Birchwood and beechwood xylans were from Sigma-Aldrich (St. Louis, MO, USA). A *Saccharomyces cerevisiae* industrial strain (purchased from Itaiquara®, Tapiratiba, SP, Brazil) was used in all SHIF experiments. All other reagents were of analytical grade.

3.2. Biocatalyst Production

3.2.1. Preparation of Chitosan-Glutaraldehyde Beads

Chitosan gel (2% or 4%, w/v) was prepared as described by Budriene et al. [44], using the coagulation of the chitosan-acetic acid solution in 0.5 M KOH. Activation of the support was carried out by the addition of glutaraldehyde (5%, v/v) in a suspension of chitosan at pH 7.0 (100 mM phosphate buffer, 1:10 $m_{support}/v_{suspension}$). After 60 min stirring at 25 °C, the support was filtered under vacuum, washed first with distilled water until neutrality, and then with ultrapure water.

3.2.2. Enzyme Immobilization

Xylose isomerase immobilization was carried out onto 2% (w/v) chitosan-glutaraldahyde according to Silva et al. [7]. The enzyme solution was prepared in 50 mM Tris-maleate buffer (pH 7.0) containing 5 mM $MgSO_4.7H_2O$ and 2.5 mM $CoCl_2.6H_2O$, in order to provide 50 $mg_{protein}/g_{support}$. The support was added to the enzyme solution at a ratio of 1:10 (w/v). After 20 h of immobilization at 25 °C under 150 rpm stirring, sodium borohydride was added (1 mg/mL) and the suspension was kept under gentle agitation for 30 min in an ice bath [45]. The derivatives were filtered and washed under vacuum, first with 200 mM Tris-maleate buffer (pH 7.0), then with ultrapure water, and finally with 50 mM Tris-maleate buffer (pH 7.0), in order to remove borohydride and adsorbed enzyme.

Xylanase complexes were immobilized onto 4% (w/v) chitosan-glutaraldehyde according to Milessi et al. [28]. The immobilizations were performed at pH 7.0 (100 mM phosphate buffer), 25 °C, and under constant stirring. A load of 20 $mg_{protein}/g_{support}$ was offered, maintaining 1:10 (w/v) ratio of mass of support to volume of enzymatic solution. After the completion of immobilization, sodium borohydride was added (1 mg/mL) and the reduction reaction proceeded for 30 min at 4 °C. The derivatives were extensively washed with 50 mM citrate buffer pH 4.8 and stored until use.

3.2.3. Biocatalyst Co-immobilization

The biocatalyst preparation was carried out through Ca-alginate gel entrapment according to the methodology described in Silva et al. [7]. The industrial strain, supplied as freshly compressed yeast cells, was used as purchased, without previous propagation or activation [46]. A solution of sodium alginate (1% w/v), immobilized XI (15% w/v), immobilized xylanase (5% w/v), and fresh yeast (10% w/v) was gently dropped into a 0.25 M $CaCl_2$/0.25 M $MgCl_2$ solution. Spherical particles (Ø = 1–1.5 mm) were produced using a pneumatic extruder [47]. The procedure was carried out in a laminar flow chamber (Airstream, ESCO, Horsham, PA, USA) and the sodium alginate and coagulation solutions were previously sterilized at 121 °C for 20 min. After immobilization, the beads were cured in a refrigerator for 12–16 h in cure solution (4 g/L of $MgSO_4$, 10 g/L of KH_2PO_4, 3 g/L of urea, 0.2 g/L of $CoCl_2$, and 4 g/L of $CaCl_2.2H_2O$).

3.3. Simultaneous Hydrolysis, Isomerization, and Fermentation (SHIF) of Birchwood Xylan

First, the birchwood xylan substrate was hydrolyzed in smaller xylo-oligomers by the action of recombinant endoxylanase of *B. subtilis* (XynA) in order to make xylan more similar to lignocellulosic hydrolysates, since after a pretreatment step the xylo-oligomers present in the medium are smaller than those in commercial xylan [28]. Xylan pre-hydrolysis was carried out by XynA (150 IU/g_{xylan}, 3.8 IU/mL) immobilized on chitosan-glutaraldehyde (35.0 ± 0.8 IU/g) for 24 h and 50 °C under 150 rpm stirring. At the end, the immobilized enzyme was recovered by filtration and the dried xylan mass retained on the filter was quantified (xylan insoluble fraction). The pH of the xylan solubilized fraction (SHIF substrate) was adjusted to 5.6 with HCl or NaOH 1M and the medium was sterilized by filtration through a 0.22 μm membrane. SHIF experiments were carried out in a shaker incubator (32 °C and 150 rpm), using sealed tubes with a total reaction volume of 2.4 mL (bead ratio of 1:1, 1.2 g of beads and 1.2 mL of medium, bead density was 1 g/cm^3). The composition of SHIF medium was 108 g/L of birchwood xylan supplemented with $MgSO_4$ (4 g/L), KH_2PO_4 (10 g/L), urea (3 g/L), $CoCl_2.6H_2O$ (0.2 g/L), and $CaCl_2.2H_2O$ (4 g/L). Samples were collected at regular intervals for determination of pH, substrate consumption, and product formation.

3.4. Influence of pH, Ca^{2+}, and Mg^{2+} on XI Activity

In order to study the influence of Ca^{2+} and Mg^{2+} ions on the catalytic activity of XI, the enzyme activity was measured at different pHs (5.0 to 8.0) and different ion concentrations. Nine medium compositions were tested for each studied pH. The standard medium was constituted of 2 M fructose at pH 8.0 (50 mM Tris-maleate buffer), 50 mM $MgSO_4$, and 2.5 mM $CoCl_2$.

3.5. *Enzymatic Xylan Hydrolysis for Xylanase Selection*

Xylan hydrolysis was carried out using the soluble fraction of the birchwood xylan obtained by adding 8 g of commercial birchwood xylan in 100 mL of 50 mM citrate buffer, pH 5.5 at 50 °C. After 1 h at 150 rpm stirring, the solution was centrifuged for 20 min at 9500× *g* and 5 °C. The supernatant was then recovered for further use at the concentration of 25 g/L of xylan. It was offered 150 IU/g_{xylan} (3.8 IU/$mL_{reactor}$). The reaction was conducted at 50 °C under mechanical stirring for 24 h.

3.6. *Analytical Methods*

3.6.1. Xylanase Activity

Xylanase activity was determined according to International Union of Pure and Applied Chemistry (IUPAC) [48] by calculating the initial velocity of xylan hydrolysis catalyzed by a known amount of enzyme. The standard substrate was birchwood xylan (1% w/v) in 50 mM citrate buffer pH 5.5. Enzyme was added to the reaction medium and incubated at 50 °C for 10 min under 250 rpm stirring. Aliquots were withdrawn at 2 min intervals, and the released reducing sugars were quantified by the dinitrosalicylic (DNS) acid method [49]. One unit of activity (IU) was defined as the amount of enzyme required to release 1 µmol of xylose per minute under the assayed conditions.

3.6.2. Xylose Isomerase Activity

Xylose isomerase activity was determined according to Giordano et al. [13], by measuring the initial velocity of fructose isomerization to glucose, under the following conditions: 2 M fructose solution prepared in 50 mM Tris-maleate buffer containing 50 mM $MgSO_4.7H_2O$ and 2.5 mM $CoC1_2.6H_2O$, at pH 7.0 and 60 °C. The glucose concentration was determined colorimetrically using the commercial enzyme kit containing glucose oxidase and peroxidase (GOD-PAP®, Bioclin, Belo Horizonte, Mg, Brazil). One international unit (IU) of xylose isomerase was defined as the amount of enzyme that released 1 µmol of glucose per minute under the assayed conditions.

3.6.3. Substrate and Product Quantification

The concentrations of XOs, xylose, xylulose, xylitol, and ethanol were determined by high performance liquid chromatography (HPLC), equipped with a Waters Sugar-Pak I column (Milford, MA, USA) (300 × 6.5 mm) coupled to a refractive index detector (W410 Waters) (Milford, MA, USA). Ultrapure water was used as eluent at a flow rate of 0.5 mL/min. The column temperature was 80 °C, the detector was set at 40 °C, and the injected volume was 20 µL. Before the analysis, the samples were filtered using a 0.22 µm filter.

4. Conclusions

A new biocatalyst composed of co-immobilized xylanases, xylose isomerase, and unmodified *S. cerevisiae* was able to produce 2G ethanol from birchwood xylan in a one-spot multi-enzymatic simultaneous hydrolysis, isomerization, and fermentation (SHIF) process. Although more studies are required to increase ethanol productivity, the SHIF process showed to be promising from the point of view of its technical viability. The SHIF process brings advantages to industrial applications for easing the integration of 1G and 2G ethanol production processes (because both are based on the same native yeast strains), while reducing the contamination risk due to the use of xylo-oligomers as substrate instead of xylose.

Author Contributions: T.S.M. and F.A.S.C. performed the experiments, analyzed the data, and wrote the paper; W.K. performed the experiments; P.W.T and T.C.Z. analyzed the data and revised the paper; R.C.G. and R.L.C.G. supervised the work and reviewed the final manuscript.

#140982/2013-2, and in part by the Coordenação de Aperfeiçoamento de Pessoal de Nível Superior - Brazil (CAPES), Finance Code 001.

Acknowledgments: The authors thank DuPont™ Genencor® (USA) for the donation of xylose isomerase, Accellerase, and Multifect xylanases; and Verdartis for the donation of xylanase from *B. subtilis*.

References

1. Antunes, F.A.F.; Chandel, A.K.; Milessi, T.S.S.; Santos, J.C.; Rosa, C.A.; da Silva, S.S. Bioethanol Production from Sugarcane Bagasse by a Novel Brazilian Pentose Fermenting Yeast Scheffersomyces shehatae UFMG-HM 52.2: Evaluation of Fermentation Medium. *Int. J. Chem. Eng.* **2014**, *2014*, 1–8. [CrossRef]
2. UNFCC Conference of the Parties. Adoption of the Paris Agreement. *Report No. FCC/CP/2015/L.9/Rev1. United Nations Framew. Conv. Clim. Chang.* 2015. Available online: http://unfccc.int/resource/docs/2015/cop21/eng/l09r01.pdf (accessed on January 2019).
3. Nigam, P.S.; Singh, A. Production of liquid biofuels from renewable resources. *Prog. Energy Combust. Sci.* **2011**, *37*, 52–68. [CrossRef]
4. Canilha, L.; Chandel, A.K.; Suzane dos Santos Milessi, T.; Antunes, F.A.F.; Luiz da Costa Freitas, W.; das Graças Almeida Felipe, M.; da Silva, S.S. Bioconversion of Sugarcane Biomass into Ethanol: An Overview about Composition, Pretreatment Methods, Detoxification of Hydrolysates, Enzymatic Saccharification, and Ethanol Fermentation. *J. Biomed. Biotechnol.* **2012**, *2012*, 1–15. [CrossRef] [PubMed]
5. Dos Santos, L.V.; de Barros Grassi, M.C.; Gallardo, J.C.M.; Pirolla, R.A.S.; Calderón, L.L.; de Carvalho-Netto, O.V.; Parreiras, L.S.; Camargo, E.L.O.; Drezza, A.L.; Missawa, S.K.; et al. Second-Generation Ethanol: The Need is Becoming a Reality. *Ind. Biotechnol.* **2016**, *12*, 40–57. [CrossRef]
6. Hahn-Hägerdal, B.; Jeppsson, H.; Skoog, K.; Prior, B.A. Biochemistry and physiology of xylose fermentation by yeasts. *Enzyme Microb. Technol.* **1994**, *16*, 933–943. [CrossRef]
7. Silva, C.R.; Zangirolami, T.C.; Rodrigues, J.P.; Matugi, K.; Giordano, R.C.; Giordano, R.L.C. An innovative biocatalyst for production of ethanol from xylose in a continuous bioreactor. *Enzyme Microb. Technol.* **2012**, *50*, 35–42. [CrossRef] [PubMed]
8. Demeke, M.M.; Dietz, H.; Li, Y.; Foulquié-Moreno, M.R.; Mutturi, S.; Deprez, S.; Den Abt, T.; Bonini, B.M.; Liden, G.; Dumortier, F.; et al. Development of a D-xylose fermenting and inhibitor tolerant industrial Saccharomyces cerevisiae strain with high performance in lignocellulose hydrolysates using metabolic and evolutionary engineering. *Biotechnol. Biofuels* **2013**, *6*, 89. [CrossRef] [PubMed]
9. Zhang, T.; Chi, Z.; Zhao, C.H.; Chi, Z.M.; Gong, F. Bioethanol production from hydrolysates of inulin and the tuber meal of Jerusalem artichoke by Saccharomyces sp. W0. *Bioresour. Technol.* **2010**, *101*, 8166–8170. [CrossRef] [PubMed]
10. Hector, R.E.; Mertens, J.A.; Bowman, M.J.; Nichols, N.N.; Cotta, M.A.; Hughes, S.R. Saccharomyces cerevisiae engineered for xylose metabolism requires gluconeogenesis and the oxidative branch of the pentose phosphate pathway for aerobic xylose assimilation. *Yeast* **2011**, *28*, 645–660. [CrossRef] [PubMed]
11. Kuyper, M.; Winkler, A.A.; van Dijken, J.P.; Pronk, J.T. Minimal metabolic engineering of Saccharomyces cerevisiae for efficient anaerobic xylose fermentation: A proof of principle. *FEMS Yeast Res.* **2004**, *4*, 655–664. [CrossRef] [PubMed]
12. Ohgren, K.; Bengtsson, O.; Gorwa-Grauslund, M.F.; Galbe, M.; Hahn-Hägerdal, B.; Zacchi, G. Simultaneous saccharification and co-fermentation of glucose and xylose in steam-pretreated corn stover at high fiber content with Saccharomyces cerevisiae TMB3400. *J. Biotechnol.* **2006**, *126*, 488–498. [CrossRef] [PubMed]
13. Giordano, R.L.C.; Giordano, R.C.; Cooney, C.L. A study on intra-particle diffusion effects in enzymatic reactions: Glucose-fructose isomerization. *Bioprocess Eng.* **2000**, *23*, 0159–0166. [CrossRef]
14. Gong, C.S. Recent Advances in D-Xylose Conversion by Yeasts. In *Annual Reports on Fermentation Processes*; Tsao, G.T., Ed.; Academic Press: New York, NY, USA, 1983; Volume 6, pp. 253–297.
15. Milessi, T.S.; Aquino, P.M.; Silva, C.R.; Moraes, G.S.; Zangirolami, T.C.; Giordano, R.C.; Giordano, R.L.C. Influence of key variables on the simultaneous isomerization and fermentation (SIF) of xylose by a native Saccharomyces cerevisiae strain co-encapsulated with xylose isomerase for 2G ethanol production. *Biomass Bioenergy* **2018**, *119*, 277–283. [CrossRef]

16. Silva, S.S.; Mussatto, S.I.; Santos, J.C.; Santos, D.T.; Polizel, J. Cell immobilization and xylitol production using sugarcane bagasse as raw material. *Appl. Biochem. Biotechnol.* **2007**, *141*, 215–227. [CrossRef] [PubMed]
17. Guisan, J.M. New Opportunities for Immobilization of Enzymes. In *Methods in Molecular Biology*; Guisan, J.M., Ed.; Springer: Clifton, NJ, USA, 2013; pp. 1–13.
18. DiCosimo, R.; McAuliffe, J.; Poulose, A.J.; Bohlmann, G. Industrial use of immobilized enzymes. *Chem. Soc. Rev.* **2013**, *42*, 6437–6474. [CrossRef] [PubMed]
19. Choi, J.-M.; Han, S.-S.; Kim, H.-S. Industrial applications of enzyme biocatalysis: Current status and future aspects. *Biotechnol. Adv.* **2015**, *33*, 1443–1454. [CrossRef] [PubMed]
20. Hernández, R.M.A.; Orive, G.; Murua, A.; Pedraz, J.L. Microcapsules and microcarriers for in situ cell delivery. *Adv. Drug Deliv. Rev.* **2010**, *62*, 711–730. [CrossRef] [PubMed]
21. Carvalho, A.F.A.; de Oliva Neto, P.; da Silva, D.F.; Pastore, G.M. Xylo-oligosaccharides from lignocellulosic materials: Chemical structure, health benefits and production by chemical and enzymatic hydrolysis. *Food Res. Int.* **2013**, *51*, 75–85. [CrossRef]
22. Gray, M.C.; Converse, A.O.; Wyman, C.E. Solubilities of Oligomer Mixtures Produced by the Hydrolysis of Xylans and Corn Stover in Water at 180 °C. *Ind. Eng. Chem. Res.* **2007**, *46*, 2383–2391. [CrossRef]
23. Biely, P.; Singh, S.; Puchart, V. Towards enzymatic breakdown of complex plant xylan structures: State of the art. *Biotechnol. Adv.* **2016**, *34*, 1260–1274. [CrossRef] [PubMed]
24. Giordano, R.L.C.; Giordano, R.C.; Zangirolami, T.C.; Tardioli, P.W.; Kopp, W.; Milessi, T.; Silva, C.R.; Cavalcanti-Montano, I.D.; Galeano-Suarez, C.A.; Rojas, M.J.; et al. Sistema catalítico e processo de obtenção de bioetanol 2g a partir de xilana/oligômeros de xilose. Brazilian Patent BR102014023394, 19 September 2014.
25. Macrelli, S.; Galbe, M.; Wallberg, O. Effects of production and market factors on ethanol profitability for an integrated first and second generation ethanol plant using the whole sugarcane as feedstock. *Biotechnol. Biofuels* **2014**, *7*, 26. [CrossRef] [PubMed]
26. Ricca, E.; Brucher, B.; Schrittwieser, J.H. Multi-Enzymatic Cascade Reactions: Overview and Perspectives. *Adv. Synth. Catal.* **2011**, *353*, 2239–2262. [CrossRef]
27. Rao, K.; Chelikani, S.; Relue, P.; Varanasi, S. A Novel Technique that Enables Efficient Conduct of Simultaneous Isomerization and Fermentation (SIF) of Xylose. *Appl. Biochem. Biotechnol.* **2008**, *146*, 101–117. [CrossRef] [PubMed]
28. Milessi, T.S.S.; Kopp, W.; Rojas, M.J.; Manrich, A.; Baptista-Neto, A.; Tardioli, P.W.; Giordano, R.C.; Fernandez-Lafuente, R.; Guisan, J.M.; Giordano, R.L.C. Immobilization and stabilization of an endoxylanase from Bacillus subtilis (XynA) for xylooligosaccharides (XOs) production. *Catal. Today* **2016**, *259*, 130–139. [CrossRef]
29. Fuxreiter, M.; Böcskei, Z.; Szeibert, A.; Szabó, E.; Dallmann, G.; Naray-Szabo, G.; Asboth, B. Role of electrostatics at the catalytic metal binding site in xylose isomerase action: Ca(2+)-inhibition and metal competence in the double mutant D254E/D256E. *Proteins* **1997**, *28*, 183–193. [CrossRef]
30. Kovalevsky, A.Y.; Hanson, L.; Fisher, S.Z.; Mustyakimov, M.; Mason, S.A.; Trevor Forsyth, V.; Blakeley, M.P.; Keen, D.A.; Wagner, T.; Carrell, H.L.; et al. Metal Ion Roles and the Movement of Hydrogen during Reaction Catalyzed by D-Xylose Isomerase: A Joint X-Ray and Neutron Diffraction Study. *Structure* **2010**, *18*, 688–699. [CrossRef] [PubMed]
31. Bhosale, S.H.; Rao, M.B.; Deshpande, V. V Molecular and industrial aspects of glucose isomerase. *Microbiol. Rev.* **1996**, *60*, 280–300. [PubMed]
32. Asboth, B.; Naray-Szabo, G. Mechanism of Action of D-Xylose Isomerase. *Curr. Protein Pept. Sci.* **2000**, *1*, 237–254. [CrossRef] [PubMed]
33. Allen, K.N.; Lavie, A.; Glasfeld, A.; Tanada, T.N.; Gerrity, D.P.; Carlson, S.C.; Farber, G.K.; Petsko, G.A.; Ringe, D. Role of the divalent metal ion in sugar binding, ring opening, and isomerization by D-xylose isomerase: Replacement of a catalytic metal by an amino acid. *Biochemistry* **1994**, *33*, 1488–1494. [CrossRef] [PubMed]
34. Kasumi, T.; Hayashi, K.; Tsumura, N. Roles of Magnesium and Cobalt in the Reaction of Glucose Isomerase from Streptomyces griseofuscus S-41. *Agric. Biol. Chem.* **1982**, *46*, 21–30. [CrossRef]
35. Jacobsen, S.E.; Wyman, C.E. Xylose Monomer and Oligomer Yields for Uncatalyzed Hydrolysis of Sugarcane Bagasse Hemicellulose at Varying Solids Concentration. *Ind. Eng. Chem. Res.* **2002**, *41*, 1454–1461. [CrossRef]
36. Dodd, D.; Cann, I.C.K.O. Enzymatic deconstruction of xylan for biofuel production. *GCB Bioenergy* **2009**, *1*, 2–17. [CrossRef] [PubMed]

37. Akpinar, O.; Erdogan, K.; Bostanci, S. Production of xylooligosaccharides by controlled acid hydrolysis of lignocellulosic materials. *Carbohydr. Res.* **2009**, *344*, 660–666. [CrossRef] [PubMed]
38. Aragon, C.C.; Santos, A.F.; Ruiz-Matute, A.I.; Corzo, N.; Guisan, J.M.; Monti, R.; Mateo, C. Continuous production of xylooligosaccharides in a packed bed reactor with immobilized–stabilized biocatalysts of xylanase from Aspergillus versicolor. *J. Mol. Catal. B Enzym.* **2013**, *98*, 8–14. [CrossRef]
39. Biely, P. Xylanolytic Enzymes. In *Handbook of Food Enzymology*; Whitaker, J., Voragen, A., Wong, D., Eds.; CRC Press: Boca Raton, FL, USA, 2003; pp. 879–916.
40. Rahim, S.N.A.; Alawi, S.; Hamid, K.H.K.; Edama, N.A.; Baharuddin, A.S. Effect of Agitation Speed for Enzymatic Hydrolysis of Tapioca Slurry Using Encapsulated Enzymes in an Enzyme Bioreactor. *Int. J. Chem. Eng. Appl.* **2015**, *6*, 38–41.
41. Carvalho, W.; Canilha, L.; Silva, S.S. Semi-continuous xylose-to-xylitol bioconversion by Ca-alginate entrapped yeast cells in a stirred tank reactor. *Bioprocess Biosyst. Eng.* **2008**, *31*, 493–498. [CrossRef] [PubMed]
42. Nakata, T.; Miyafuji, H.; Saka, S. Ethanol production with β-xylosidase, xylose isomerase, and *Saccharomyces cerevisiae* from the hydrolysate of Japanese beech after hot-compressed water treatment. *J. Wood Sci.* **2009**, *55*, 289–294. [CrossRef]
43. Lastick, S.M.; Tucker, M.Y.; Beyette, J.R.; Noll, G.B.; Grohmann, K. Simultaneous fermentation and isomerization of xylose. *Appl. Microbiol. Biotechnol.* **1989**, *30*, 574–579. [CrossRef]
44. Budriene, S.; Gorochovceva, N.; Romaskevic, T.; Yugova, L.V.; Miezeliene, A.; Dienys, G.; Zubriene, A. β-Galactosidase from Penicillium canescens. Properties and immobilization. *Cent. Eur. J. Chem.* **2005**, *3*, 95–105. [CrossRef]
45. Guisán, J. Aldehyde-agarose gels as activated supports for immobilization-stabilization of enzymes. *Enzyme Microb. Technol.* **1988**, *10*, 375–382. [CrossRef]
46. Basso, L.C.; de Amorim, H.V.; de Oliveira, A.J.; Lopes, M.L. Yeast selection for fuel ethanol production in Brazil. *FEMS Yeast Res.* **2008**, *8*, 1155–1163. [CrossRef] [PubMed]
47. Trovati, J.; Giordano, R.C.; Giordano, R.L.C. Improving the Performance of a Continuous Process for the Production of Ethanol from Starch. *Appl. Biochem. Biotechnol.* **2009**, *156*, 76–90. [CrossRef] [PubMed]
48. Ghose, T.K.; Bisaria, V.S. Measurement of hemicellulase activities: Part I Xylanases. *Pure Appl. Chem.* **1987**, *59*, 1739–1751. [CrossRef]
49. Miller, G.L. Use of dinitrosalicylic acid reagent for determination of reducing sugar. *Anal. Chem.* **1959**, *31*, 426–428. [CrossRef]

7

Surfactant Imprinting Hyperactivated Immobilized Lipase as Efficient Biocatalyst for Biodiesel Production from Waste Cooking Oil

Huixia Yang and Weiwei Zhang *

State Key Laboratory of High-efficiency Utilization of Coal and Green Chemical Engineering, School of Chemistry and Chemical Engineering, Ningxia University, Yinchuan 750021, China; yhx6668297@sina.com
* Correspondence: zhangww@nxu.edu.cn

Abstract: Enzymatic production of biodiesel from waste cooking oil (WCO) could contribute to resolving the problems of energy demand and environment pollutions.In the present work, *Burkholderia cepacia* lipase (BCL) was activated by surfactant imprinting, and subsequently immobilized in magnetic cross-linked enzyme aggregates (mCLEAs) with hydroxyapatite coated magnetic nanoparticles (HAP-coated MNPs). The maximum hyperactivation of BCL mCLEAs was observed in the pretreatment of BCL with 0.1 mM Triton X-100. The optimized Triton-activated BCL mCLEAs was used as a highly active and robust biocatalyst for biodiesel production from WCO, exhibiting significant increase in biodiesel yield and tolerance to methanol. The results indicated that surfactant imprinting integrating mCLEAs could fix BCL in their active (open) form, experiencing a boost in activity and allowing biodiesel production performed in solvent without further addition of water. A maximal biodiesel yield of 98% was achieved under optimized conditions with molar ratio of methanol-to-WCO 7:1 in one-time addition in hexane at 40 °C. Therefore, the present study displays a versatile method for lipase immobilization and shows great practical latency in renewable biodiesel production.

Keywords: biodiesel; waste cooking oil; lipase immobilization; interfacial activation; functionalized magnetic nanoparticles

---

1. Introduction

Over the past decades, biodiesel has attracted great interest as a sustainable alternative for fossil fuels in virtue of the depletion of fossilized fuel resources and their environmental impacts [1]. Biodiesel is a renewable and clean energy, and possess favorable advantages in combustion emission like low emissions of CO, sulfur free, low hydrocarbon aroma, high cetanenumber, and high flash point [2].

The conventional chemical technologies for biodiesel production involve the use of acid or basic catalysts (i.e., NaOH, KOH, and H_2SO_4), thus numerous disadvantages are inescapable, for example, high corrosive procedure, high energy consumption, high quantities of waste pollution, and costly in efficient product separation processes [3]. Furthermore, in order to prevent the hydrolysis reaction and saponification, high quality oils are required, with low contents of water and free fatty acids [4].

Feedstocks used for biodiesel can be allocated five categories, including edible vegetable oils, non-edible plant oils, animal fats, microalgae oils, and waste oils [5]. The global application of first-generation biodiesel produced by using edible oils, was restricted due to food scarcity and high cost of the edible oils [6]. Biodiesel production from waste cooking oils (WCO) could be a promising and cost effective candidate in handling issues associated with energy crisis, environmental concerns, and total cost reduction of biodiesel production [7]. Moreover, 15 million tons of WCO are produced annually throughout the world [8], bringing great challenge in reasonable management of such oils on account of environment concerns [9]. However, using WCO as raw material is quite challenging as it contains a high amount of free fatty acids (FFAs) and water which could hinder the homogeneous alkaline-catalyzed transesterification in conventional biodiesel production processes [10]. Complete conversion of these low-quality feedstocks like WCO could be accomplished in enzymatic biodiesel production without saponification. Therefore, enzyme-catalyzed transesterification has become a laudable potential alternative for biodiesel synthesis.

Particularly, lipases are foremost and efficient enzymes implemented in biodiesel production. Lipase-catalyzed process exhibits key advantages such as no soap formation, high-purity products, easy product removal, adaptable to different biodiesel feedstock, environmentally friendly, and mild operating conditions [5]. However, the commercialization of enzymatic biodiesel production remains complicated, because of high price and low stability of lipases as well as low reaction rate of biocatalysis. Heterogeneous enzyme-catalyzed transesterification using immobilized lipases is a possible solution to these problems [11].

Immobilization of enzymes has been investigated for many years, and lipase can generally be immobilized by various techniques such as cross-linking, adsorption, entrapment and encapsulation [12,13]. Thereinto, cross-linked enzyme aggregates (CLEAs) is a cheap and efficient strategy for enzyme immobilization, which has broad applicability over numerous enzyme classes. Owing to its outstanding resistance to organic solvents, extreme pH, and high temperatures, CLEAs has attracted growing attention in cost effective biocatalysis [14]. Nevertheless, small particle size and low mechanical stability of CLEAs could directly affect mass transfer and stability under operational conditions, thus accordingly cause problems in practical use [15]. An alternative approach for circumventing compressed construction of CLEAs is to use "smart" magnetic CLEAs (mCLEAs). Magnetic nanoparticles (MNPs) could provide enhanced stability over repeated uses, especially for enzymes having low amount of lysine residues on their surface. Besides, mCLEAs could perform easily separation using a permanent magnet, affording novel combinations of bioconversions and down-streaming processes, thus provide the necessary reduction in enzymecosts to enable commercial viability.

Among various types of nanomaterials, MNPs have attracted substantial attention in enzyme immobilizations. However, bare MNPs tend to aggregate due to their high surface energy and are easily oxidized in the air leading to loss of magnetism and dispersibility, thus limiting their exploitation in practical applications [16]. The surface modification with an organic or an inorganic shell is an appropriate strategy to address these issues. Due to their excellent biocompatibility, slow biodegradation, high surface area-to-volume-ratio, and unique mechanical stability, Hydroxyapatite (HAP) could be a proper inorganic surface coating for MNPs [17]. Moreover, HAP-coated MNPs can be easily functionalized with organosilanes, and consequently has great application potential in enzyme immobilization.

Burkholderia cepacia lipase (BCL)is one of the most widely used lipases in biocatalysis [18]. On account of its versatility to accommodate a wide variety of substrates, high heat resistance, and good tolerance to polar organic solvents, BCL has been extensively used in various biotechnological processes, especially for biodiesel production. The active site of BCL is shielded by a mobile element, called the lid or flap [19]. The displacement of lid or flap to closed or open position, which directly impacts the accessibility of active site, determines the enzyme in an in active or active conformation. In general, substrate access to the underlying active site is prohibited in its closed configuration. As the stabilization of the open conformation of all lipases could remarkably increase their catalytic activity, a favorable method to obtain highly active biocatalysts should try to immobilize lipases in their most active form (open conformation).

Generally, the preparation of immobilized enzyme with enhanced activity and stability is a persistent goal of the biotechnology industry to seek maximum profit. Therefore, developing a simple and efficient approach for lipase interfacial activation in immobilization is highly desirable. Bioimprinting is a commonly used method for achieving hyperactivation of lipases in organic media.

The principle of bioimprinting is to "anchor" the enzyme in its active form, which could be achieved by binding with imprint molecules (such as surfactants, natural substrates, substrate analogs etc.). From an applied point of view, the dramatic hyperactivation of lipases by low concentrations of surfactants is an expeditious and facile method for lipase interfacial activation [20].

To develop an efficient and environmentally benign process for the biodiesel production from waste cooking oils, in the present study surfactant imprinting strategy on BCL was implemented in combination with mCLEAs immobilization using HAP-coated MNPs. Subsequent cross-linking could "lock" BCL in its favorable conformation, while HAP-coated MNPs could facilitate the recovery of immobilized BCL and simplify the biodiesel purification. To the best of our knowledge, this is the first report on BCL immobilization integrating surfactant imprinting and mCLEAs. The optimal conditions for mCLEAs preparation, along with the effect of different surfactants (anionic, cationic, and non-ionic) on the catalytic activity of BCL mCLEAs in transesterification were studied. The optimized surfactant-activated BCL mCLEAs was further used in transesterification of waste cooking oils to biodiesel. In addition, a detailed analysis of solvents, methanol-to-oil molar ratio, and temperatures on the yield of biodiesel production was presented. The results obtained in the research are expected to provide a reliable basis for further exploration of lipase immobilization and efficient biodiesel production in industry.

2. Results and Discussion

2.1. Preparation and Characterization of Immobilized Lipase

In this study, the prepared MNPs encapsulated by hydroxyapatite (HAP) were used as immobilization supports. The amino functionalization of HAP-coated MNPs was carried out using 3-aminopropyltrimethoxysilane (APTES) for efficient enzyme attachment. Typically, the preparation procedure of immobilization supports and surfactant-activated BCL mCLEAs were performed according to the scheme shown in Scheme 1. The prepared magnetic supports and immobilized BCL were characterized by fourier transform infrared spectroscopy (FT-IR), transmission electron microscope (SEM) and vibrating sample magnetometer (VSM).

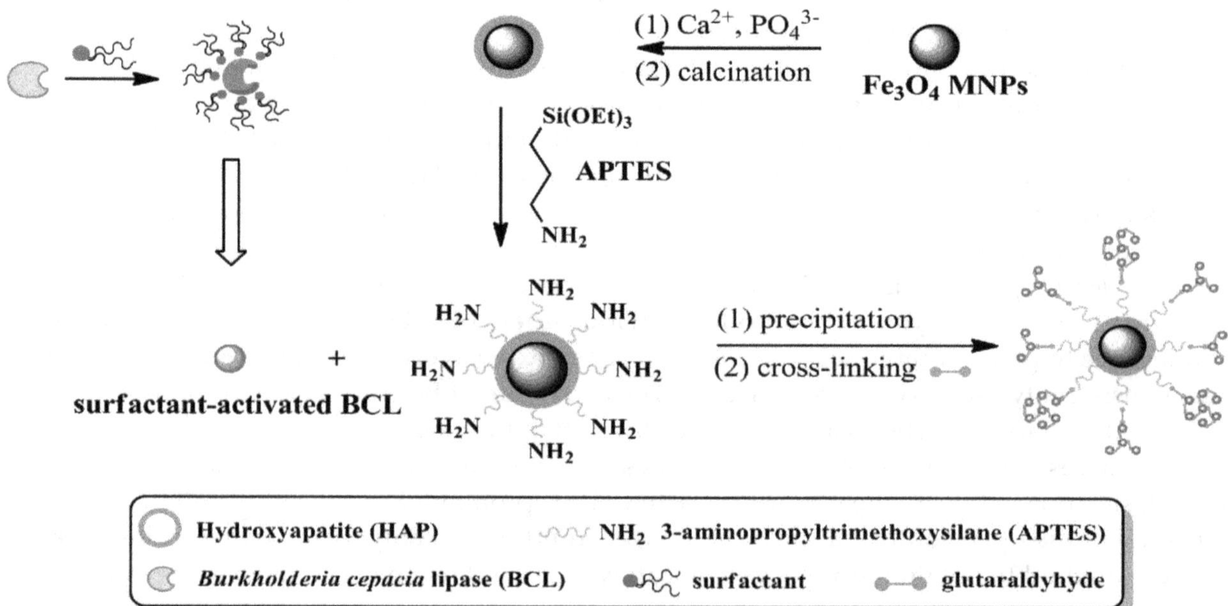

Scheme 1. Preparation procedure of immobilization supports and surfactant-activated *Burkholderia cepacia* lipase (BCL) magnetic cross-linked enzyme aggregates (mCLEAs).

FTIR characterization was performed to investigate the chemical composition of functionalized MNPs and immobilized BCL. Spectra were recorded on over the region from 4000 to 400 cm^{-1}.

As shown in Figure 1, the strong peak at 588 and 639 cm^{-1} corresponds to the stretching vibration of Fe-O bond. The characteristic absorption bands related to the HAP appease at 565 and 1044 cm^{-1}, which are assigned to phosphate groups [21]. In the IR spectrum of modified MNPs and BCL mCLEAs, the characteristic absorption bands related to the functional groups of HAP emerged clearly, which demonstrated the successful incorporation of MNPs with HAP. For all immobilized lipases, including BCL CLEAs, BCL mCLEAs and surfactant-activated BCL mCLEAs, the typical IR bands responsible for the lipase that were chemically covalent-bonded to the functionalized MNPs were observed at 1642 cm^{-1} for amide I (C=O stretching vibration) and at 1539 cm^{-1} for amide II (N-H bending vibration), respectively. Besides, compared with the results shown in Figure 1, aliphatic C-H stretch band at 2859 and 2927 cm^{-1}, corresponding to C-H stretching vibrations, are clearly observed in all immobilized lipases, which also indicated the successful loading of lipase.

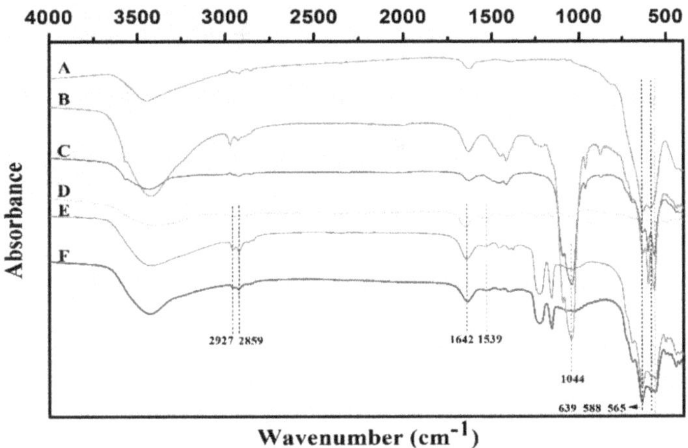

Figure 1. Spectra of (**A**) Fe_3O_4 MNPs, (**B**) hydroxyapatite coated magnetic nanoparticles (HAP-coated MNPs), (**C**) 3-aminopropyltrimethoxysilane (APTES)-HAP-coated MNPs, (**D**) BCL CLEAs, (**E**) BCL mCLEAs, (**F**) Triton-activated BCL mCLEAs.

In order to assess morphology, size and composition of functionalized MNPs and immobilized BCL, SEM images were collected and illustrated in Figure 2. As seen in Figure 2, bare Fe_3O_4 MNPs formed significantly dense agglomeration, because of their high surface energy and strong dipole-dipole interactions. It is obvious that the structure of Fe_3O_4 MNPs becomes looser and more evenly distributed after being functionalized with HAP (Figure 2B) and APTES (Figure 2C), suggesting that surface modification is favorable for preventing aggregation of Fe_3O_4 MNPs. At the same time, the rough surface of Fe_3O_4 MNPs also increased the surface area for attachment of enzyme.

The crucial structure factors in aggregated-based enzyme immobilization, including morphological topographies, structural arrangement and size, play an important role in affecting substrate affinity and stability of biocatalyst [22]. Besides, the particle size of enzymes is an important property of any heterogeneous catalysis since it can directly affect the diffusion of substrates and catalytic efficiency, especially in the internal enzymes of highly compact aggregates [23]. SEM images (Figure 2D) of standard BCL CLEAs revealed no defined morphologies and large size particles. Moreover, standard BCL CLEAs presented a uniform and compact surface with the presence of few tiny pores. On the contrary, after the incorporation of functionalized MNPs, BCL mCLEAs formed spherical structures and small particle sizes, which could reduce inner steric hindrance in closely packed CLEAs. It is noteworthy that the presence of functionalized MNPs displayed large active surface available for lipase immobilization, therefore were important for development of a stabilized enzyme-matrix. Furthermore, a loose and homodispersed structure of Triton-activated BCL mCLEAs was found in Figure 2F, suggesting that the formation of large aggregates were forbidden by the imprinting of surfactants. From the SEM outcomes, it can be discerned that, thanks to the coating of surfactants, lipase could be uniformly dispersed on functionalized MNPs, which could contribute to a wider surface area with more catalytic sites and decrease the diffusion limit. Consequently, compared with standard BCL CLEAs, Triton-activated BCL mCLEAs could perform superior catalytic efficiency.

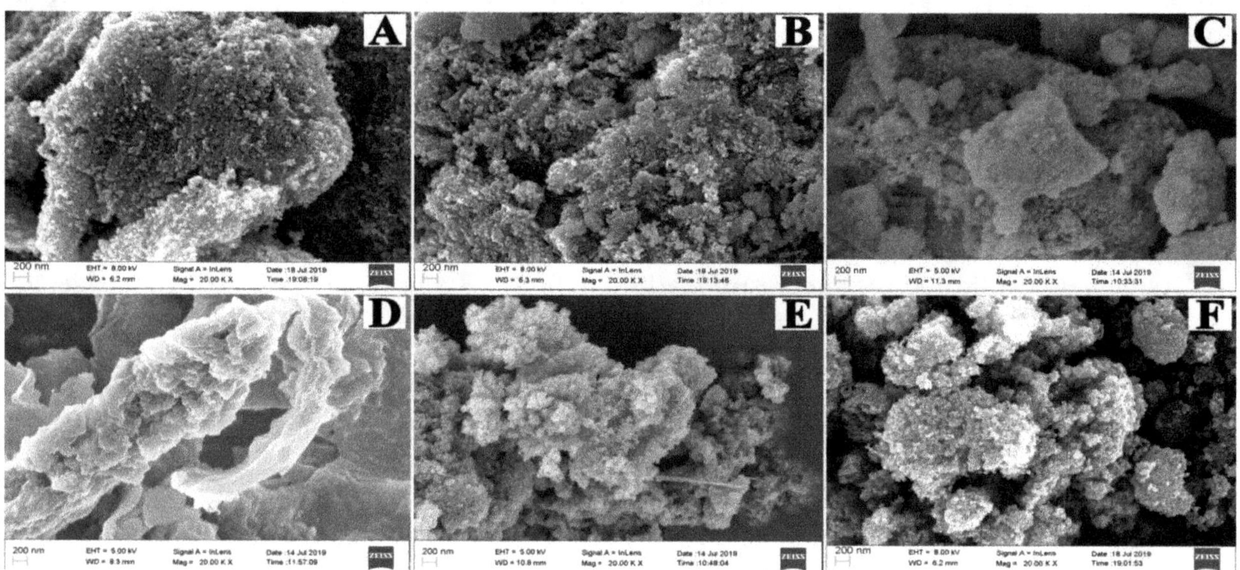

Figure 2. Images of(**A**) Fe_3O_4 MNPs, (**B**) HAP-coated MNPs, (**C**) APTES-HAP-coated MNPs, (**D**) BCL CLEAs, (**E**) BCL mCLEAs, (**F**) Triton-activated BCL mCLEAs.

The magnetic property of functionalized MNPs and immobilized BCL were measured using VSM. The hysteresis curves of the Fe_3O_4 MNPs, HAP-coated MNPs, APTES-HAP-coated MNPs, BCL mCLEAs and Triton-activated BCL mCLEAs shown in Figure 3, exhibited a perfect sigmoidal behavior, corresponding to a typical superparamagnetism.

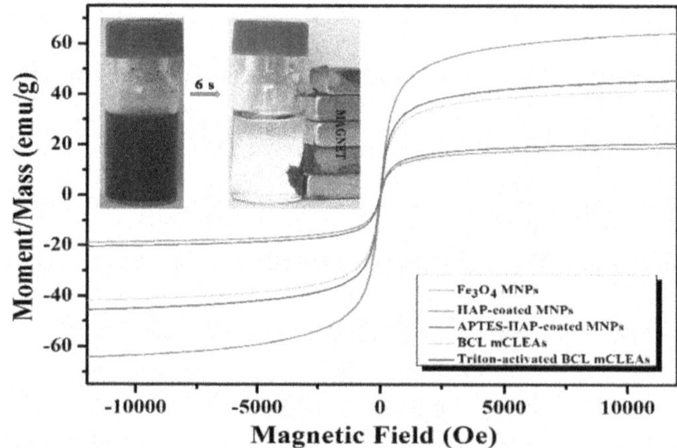

Figure 3. Hysteresis loops of Fe_3O_4 MNPs, HAP-coated MNPs, APTES-HAP-coated MNPs, BCL mCLEAs and Triton-activated BCL mCLEAs. The inner shows the easy magnetic separation of Triton-activated BCL mCLEAs in reaction mixture.

With further functionalization of MNPs, the saturation magnetization value decreased and correlated with the increase of the core-shell layer. Interestingly, it is obviously observed that the saturation magnetization value of Triton-activated BCL mCLEAs increased visibly compared to BCL mCLEAs. It might be due to the uniform dispersion of lipase on MNPs and availability of large surface area which decreased the shielding-effect of the out layer substances. As seen in Figure 3 (inner), Triton-activated BCL mCLEAs showed fast response (6s) to the external magnetic field andcould be easily recovered from the reaction mixture. After removing the external magnetic field, the magnetic immobilized BCL redispersed rapidly by a slight shake, indicating good dispersion and efficient recyclability in industrial application.

2.2. Optimization of the Immobilization Conditions

In this study, the enzymes were precipitated by adding water-miscible organic solvents (acetone, ethanol and 2-propanol), PEG 800 and ammonium sulfate. The optimum precipitant was selected by measuring the transesterification activity of the corresponding BCL mCLEAs. Compared with free BCL, all BCL mCLEAs prepared using different precipitants performed higher transesterification activity in organic solvent. Among the protein precipitants evaluated, ammonium sulfate showed maximum recovery of activity (Figure 4a), therefore was further used in BCL immobilization.

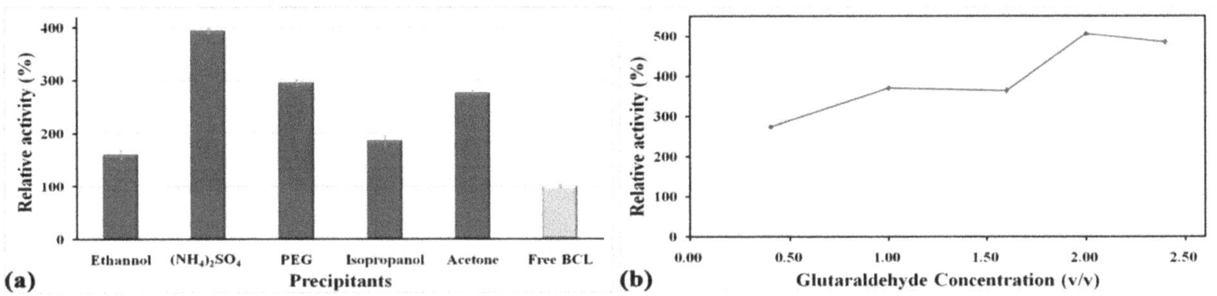

Figure 4. (**a**) Precipitant type and (**b**) glutaraldehyde concentration on the activities of BCL mCLEAs.

Traditionally, glutaraldehyde has been extensively used as the cross-linking agent to prepare CLEAs of various enzymes and exhibited a strong effect on activity and particle size of enzyme aggregates. The activity recovery of CLEAs greatly depends on the type of enzyme and the concentration of glutaraldehyde [24]. Lower glutaraldehyde concentration affects the cross-linking efficiency, which might result in enzyme leakage in immobilization, while excessive glutaraldehyde can induce the flexibility of enzymes and the active site availability, consequently, decreasing the activity recovery of CLEAs [25,26]. In this study, the influence of glutaraldehyde concentration on activity of BCL mCLEAs was investigated by using various concentrations of glutaraldehyde in cross-linking. As shown in Figure 4b, the optimum glutaraldehyde concentration of BCL mCLEAs was 2.0% (v/v).

2.3. Hyperactivation of BCL mCLEAs with Surfactants

A pivotal challenge in lipase immobilization is to open the lid of lipases and fix their open form for the exposure of active site. Surfactant imprinting is an efficient approach to activate lipases by facilitating lid-opening. Like other lipases, BCL also consists of a mobile element at the surface, which composed of two helical elements (a5- and a9-helices) and covers the active site [18]. To improve the catalytic performance of BCL mCLEAs, BCL was imprinted in the presence of surfactants prior to immobilization. Thus, four different surfactants with different properties (cationic, anionic and non-ionic) were investigated for modulating the activity of BCL mCLEAs in biodiesel production. As seen in Figure 5, Triton X-100 exhibited maximum effect on the enhancement of lipase activity in low surfactant concentration, while the addition of sodium bis-2-(ethylhexyl) sulfosuccinate (AOT) showed the least influence. The optimal surfactant in proper concentration acting as a bipolar agent, could simulate the amphiphilic environment to benefit the exposure of hydrophobic regions in the active site. Meanwhile, surfactants may also promote the dissociation of large aggregates formed, thus slightly increase the enzymatic activity of lipase (Figure 2).

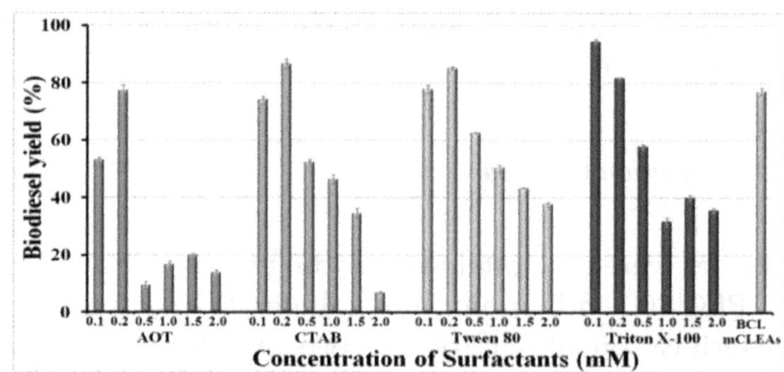

Figure 5. Four different surfactants activation on activity of surfactant-activated BCL mCLEAs in biodiesel production.

However, the increase of surfactant concentration led to gradual decrease of biodiesel yield in all cases, indicating that surfactants showed positive and negative effect on the activity of lipase. Additional detergent molecules may bind to the active site region of lipase, blocking the substrate access, inducing inhibition [27]. Compared with ionic surfactants (AOT and cetrimonium bromide (CTAB)), nonionic surfactants (Triton X-100 and Tween 80) were preferred aiming at regulating the activity of BCL (Figure 5). As the main interaction between the enzyme and nonionic surfactants is hydrophobic interaction while anionic or cationic surfactants perform electrostatic interactions [28], mild hydrophobic interaction between BCL and the surfactant might be important to trigger the interfacial activation mechanism. Therefore, nonionic Triton X-100 and Tween 80 were further studied to confirm the optimal amphiphile and surfactant concentration. As performed in Figure 6, the maximum hyperactivation of BCL mCLEAs was observed in the pretreatment of BCL with 0.1 mM Triton X-100, and the optimal Triton-activated BCL mCLEAs were used for further experiments.

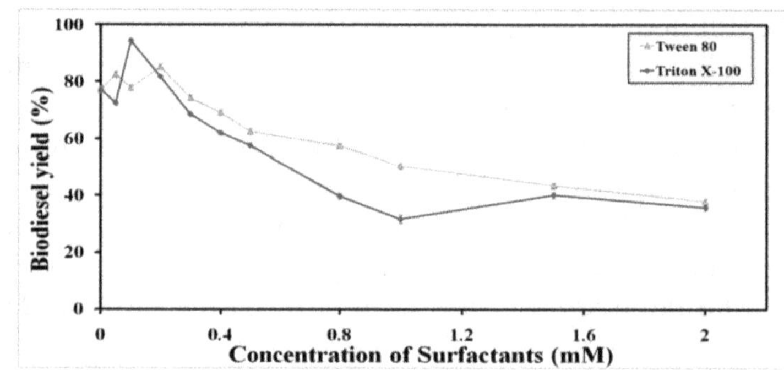

Figure 6. Surfactants (Triton X-100 and Tween 80) concentration in surfactant-activated BCL mCLEAs preparation.

2.4. Biodiesel Production

For the economic feasibility of biodiesel production, solvents, methanol-to-oil molar ratio, and reaction temperature are important variables to optimize for transesterification step. As a result of oxidative reactions occurring during cooking and long-term storage in air, WCO generally exhibits a dramatic increase in viscosity and saponification value [7]. Compared to the fresh oil, high viscose WCO is not favored in biodiesel production. Using solvents could reduce the viscosity of the reaction medium and decrease the diffusion limitations, while it might also directly affect the enzyme structure and activity. In general, hydrophobic solvents could promote the interface and stabilize lipases on their open assembly, causing the hyperactivation of these enzymes. To select the most suitable medium, five hydrophobic solvents commonly used in transesterification were tested in biodiesel production (Figure 7a). It can be clearly seen that biodiesel yield is remarkably dependent on the type of solvent. Overall, Triton-activated BCL mCLEAs exhibited higher activity than BCL mCLEAs and free BCL in all the solvents tested, and the changing trend of their activity in various solvents was accord with

BCL mCLEAs and free BCL. In case of Triton-activated BCL mCLEAs, the best results were achieved using n-hexane with a yield of up to 94% biodiesel, which was 3.3-fold higher than that in free BCL catalyzed reaction. Interestingly, surfactant hyperactivation in combination with immobilization could fasten lipase in their active conformation, allowing biodiesel production performed in solvent without further addition of water, which was in accordance with earlier reports [29,30].

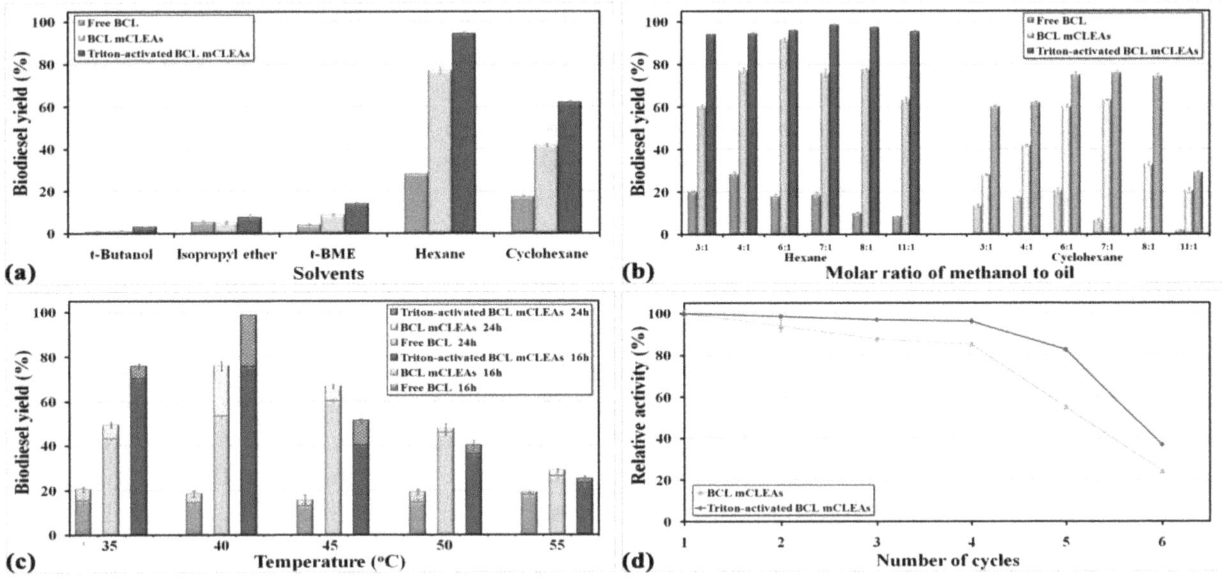

Figure 7. Reaction parameters on biodiesel production catalyzed by free BCL, BCL mCLEAs and Trion-activated BCL mCLEAs and reusability of immobilized BCL. (**a**) Solvents, (**b**) molar ratio of methanol to oil, (**c**) temperature, (**d**) reusability.

The methanol:oil molar ratio can have a significant effect on the reaction yield because excess methanol increases the reaction rate and drives high yield of biodiesel, while a high concentration of methanol leads to inactivation of lipases. In this study, experiments were performed at different molar ratios of methanol to WCO ranging from 3:1 (stoichiometric ratio) to 11:1 both in hexane and cyclohexane with methanol added only once. As shown in Figure 7b, Triton-activated BCL mCLEAs exhibited higher biodiesel yields in one-time addition of methanol under all the experimental conditions, especially when hexane was used as solvent. Meanwhile, owing to the significant inhibitory effect of excess methanol in one time addition, free BCL showed low activity in biodiesel production. It is worth noting that yields of Triton-activated BCL mCLEAs catalyzed reactions in hexane exceeded 90% in a wide range of methanol-to-WCO ratio, while the reaction yield of BCL mCLEAs catalyzed reaction decreased with methanol-to-WCO ratio exceeding 6:1. Consequently, it can be confirmed that surfactants pretreatment provided not only hyperactivation but also protection to lipases from denaturation in excess methanol. In addition, the maximum biodiesel yield was observed at a methanol-to-WCO ratio of 7:1 for Triton-activated BCL mCLEAs. Consequently, the minimal stoichiometric methanol-to-WCO ratio of 7:1 was chosen in further experiments.

Most of the enzymatic transesterification depends on temperature, which could enhance reaction rate and improve the dispersion of immobilized particles in reaction medium with better mass transfer between the reactants [31]. However, thermal denaturation of the enzyme might occur with elevation of temperature, typically according to the property of enzyme and immobilized methods. The effect of temperature on the yield of biodiesel during the transesterification of WCO has been investigated over a temperature range from 35 to 55 °C. According to Figure 7c, the optimum operational temperature for Triton-activated BCL mCLEAs and BCL mCLEAs was 40 °C, with biodiesel yields of 98% and 76% respectively after 24 h, and further increase of temperature will result in decrease of biodiesel yields simultaneously. Besides, Triton-activated BCL mCLEAs showed better activity below 40 °C, while BCL mCLEAs performed higher biodiesel yield over 40 °C. The suitable covalent cross-linking with functionalized MNPs provided extra structure stabilization in mCLEAs, requiring much more energy to the disruption of this stable structure than free enzyme [32]. Nevertheless, the accessible active site of lipases achieved by surfactants pretreatment might be more sensitive to high temperature denaturation [33].

In summary, the optimal reaction conditions for Triton-activated BCL mCLEAs catalyzed transesterification of WCO are as follows: hexane used as solvent, molar ratio of methanol-to-WCO 7:1 in one-time addition, reaction temperature 40 °C. To verify the feasibility of the whole process at a larger scale, transesterification of WCO were performed under optimal conditions adding proper amount of Triton-activated BCL mCLEAs (the initial content of BCL was 240 mg in immobilization) to a mixture of 1 g WCO in 20 mL hexane. The biodiesel yield reached 94% after shaking at 40 °C for 48 h. Triton-activated BCL mCLEAs showed good activity and stability under higher oil content, indicating the possibility of its scale-up application in bioreactor systems.

2.5. *Reusability*

Reusability of immobilized enzyme is a chief criterion for its cost-effective use for potential industry applications. The utilization of functionalized MNPs facilitates the consequent reuse of immobilized enzyme. To investigate the reusability of BCL mCLEAs and Triton-activated BCL mCLEAs, the immobilized lipases were recovered by magnetic separation, and applied in the consecutive batches of biodiesel reactions under optimized conditions. Assessments of the operational stability were analyzed for 6 cycles and presented in Figure 7d. As observed, Triton-activated BCL mCLEAs showed no significant loss in the catalytic activity after subsequent consecutive reuse for 4 cycles, and kept 82% relative activity after continuous running 5 cycles. Meanwhile, the relative activity of BCL mCLEAs was 55% after 5 cycles, implying that BCL could possess good long-term stability with surfactant pretreatment. The protein denaturation in one time addition of methanol and byproduct inhibition might be account for the decrease in biodiesel yield in long-term reuses [34].

3. Materials and Methods

3.1. *Materials*

Burkholderia cepacia lipase (powder, Amano Lipase PS, ≥3000 U/g) and fatty acid methyl ester standards were purchased from Sigma-Aldrich (St. Louis, MO, USA). Also, 3-aminopropyl triethoxysilane (APTES), glutaraldehyde (25%, v/v) and 2-phenyl ethanol (>98%, CP) were supplied by Aladdin (Shanghai, China). Sodium bis-2-(ethylhexyl) sulfosuccinate (AOT) were procured from Acros (USA). Waste cooking oil (WCO) was obtained from local restaurant around Ningxia University campus (Yinchuan, China) with the following fatty acid compositions: 10.48% palmitic acid, 15.04% stearicacid, 38.44% oleic acid, 23.76% linoleic acid, and 1.72% linolenic acid. The WCO sample was filtered to separate impurities and solids in the oil. The physical properties of WCO are saponification value of 197.3 mg KOH/g, acid value of 4.37 mg KOH/g, and average molecular weight of 870.9 g/mol. All other chemicals were of analytical or chromatographical grade and used as purchased.

3.2. *Preparation of Magnetic Support*

Preparation of HAP-coated MNPs was carried out according to the previously reported method [35]. Initially, MNPs cores were prepared by the conventional co-precipitation method. Typically, $FeCl_2{\cdot}4H_2O$ (1.1 g) and (3.0 g) of $FeCl_3{\cdot}6H_2O$ were dissolved in 90 mL deionized water under the protection of argon, with subsequent addition of 25% ammonia solution (30 mL) under vigorous stirring at room temperature. After stirring for 30 min, a 60 mL aqueous solution composed of $Ca(NO_3)_2{\cdot}4H_2O$ (7.1 g) and $(NH_4)_2HPO_4$ (2.3 g) adjusted to pH=11 was added drop wise to the above suspension under continuous stirring. Subsequently, the resultant mixture was heated to 90 °C and stirred for 2 h. After cooling to room temperature and aging in the mother solution overnight, the obtained precipitates were washed several times with deionized water until neutral and lyophilized for 12 h. The HAP-coated MNPs were obtained by calcining the materials in air at 300 °C for 3 h.

To obtain 3-aminopropyl trimetoxysilane functionalized HAP-coated MNPs (APTES-HAP-coated MNPs), HAP-coated MNPs (1.0 g) were suspended in a solution composed of 30 mL anhydrous toluene and 0.44 g of APTES. The mixture was refluxed under Ar atmosphere for 12 h, and then washed several times with ethanol, magnetically separated, and subsequently lyophilized prior to use.

3.3. Lipase Immobilization

BCL mCLEAs were produced according to the procedure described in Scheme 1. Firstly, 10 mg of APTES-HAP-MNPs were dispersed in 1 mL of BCL solution (10 mg/mL, 0.1 M phosphate buffer, pH 7.0) and shaken for 15 min at 30 °C. Then 5 mL of precipitant was added with stirring at 4 °C for 30 min. After precipitation, glutaraldyhyde was added drop wise into the suspension and stirred for 3 h at 30 °C. Afterwards, BCL mCLEAs were collected by centrifugation and washed thrice with phosphate buffer and deionized water, lyophilized and finally stored at 4 °C.

During optimization of the immobilization conditions, the effects of precipitants (acetone, ethanol, isopropanol, PEG 800 (1 g/mL), and saturated ammonium sulfate solution) and concentration of glutaraldehyde on the activity recovery of BCL mCLEAs were investigated.

The surfactant-activated BCL mCLEAs was prepared using cationic (CTAB), anionic (AOT) and nonionic (Tween 80 and Triton Triton X-100) surfactants at various concentrations. Then, 1 mL of BCL solution and appropriate amount of surfactant were mixed and stirred at 4 °C for 30 min. After incubated for 24 h at 4 °C, the suspended solution was sequentially used for BCL mCLEAs preparation under optimal conditions.

3.4. Characterization

The prepared support matrix and immobilized lipase described above were characterized using FTIR, SEM and VSM. The Fourier transform infrared (FTIR) spectra were acquired using a Perkin Elmer Frontier spectrometer (Spectrum Two, Waltham, MA, USA) equipped with an Attenuate Total Reflection (ATR) accessory. Samples were analyzed as KBr pellets in the range of 400 to 4000 cm^{-1} at a resolution of 0.5 cm^{-1}. The morphology of the particle surface was observed using a scanning electron microscope (SEM, Sigma HD, ZEISS, Germany), with deposition of a thin coating of gold onto the samples prior to analyses. The magnetic properties were detected by a vibrating sample magnetometer (VSM, MicroSense EZ9, Lowell, MA, USA) at room temperature.

3.5. Activity Assay

In studying the optimal conditions for BCL mCLEAs preparation, the enzymatic transesterification activities of free lipase and immobilized BCLs were assayed via transesterification reaction of 2-phenyl ethanol with vinyl acetate according to the method introduced previously [36]. The reaction mixture contained 10 mg of 2-phenylethanol, 1 mL of vinyl acetate and 10 mg of lipase (the initial content of BCL was 10 mg in preparing BCL CLEAs and BCL mCLEAs), and the reactions were carried out at 30 °C with continuous shaking at 220 rpm. After 24 h of reaction, samples were withdrawn and analyzed by high-performance liquid chromatography (HPLC). All experiments were repeated at least three times. The relative activity of BCL mCLEAs was calculated with the following equation:

$$\text{Relative activity }(\%) = \frac{\text{Transesterification yield of immobilized BCL}}{\text{Trasesterification yield of free BCL}} \times 100$$

3.6. Enzymatic Transesterification for Biodiesel Production

The transesterification of WCO were carried out at 40 °C in a 10 mL screw-capped vessel for 24 h with continuous shaking at 220 rpm. Unless otherwise stated, a typical reaction mixture consisted of 50 mg WCO, 2.0 mL hexane, 10 mg of lipase (the initial content of BCL was 10 mg in preparing

BCL CLEAs and BCL mCLEAs) and methanol using methanol: oil molar ratio of 4:1. Single factor optimization was conducted to determine optimal reaction parameters for transesterification of WCO to biodiesel. Various conditions including kinds and concentration of surfactants, solvents, molar ratio of methanol to oil and temperature (°C) were investigated. The transesterification reaction of large scale with 1 g WCO were carried out as described in Section 2.4. All biodiesel reactions were performed in dried solvents without any water added. The yield of biodiesel (20 μL) was analyzed in different time intervals using gas chromatography.

3.7. *Analytical Methods*

HPLC was conducted with Shimadzu LC-2010A HT apparatus using C18 column (UltimateXB-C18, 5 μm, 4.6 × 150 mm, Welch). The samples were analyzed with a mixture of MeOH/water = 80:20 (v/v) as eluent at 0.8 mL/min for 9 min at 254 nm.

Fatty acid methyl esters (FAMEs) were analyzed by a Fuli9790 plus gas chromatography (Fuli, Zhejiang, China) fitted with a flame ionization detectorcity (FID, Zhejiang, China), and a KB-FFAP column (30 m × 0.32 mm × 0.25 μm). Nitrogen gas was a carrier at continues flow of 1.0 mL/min. The oven (Zhejiang, China) temperature was set and at 160 °C maintained for 2 min, then a heating ramp was applied up to 240 °C at a rate of 10 °C /min, and the temperature of the oven was maintained at 240 °C for 15 min. The temperatures of the injector (Zhejiang, China) and the detector (Zhejiang, China) were set at 270 and 280 °C, respectively. Methyl tridecanoate was used as internal standard, and the biodiesel yield (%) was calculated by peaks area of standard FAME peaks.

3.8. *Reusability*

The reusability of Triton-activated BCL mCLEAs and BCL mCLEAs for the transesterification of WCO were also investigated under optimal conditions. After each batch reaction, immobilized BCL was recovered by magnetic separation and washed with n-hexane. The washed biocatalyst was reused consecutively in repetitive cycles. The biodiesel yield of the first reaction was set as 100% and the FAMEs yield in the subsequent reactions was calculated accordingly.

4. Conclusions

A facile and effectual surfactant imprinting method to expose the lipase active site integrating amino functionalized HAP-coated MNPs was established to immobilize CLEAs of BCL attaining enhanced activity and stability. The as-prepared Triton-activated BCL mCLEAs was subsequent processed in enzymatic transesterification of waste cooking oil for biodiesel production, and showed 98% biodiesel yield under optimal conditions, which was 5.3-fold higher than the free lipase. This study proved that hyperactivation with surfactant could significantly improve the resistance of lipase to methanol in one-time addition, when compared to BCL mCLEAs and free BCL. In addition, surfactant imprinting in combination with immobilization could fasten lipase in their active conformation, allowing biodiesel production performed in solvent without further addition of water, and thus displayed priority in downstream purification of biodiesel over ordinary immobilization methods. Besides, the green immobilization with functionalized MNPs facilitates fast and easy recovery of lipase, and the corresponding immobilized BCL was reused for 4 cycles without significant loss in the catalytic activity. Furthermore, this work provides a promising approach for immobilization of other lipases, which can be used with success in green and clean production processes.

Author Contributions: Conceptualization, W.Z.; Investigation, H.Y.; supervision, funding acquisition, writing—original draft preparation and review and editing, W.Z.

Acknowledgments: The help from Fu Zheng (School of Physics & Electronic-Electrical Engineering, Ningxia University) for magnetization measurements is gratefully recognized.

References

1. Mahmudul, H.M.; Hagos, F.Y.; Mamat, R.; AbdulAdam, A.; Ishak, W.F.W.; Alenezi, R. Production, characterization and performance of biodiesel as an alternative fuel in diesel engines-a review. *Renew. Sustain. Energy Rev.* **2017**, *72*, 497–509. [CrossRef]
2. Laesecke, J.; Ellis, N.; Kirchen, P. Production, analysis and combustion characterizationof biomass fast pyrolysis oil-biodiesel blends for use in diesel engines. *Fuel* **2017**, *199*, 346–357. [CrossRef]
3. Jamil, F.; Alhaj, L.; Almuhtaseb, A.H.; Baawain, M.; Rashid, U.; Ahmad, M.N.M. Current scenario of catalysts for biodieselproduction: A critical review. *Rev. Chem. Eng.* **2018**, *34*, 267–297. [CrossRef]
4. Banković-Ilić, I.B.; Stamenković, O.S.; Veljković, V.B. Biodiesel production fromnon-edible plant oils. *Renew. Sustain. Energy Rev.* **2012**, *16*, 3621–3647. [CrossRef]
5. Hama, S.; Noda, H.; Kondo, A. How lipase technology contributes to evolution ofbiodiesel production using multiple feedstocks. *Curr. Opin. Biotechnol.* **2018**, *50*, 57–64. [CrossRef]
6. Gebremariam, S.N.; Marchetti, J.M. Economics of biodiesel production: Review. *Energy Convers. Manag.* **2018**, *168*, 74–84. [CrossRef]
7. Moazeni, F.; Chen, Y.-C.; Zhang, G. Enzymatic transesterification for biodiesel production from usedcooking oil, a review. *J. Clean. Prod.* **2019**, *216*, 117–128. [CrossRef]
8. Lopresto, C.; Naccarato, S.; Albo, L.; De Paola, M.; Chakraborty, S.; Curcio, S.; Calabro, V. Enzymatic transesterification of waste vegetable oil to producebiodiesel. *Ecotoxicol. Environ. Saf.* **2015**, *121*, 229–235. [CrossRef]
9. Chhetri, A.; Watts, K.; Islam, M. Waste cooking oil as an alternate feedstock forbiodiesel production. *Energies* **2008**, *1*, 3–18. [CrossRef]
10. Sabudak, T.; Yildiz, M. Biodiesel production from waste frying oils and itsquality control. *Waste Manag.* **2010**, *30*, 799–803. [CrossRef]
11. Kim, K.H.; Lee, O.K.; Lee, E.Y. Nano-immobilized biocatalysts for biodiesel production from renewable and sustainable resources. *Catalysts* **2018**, *8*, 68. [CrossRef]
12. Facin, B.R.; Melchiors, M.S.; Valério, A.; Oliveira, J.V.; Oliveira, D. Driving immobilized lipases as biocatalysts: 10 years state of the art and future prospects. *Ind. Eng. Chem. Res.* **2019**, *58*, 5358–5378. [CrossRef]
13. Filho, D.G.; Silva, A.G.; Guidini, C.Z. Lipases: Sources, immobilization methods, and industrial applications. *Appl. Microbiol. Biotechnol.* **2019**, *103*, 7399–7423. [CrossRef]
14. Xu, M.-Q.; Wang, S.-S.; Li, L.-N.; Gao, J.; Zhang, Y.-W. Combined cross-linked enzyme aggregatesas biocatalysts. *Catalysts* **2018**, *8*, 460. [CrossRef]
15. Sheldon, R.A. CLEAs, combi-CLEAs and 'smart' magnetic CLEAs: Biocatalysis in a bio-based economy. *Catalysts* **2019**, *9*, 261. [CrossRef]
16. Liu, D.-M.; Chen, J.; Shi, Y.-P. Advances on methods and easy separated support materials forenzymes immobilization. *TrAC-Trends Anal. Chem.* **2018**, *102*, 332–342. [CrossRef]
17. Izadia, A.; Meshkinia, A.; Entezari, M.H. Mesoporous superparamagnetic hydroxyapatite nanocomposite: Amultifunctional platform for synergistic targeted chemo-magnetotherapy. *Mater. Sci. Eng. C Mater.* **2019**, *101*, 27–41. [CrossRef]
18. Sánchez, D.A.; Tonetto, G.M.; Ferreira, M.L. Burkholderia cepacia lipase: A versatile catalyst in synthesisreactions. *Biotechnol. Bioeng.* **2018**, *115*, 6–24. [CrossRef]
19. Barbe, S.; Lafaquiere, V.; Guieysse, D.; Monsan, P.; Remaud-Siméon, M.; Andre, I. Insights into lid movements of Burkholderia cepacialipase inferred from molecular dynamics simulations. *Proteins* **2009**, *77*, 509–523. [CrossRef]
20. Gabriele, F.; Spreti, N.; Giacco, T.D.; Germani, R.; Tiecco, M. Effect of surfactant structure on the superactivity of Candida rugosalipase. *Langmuir* **2018**, *34*, 11510–11517. [CrossRef]
21. Zhang, N.; Gao, T.; Wang, Y.; Wang, Z.; Zhang, P.; Liu, J. Environmental pH-controlledloading and release of protein on mesoporous hydroxyapatite nanoparticles forbone tissue engineering. *Mater. Sci. Eng. C Mater.* **2015**, *46*, 158–165. [CrossRef]

22. Kannan, S.; Marudhamuthu, M. Development of chitin cross-linked enzyme aggregates of L-methioninase for upgraded activity, permanence andapplication as efficient therapeutic formulations. *Int. J. Biol. Macromol.* **2019**, *141*, 218–231. [CrossRef]
23. Shaarani, S.M.; Jahim, J.M.; Rahman, R.A.; Idris, A.; Murad, A.M.A.; Illias, R.M. Silanized maghemite for cross-linked enzyme aggregates of recombinantxylanase from Trichoderma reesei. *J. Mol. Catal. B Enzym.* **2016**, *133*, 65–76. [CrossRef]
24. Rehman, S.; Bhatti, H.N.; Bilal, M.; Asgher, M. Cross-linked enzyme aggregates (CLEAs) of Pencilluim notatum lipaseenzymewith improved activity, stability and reusabilitycharacteristics. *Int. J. Biol. Macromol.* **2016**, *91*, 1161–1169. [CrossRef]
25. Cui, J.; Lin, T.; Feng, Y.; Tan, Z.; Jia, S. Preparation of spherical cross-linked lipaseaggregates with improved activity, stabilityand reusability characteristic in water-in-ionicliquid microemulsion. *J. Chem. Technol. Biotechnol.* **2017**, *92*, 1785–1793. [CrossRef]
26. Yamaguchi, H.; Kiyota, Y.; Miyazaki, M. Techniques for preparation of cross-linked enzyme aggregates and their applications in bioconversions. *Catalysts* **2018**, *8*, 174. [CrossRef]
27. Bañó, M.C.; González-Navarro, H.; Abad, C. Long-chain fatty acyl-CoA esters induce lipase activation in the absence of a water-lipid interface. *BBA Mol. Cell Biol. Lipids* **2003**, *1632*, 55–61. [CrossRef]
28. Delorme, V.; Dhouib, R.; Canaan, S.; Fotiadu, F.; Carriere, F.; Cavalier, J.F. Effects of surfactants on lipase structure, activity, and inhibition. *Pharm. Res.* **2011**, *28*, 1831–1842. [CrossRef]
29. Mukherjee, J.; Gupta, M.N. Molecular bioimprinting of lipases with surfactantsand its functional consequences in low water media. *Int. J. Biol. Macromol.* **2015**, *81*, 544–551. [CrossRef]
30. Zhang, W.-W.; Yang, X.-L.; Jia, J.Q.; Wang, N.; Hu, C.-L.; Yu, X.-Q. Surfactant-activated magnetic cross-linked enzyme aggregates (magnetic CLEAs) of Thermomyces lanugunosus lipase for biodiesel production. *J. Mol. Catal. B Enzym.* **2015**, *115*, 83–89. [CrossRef]
31. Lee, H.V.; Yunus, R.; Juan, J.C.; Taufiq-Yap, Y.H. Process optimization design forjatropha-based biodiesel production using response surface methodology. *Fuel Process. Technol.* **2011**, *92*, 2420–2428. [CrossRef]
32. Dong, T.; Zhao, L.; Huang, Y.; Tan, X. Preparation of cross-linked aggregates ofaminoacylase from Aspergillus melleus by using bovine serum albumin as aninert additive. *Bioresour. Technol.* **2010**, *101*, 6569–6571. [CrossRef]
33. Rehm, S.; Trodler, P.; Pleiss, J. Solvent-induced lid opening in lipases: A molecular dynamics study. *Protein Sci.* **2010**, *19*, 2122–2130. [CrossRef]
34. Xie, W.; Huang, M. Immobilization of Candida rugosa lipase onto graphene oxide Fe3O4nanocomposite: Characterization and application for biodiesel production. *Energy Convers. Manag.* **2018**, *159*, 42–53. [CrossRef]
35. Xie, W.; Zang, X. Covalent immobilization of lipase onto aminopropyl-functionalizedhydroxyapatite-encapsulated-γ-Fe_2O_3 nanoparticles: A magneticbiocatalyst for interesterification of soybean oil. *Food Chem.* **2017**, *227*, 397–403. [CrossRef]
36. Zhang, W.-W.; Yang, H.-X.; Liu, W.-Y.; Wang, N.; Yu, X.-Q. Improved performance of magnetic cross-linked lipase aggregates by interfacial activation: A robustand magnetically recyclable biocatalyst for transesterification of Jatropha oil. *Molecules* **2017**, *22*, 2157. [CrossRef]

8

Bacillus subtilis Lipase A—Lipase or Esterase?

Paula Bracco [1], Nelleke van Midden [1], Epifanía Arango [1], Guzman Torrelo [1], Valerio Ferrario [2], Lucia Gardossi [2] and Ulf Hanefeld [1,*]

1 Biokatalyse, Afdeling Biotechnologie, Technische Universiteit Delft, Van der Maasweg 9, 2629 HZ Delft, The Netherlands; paulabracco@gmail.com (P.B.); nellekevanmidden@gmail.com (N.v.M.); epifaniarango@gmail.com (E.A.); guzman.torrelo@hotmail.com (G.T.)

2 Dipartimento di Scienze Chimiche e Farmaceutiche, Università degli Studi di Trieste, Via Licio Giorgieri 1, 34127 Trieste, Italy; valerio.ferrario@gmail.com (V.F.); gardossi@units.it (L.G.)

* Correspondence: u.hanefeld@tudelft.nl

Abstract: The question of how to distinguish between lipases and esterases is about as old as the definition of the subclassification is. Many different criteria have been proposed to this end, all indicative but not decisive. Here, the activity of lipases in dry organic solvents as a criterion is probed on a minimal α/β hydrolase fold enzyme, the *Bacillus subtilis* lipase A (BSLA), and compared to *Candida antarctica* lipase B (CALB), a proven lipase. Both hydrolases show activity in dry solvents and this proves BSLA to be a lipase. Overall, this demonstrates the value of this additional parameter to distinguish between lipases and esterases. Lipases tend to be active in dry organic solvents, while esterases are not active under these circumstances.

Keywords: hydrolase; lipase; esterase; *Bacillus subtilis* lipase A; transesterification; organic solvent; water activity

1. Introduction

Lipases and esterases both catalyze the hydrolysis of esters. This has led to the longstanding question: how can we distinguish between a lipase and an esterase? As the simple hydrolysis of an ester does not suffice, a range of different criteria has been suggested [1–5]. (1) The oldest distinction is the kinetic and structural criterion of interfacial activation, which was already described in 1936 [6]. However, several lipases do not fulfill this; in particular, the much-used *Candida antarctica* lipase B (CALB) does not [7]. (2) Directly linked to the interfacial activation is the lid that covers the active site of many lipases and, via a conformational change, makes the active site more accessible once an interface is present. Again, this is not the case for all lipases [1–3,7,8]. (3) Primary sequence data were shown not to be distinctive enough [2]. (4) Substrates and inhibitors, such as Orlistat, can be utilized to distinguish between esterases and lipases but, again, they are not precise. However, the different substrate ranges are indicative. Esterases tend to be capable of the hydrolysis of water-soluble esters and, in general, short and/or branched side chain esters, while lipases hydrolyze triglycerides, apolar esters, substituted with linear side chains, as well as waxes. This is seen as a reliable but not decisive criterion [2,9]. (5) The activity of the enzyme in the presence of (water-miscible) organic solvents has been proposed as a property of lipases, but other enzymes fulfill this criterion, too [2,10–14]. (6) A parameter already investigated some time ago is the activity of lipases in the absence of water, i.e., in modestly polar, water-non-miscible solvents at very low water activities (a_W). Out of all enzymes tested, only lipases and the closely related cutinases are active at low a_W [1–5,10,15–19]. While not all lipases display this property, it is highly distinctive [20,21].

To probe whether a_W is indeed a suitable parameter to distinguish between lipases and esterases and between lipases and other hydrolases in general, we studied the behavior of *Bacillus subtilis* lipase A (BSLA) [9]. BSLA is a small (181 amino acids, 19 kDa) serine hydrolase (Figure 1). It is neither interfacially activated nor does it have a lid (criteria one and two) [9,22–24] and sequence data are not conclusive, but it is a minimal α/β hydrolase fold enzyme [9,23,25]. The substrate range clearly qualifies BSLA as a lipase, as does the stability in the presence of solvents [9,26–29]. This stability has even been significantly improved in recent mutational studies and BSLA mutants can be very stable in the presence of water-miscible solvents, such as dimethyl sulfoxide (DMSO), dioxane and trifluoroethanol [30,31]. Studies on BSLA in dry organic solvents are, however, missing. As an experimental parameter, we demonstrate the activity of BSLA in dry toluene. Toluene is not water-miscible and has a logP of 2.5 [32]. It is commonly used in organic synthesis and is highly suitable for lipases and also other enzymes with an α/β hydrolase fold. To date, only lipases were shown to be active in toluene with a very low a_W [1,3].

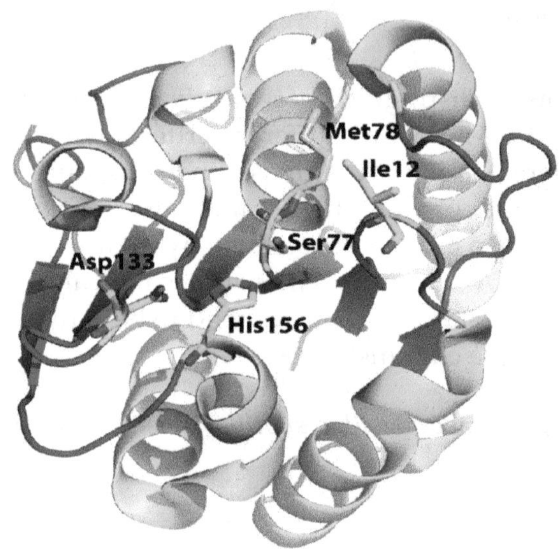

Figure 1. *Bacillus subtilis* lipase A (BSLA) is the smallest serine hydrolase with an α/β hydrolase fold. With only 181 amino acids, it has a molecular weight of 19 kDa. The depicted BSLA structure is pdb 1R50 and the catalytic triad His156, Ser77 and Asp133 and the oxyanion hole Ile12 and Met78 are highlighted. The figure was created with PyMOL.

Additionally, we extend the structural assignment of the hydrolase character with the GRID-based (Fortran program [33]) Global Positioning System in Biological Space (Bio GPS) investigation [33]. BioGPS utilizes surface shape and polarity as criteria. It is neither based on direct sequence comparison, nor on structure superimposition [33,34]. Earlier studies with this method had placed CALB in both the esterase and lipase group. CALB works extremely well in dry solvents and is, therefore, often applied in reactions that require these conditions, such as dynamic kinetic resolutions [1,3,7,35]. On the other hand, it misses interfacial activation and major conformational changes do not take place when CALB comes into contact with an apolar second phase (see above). As such, BioGPS recognized the ambivalence in the assignment of CALB as a lipase well.

Here, we describe the investigation of BSLA by BioGPS and a comparison to other lipases, in particular CALB. We also probe the lipase character of both BSLA and CALB at different a_W. In this manner, new experimental and computational criteria for the esterases and lipases are introduced and investigated.

2. Results

2.1. BioGPS

BioGPS descriptors can be utilized to explore enzyme active site properties and to group them according to their similarities and differences. As such, they can help to explore promiscuous activities.

In an earlier study, the character of CALB was investigated in a set of 42 serine hydrolases. The set contained 11 amidases, nine proteases, 11 esterases and 11 lipases, one of them being CALB [33]. Here, we expand this set with BSLA, utilizing the pdb 1R50 with a resolution of 1.4 Å for the structural information (Table S1). Three probes were used to map specific electrostatic and geometrical active site properties. The O-probe evaluates the H-bond donor properties of the enzymes; the N1 probe, on the contrary, evaluates the H-bond acceptor properties and the DRY probe evaluates the hydrophobic interactions [33]. The DRY probe is clearly of special importance for enzymes that accept hydrophobic substrates, as is the case for lipases.

Considering each property separately, the O-probe located BSLA (pdb 1R50) not among the lipases, but in the amidases cluster, together with a number of esterases (Figure S1a). Equally, the N1-probe (Figure S1b) placed BSLA among the amidases. The DRY probe (Figure S1c), again, placed BSLA amongst the amidases and esterases. This is, in all cases, in contrast to CALB (pdb 1TCA) but it should also be noted that *Candida rugosa* lipase (CRL), a classic lipase with a prominent movement of the lid (criteria one and two) is also always outside the lipase cluster in the different analyses. The previous study ascribed this behavior to the lower hydrophobic nature of the active site of CRL (pdb 1CRL) when compared to the other lipases [8,33].

In the global score, which considers all the mapped properties of the BioGPS together, BSLA can be found firmly among the amidases and esterases (Figure 2), while CALB is in the lipase cluster in the area overlapping with the esterase cluster. CRL, again, is outside the lipase cluster and indeed seems to take a separate position.

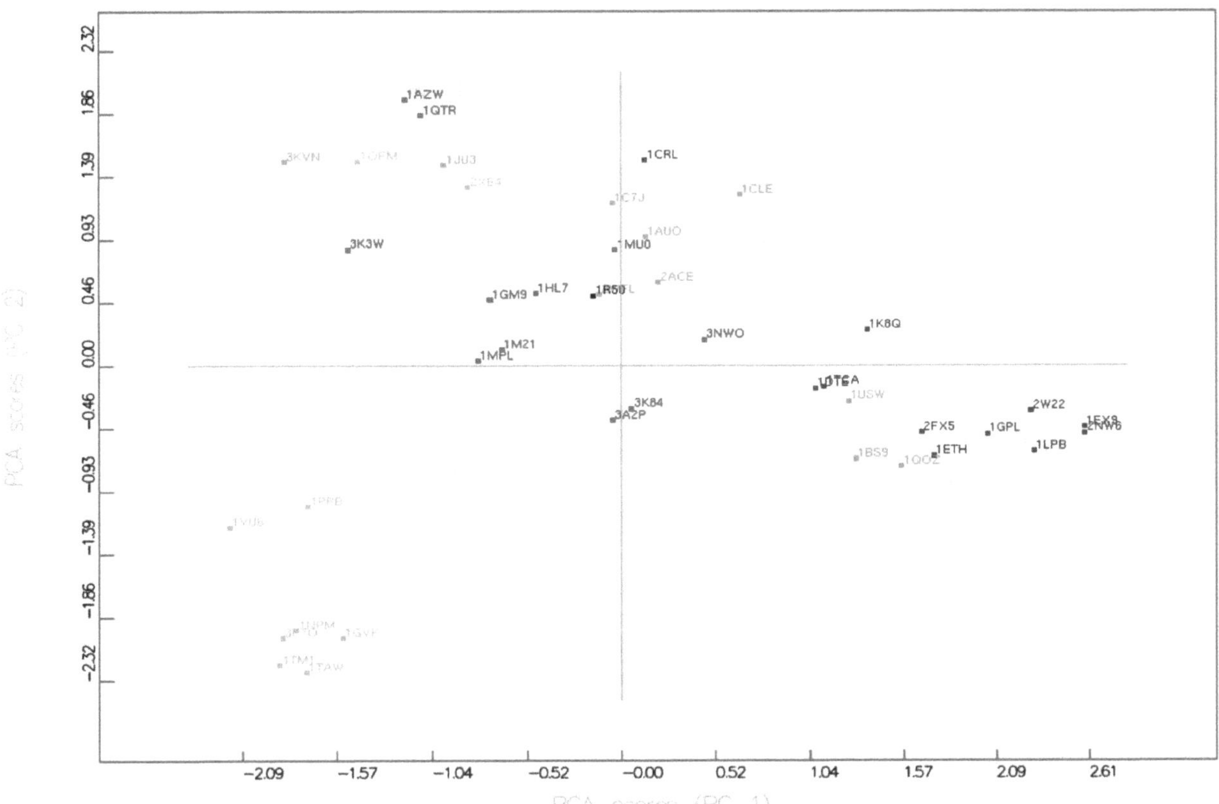

Figure 2. BioGPS of 43 serine hydrolases, for BSLA the data of pdb 1R50 were utilized (global score). Each analyzed enzyme structure is placed within a multidimensional space. Relative distances between each enzyme and all the other enzymes are determined by a statistical principal component analysis. The pdb codes of the processed enzyme structures are indicated in different colors according to their class: lipases in blue, amidases in red, proteases in cyan and esterases in green; the BSLA structure is in black.

BSLA is, according to its substrate range, very clearly a lipase and not an esterase. Amidase activity has to date not been reported for BSLA. While initially surprising, these results also indicate that the study should be extended with an activity assay for amidases.

2.2. *Amidase Activity*

To probe for amidase activity in BSLA, an amidase activity assay is utilized (Scheme 1). This assay employs benzyl chloroacetamide as standard amide. The released amine reacts with 4-nitro-7-chloro-benzo-2-oxa-1,3-diazole, yielding an adduct that can directly be quantified spectrophotometrically at 475 nm [36].

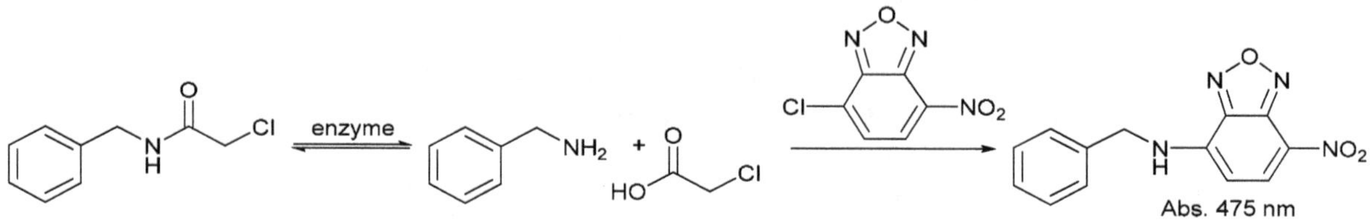

Scheme 1. Amidase activity assay [36]. The assay can be quantified spectrophotometrically. BSLA showed no activity in this assay, ruling out amidase activity.

BSLA showed no activity in this assay. A control experiment with another serine hydrolase, the acyltransferase from *Mycobacterium smegmatis* (*Ms*Act), was performed. This enzyme is an acyltransferase [37,38] and displays promiscuous amidase activity [39,40]. *Ms*Act exhibited activity in this 24 h assay (> 20% conversion of the 5 mM substrate), showing that even minor, promiscuous activities are detectable. This rules out amidase activity for BSLA and supports the earlier assignment of the enzyme as a lipase.

2.3. *BSLA Activity in Dry Organic Solvents*

To probe the activity of BSLA at low a_W, toluene was used as the solvent and the transesterification of 1-octanol with vinyl acetate was performed as a test reaction (Scheme 2). The use of 1-octanol as a long chain aliphatic compound is a good substrate for lipases [1–5,9] and vinyl acetate is a readily available and widely utilized acyl donor in lipase catalyzed acylation reactions [1,3,41,42]. All reactions were performed with lyophilized BSLA. In parallel, CALB was also tested to ensure direct comparability with one of the most-used lipases. CALB was utilized both as lyophilized enzyme and immobilized as Novozym 435. The latter preparation is most commonly employed, both in the laboratory and on industrial scale [43].

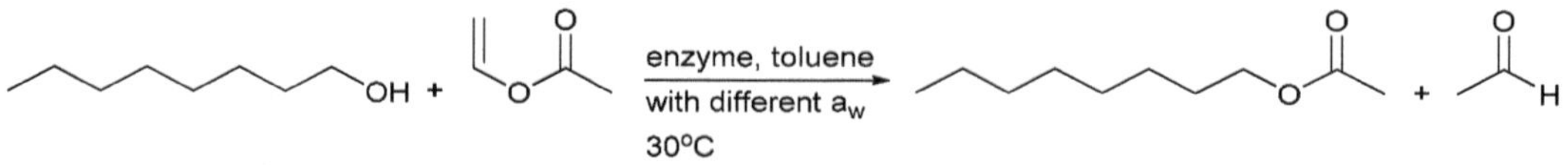

Scheme 2. Test reaction for the activity of BSLA at low a_W. The reaction was performed in toluene at 30 °C, with a ratio of 1-octanol to vinyl acetate of 1:5 and $a_W < 0.1$, 0.23 and 0.75.

BSLA and CALB were produced by expressing the codon-optimized genes in *E. coli* BL21 (DE3) within pET22b. Subsequent purification gave both enzymes a good purity (Figure 3). With this expression system, both enzymes are not glycosylated. The CALB Novozym 435 produced and immobilized by Novozymes, however, is expressed in *Aspergillus oryzae* and it is, therefore, glycosylated [44].

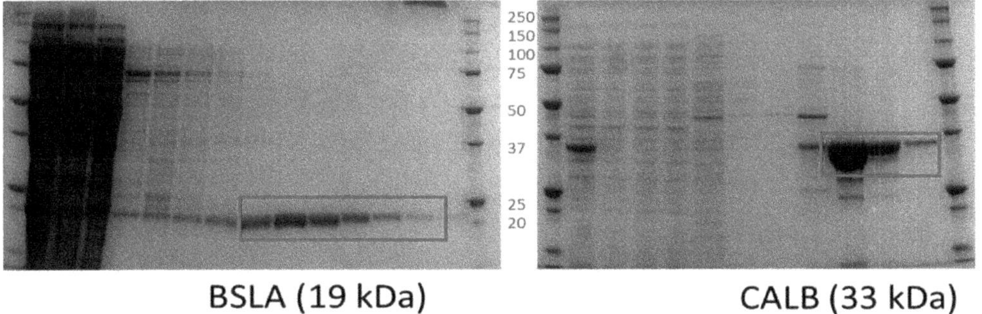

Figure 3. Sodium dodecyl sulfate–polyacrylamide gel electrophoresis (SDS-PAGE) gels of purified BSLA (19 kDa) and Candida antarctica lipase B (CALB) (33 kDa).

Three different a_W were tested < 0.1 to establish whether BSLA shows the activity in dry solvent only observed for lipases, with a_W = 0.23 as a low value at which most enzymes lose all their activity and a_W = 0.75, an activity at which most enzymes are active [10,21,45,46]. To rigorously ascertain these values of the solvent and reagents, including the internal standard, decane and the enzyme preparations were equilibrated via the vapor phase with dried molecular sieves (activated at elevated temperatures, 5 Å) for a_W < 0.1 [47]. For the other a_W, the enzyme preparations and the other components were equilibrated via the gas phase with an oversaturated solution of potassium acetate (a_W = 0.23) and sodium chloride (a_W = 0.75) [48–52]. For all components, the water content was determined by Karl Fischer titration and equilibrations were considered complete when no changes were observed any more (24–48 h, Table 1). As vinyl acetate was found to negatively affect the Karl Fischer titration, it was freshly distilled and dried with activated molecular sieves for 16 h before use. The activity of the different enzyme preparations was also followed with the tributyrin and *p*-nitrophenol acetate activity assays [2,5,53–56] during equilibration, to establish optimal equilibration times. For BSLA, a small loss of activity over time was observed, while both CALB preparations were stable.

Table 1. Equilibration to different a_W via vapor phase over a saturated solution of salt [47,51] and via the salt pair method [50]. All reaction components, except the acyl donor, were mixed and equilibrated overnight at 30 °C. Finally, dried and freshly distilled vinyl acetate was added in order to start the reaction. The water content was determined by Karl Fischer Titration after 48 h.

a_W	Agent (Vapor Phase or Salt Pair)	Moles of H_2O/mol of Salt	Water Content (ppm)
<0.1	Mol. sieves	0	~20
0.25	NaAc anhydr. (salt pair)	1.5	~180
0.57	Na_2HPO_4 anhydr. (salt pair)	5.0	~360
0.23	KAc (vapor phase)	NA [a)]	~120
0.75	NaCl (vapor phase)	NA [a)]	~400

[a)] Not applicable (NA).

Once reagents and enzymes were equilibrated, the reactions were performed with 100 mM 1-octanol and 500 mM vinyl acetate in previously equilibrated toluene at 30 °C and 1000 rpm (Figure 4). Equal activity of the enzymes (Units) was utilized as determined with the tributyrin activity assay. CALB and, in particular, the well-established commercial preparation of CALB, Novozym 435, performed very well. In both cases, full conversion to 1-octyl acetate was observed. In comparison, BSLA displayed lower conversions (Figure 4). However, the key indicator for a lipase is its activity at low a_W. Here, BSLA and Novozym 435 performed best. For the synthesis of 1-octyl acetate, the trend is a reduction in specific rate at higher a_W (Figure 5). BSLA is very active in dry solvent, as is Novozym 435. Both display lower activities at higher a_W. CALB does not follow this trend.

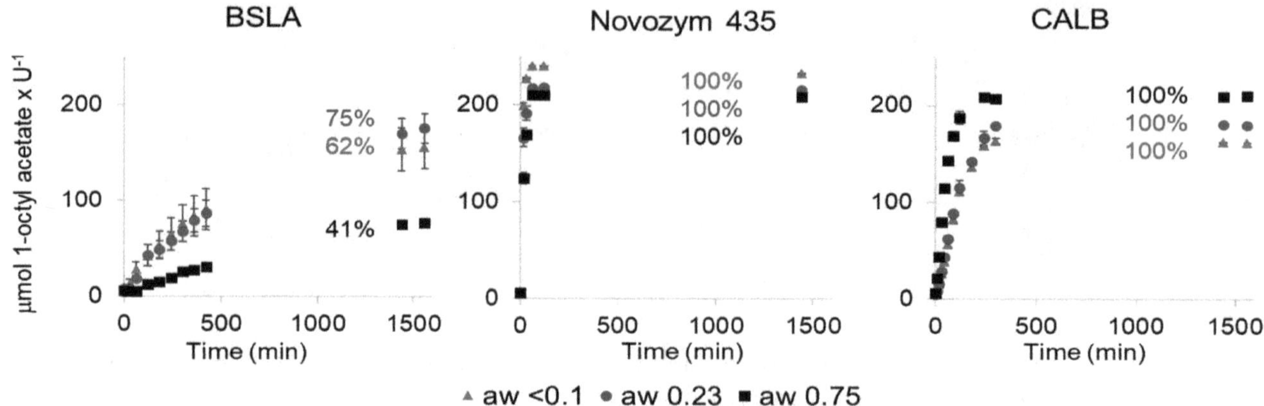

Figure 4. Activity of BSLA, CALB and Novozym 435 in toluene with different a_W. U = µmol butyric acid × min^{-1} in tributyrin activity assay, 0.5–1.2 U of catalyst, 1-octanol (100 mM), vinyl acetate (5 eq.), ISTD: Decane (500 mM), 1 mL reaction volume, 24 h, 30 °C and 1000 rpm. Blanks were performed in the absence of enzyme and showed no conversion. Final conversions are given as inset; the color corresponds to the a_W.

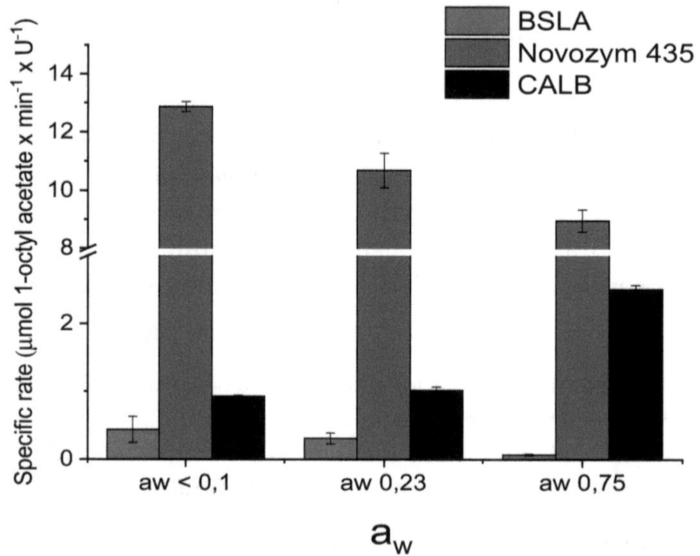

Figure 5. Activity of BSLA, CALB and Novozym 435 in toluene with different a_W. Reaction conditions: 0.5–1.2 U of catalyst, 100 mM 1-octanol, 500 mM vinyl acetate, 500 mM decane (ISTD), in dry toluene (1 mL reaction) at 30 °C and 1000 rpm. U: µmol butyric acid × min^{-1}. Blanks were performed in the absence of enzyme and showed no conversion.

In an earlier study, it had been demonstrated, for different CALB preparations, that this change in activity in the synthesis reaction to 1-octyl acetate can be due to the hydrolysis of the acyl donor vinyl acetate [47]. Therefore, the synthesis reaction at $a_W < 0.1$ was repeated for BSLA with a 1-octanol to vinyl acetate ratio of 1:1 (Figure 6). Almost the same rate and conversion was observed as with the 1:5 ratio, indicating that, at this low a_W, essentially no hydrolysis occurred, as was the case for Novozym 435, as reported earlier. Overall, these differences in performance at altered a_W can be ascribed to several influences [47,57,58]. Novozyme 435 is an immobilized enzyme and its high activity can be linked to the dispersion of the enzyme on a large surface, promoting its mass transfer and preventing particle aggregation. In contrast, the lyophilized enzymes have a reduced accessibility of the individual enzymes in the preparation. Furthermore, it is well established that immobilized enzymes are better protected against the acetaldehyde that is a side product of the acylation reaction [59]. A difference in susceptibility to acetaldehyde induced deactivation might also cause the alterations in rate between the two pure enzymes. However, similarly, ionization and water clustering can influence the activity [60,61], leading to these alterations. To demonstrate that the observed effect is general, the experiments were repeated, but this time with BSLA that was dried by co-lyophilization with a salt to establish the desired a_W [62,63]. The enzyme is now in a different environment and two different a_W were established, < 0.1 and 0.57. At < 0.1, very similar results were obtained. Equally, at higher a_W, the ester formation slowed down as before, but could be restarted by adding additional vinyl acetate (Figure 7).

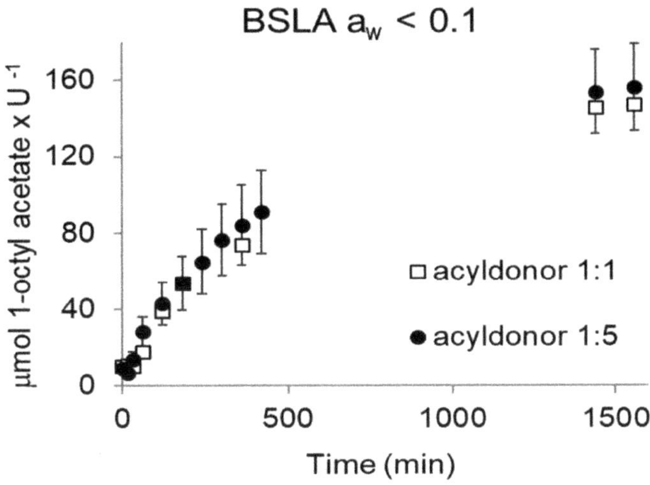

Figure 6. Activity of BSLA, toluene at a_W < 0.1. Reaction conditions: 0.5–1.2 U of catalyst, 100 mM 1-octanol, 100 mM or 500 mM vinyl acetate, 500 mM decane (ISTD), in dry toluene (1 mL reaction) at 30 °C and 1000 rpm. U: µmol butyric acid × min^{-1}. Blanks were performed in the absence of enzyme and showed no conversion.

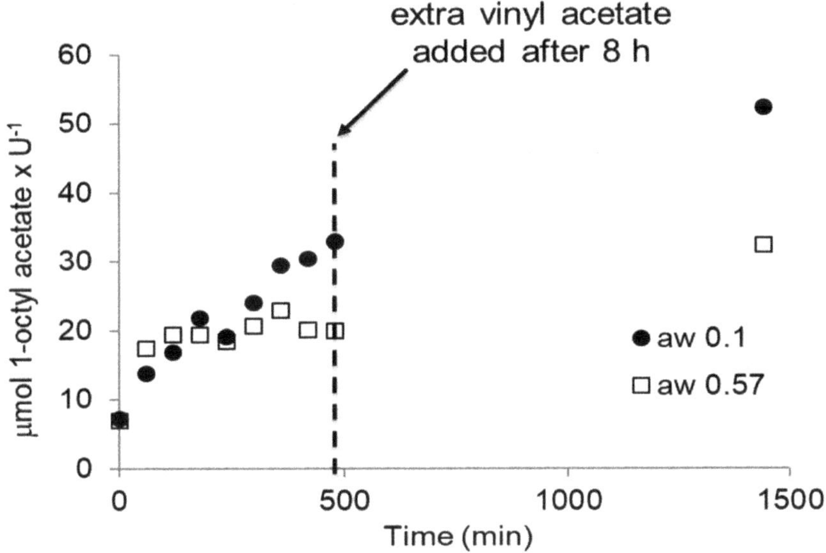

Figure 7. Activity of BSLA co-lyophilized with the appropriate salt, toluene at a_W < 0.1 or 0.57. Reaction conditions: 0.5–1.2 U of catalyst, 100 mM 1-octanol, 100 mM vinyl acetate, 500 mM decane (ISTD), in dry toluene (1 mL reaction) at 30 °C and 1000 rpm. U: µmol butyric acid × min^{-1}. Blanks were performed in the absence of enzyme, i.e., in the presence of salt, and showed no conversion. After 8 h (480 min) an additional equivalent of vinyl acetate was added.

To confirm this activity of BSLA (equilibrated via the gas phase) in dry toluene as a general property, the reaction was repeated in dry methyl-t-butyl ether (MTBE) at the same low $a_W < 0.1$. Enzymes display the same activity in organic solvents when these have the same a_W [64]. Indeed, the BSLA-catalyzed esterification displayed a very similar reaction progress in MTBE and toluene (Figure 8). This confirms the activity of BSLA at low a_W, in line with the earlier observed catalytic activity of CALB at low a_W [47] and of *Rhizomucor miehei* lipase at very low a_W [65].

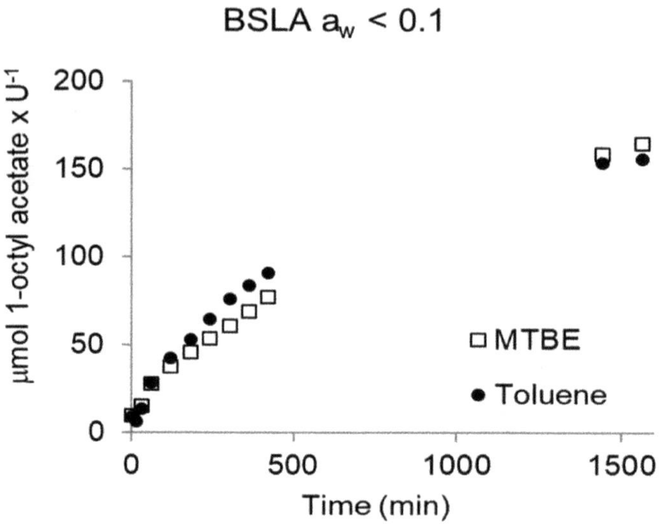

Figure 8. Activity of BSLA, in MTBE and toluene $a_W < 0.1$. Reaction conditions: 0.5–1.2 U of catalyst, 100 mM 1-octanol, 500 mM vinyl acetate, 500 mM decane (ISTD), in dry solvent (1 mL reaction) at 30 °C and 1000 rpm. U: µmol butyric acid × min^{-1}. Blanks were performed in the absence of enzyme and showed no conversion.

3. Discussion

Interestingly, the BioGPS analysis seems to identify features of the BSLA active site which are shared by other amidase enzymes. In particular, BSLA seems to share similar H-bond capabilities with amidases, as evidenced by the single-probe clustering. The possible promiscuous amidase activity of BSLA was probed with an amidase activity assay (Scheme 1) [36]. This revealed a complete absence of amidase activity. While indicative, this is not conclusive, as this might also be due to substrate specificity. Amidases are characterized by a developed network of H-bond acceptors and donors as described in previous work [33]. The aromatic moiety of the substrate molecule might thus prevent a good interaction with such H-bond/hydrophilic network. More generally, for amidases, the necessity of a hydrogen bond network that stabilizes the NH hydrogen to suppress its deprotonation was reported earlier [66]. Given the very open active site of BSLA, the minimal serine hydrolase with an α/β hydrolase fold, it is not entirely surprising that this type of hydrogen bond network has never been described for this enzyme.

The test of a_W as parameter for the assignment of a serine hydrolase as lipase gave conclusive results. BSLA and CALB displayed good activity at low a_W. In line with earlier results, the synthetic catalytic activity of CALB varies depending on the preparation. Earlier studies had shown that the observed synthetic catalytic activity competes with the hydrolytic activity—that is to say, the parallel hydrolysis of vinyl acetate [47]. This can lead to an apparent decrease in synthetic activity, as is observed for Novozym 435 and BSLA at higher a_W (Figure 5). This trend has already been reported for Novozym 435 [47]. Just like Novozym 435 [47], BSLA displays essentially no hydrolysis of vinyl acetate at low $a_W < 0.1$. This is confirmed in experiments with a ratio of 1:1 of alcohol to vinyl acetate; a similar rate of synthesis was observed (Figures 6 and 7). The fact that free CALB displays higher synthetic rates at higher a_W is also in line with the literature [47]. It had earlier been demonstrated for

CALB that the ratio of synthesis to hydrolysis depends on the preparation of the enzyme used and that it increases with a_w for purified, free CALB [47]. The activity of BSLA at low a_w was proven also with a different solvent, MTBE (Figure 8).

BioGPS is a complimentary computational tool to investigate the character of an enzyme and delivers a useful input to help us explore the scope of an enzyme more thoroughly. The parameter a_w is an indicative tool to determine whether an enzyme is a lipase or and esterase. Just like the substrate scope, it is not absolute, but is highly indicative. Essentially, a serine hydrolase that is active at low a_w is a lipase and not an esterase, while the reverse statement is not valid. Or, as it was recently summarized: "This long-standing and biased question could be compared to the search for differences between humans and mammals, which implicitly means that one does not consider humans as mammals! Obviously, lipases are a special kind of esterases like humans are a special kind of mammals." [2].

4. Materials and Methods

4.1. Materials

Chemicals and Enzymes

1-propanol, 1-octanol, toluene extra dry, decane, *p*-nitrophenylbutyrate, *p*-nitrophenol, 3-[(3-cholamidopropyl)dimethylammonio]-1-propanesulfonate (CHAPS), tributyrin and 2-methyl-2-propanol, butyric acid and caprylic acid were purchased from Sigma-Aldrich (Schnelldorf, Germany) and Acros (Geel, Belgium), and used without previous purification. Vinyl acetate was purchased from Sigma-Aldrich and distilled before use. Novozym 435 (immobilized lipase B from *Candida antarctica*) was made available by Novozymes (Bagsværd, Denmark). Bovine serum albumin protein and lysozyme from chicken egg whites were purchased from Sigma Aldrich. Bradford reagent was purchased from Biorad (Hercules, C.A., USA). Medium and buffer components were purchased from BD, Merck (Darmstadt, Germany) or J.T. Baker (Geel, Belgium).

Strains and Plasmids

Strains *Escherichia coli* (*E. coli*) HB2151 and *E. coli* HB2151 pCANTAB 5E bsla were kindly provided by Prof. Bauke Dijkstra and Prof. Wim Quax, University of Groningen, the Netherlands. Strains *Escherichia coli* (*E. coli*) BL21 (DE3), *E. coli* TOP10 and plasmid pET22b(+) were utilized for all further work.

4.2. Methods

Cloning pET22bbsla and pET22bcalb

The gene of BSLA (as confirmed by sequencing ID: CP011115.1, range from 292296 to 292841, protein AKCA5803.1) was amplified by PCR from vector pCANTAB 5E BSLA using primers BSLA F: 5′-CCTTTCTATGCGGCCCAGC–3′ and BSLA + XhoI R: 5′-CCGCTCGAGCGCCTTCGTATTCTGG-3′. Thereby, restriction site XhoI was introduced for subsequent cloning of BSLA (NcoI, XhoI) into vector pET22b+ (in frame with pelB and His-tag signals). The resulting vector was named pET22bBSLA. The wild type CALB was synthesized by BaseClear (Leiden, The Netherlands). The codon-optimized genes were cloned into pET22b(+) using the NcoI and NotI restriction sites as previously described [67], in order to be in frame with the pelB sequence and a C-terminal His-tag of the plasmid.

Expression and Purification of BSLA

This protocol was adapted from [24]. A freshly grown colony of *E. coli* HB 2151 pCANTAB 5E BSLA was used to inoculate a 1 L shake flask containing 100 mL of 2xTY medium (1.6% *w*/*v* bactotryptone, 1% *w*/*v* bacto yeast extract and 0.5% *w*/*v* sodium chloride), ampicillin (100 µg/mL final concentration) and isopropyl-β-d-galactopyranoside (IPTG, 1 mM final concentration). After 16 h at 28 °C and 150 rpm (Innova Incubator, Hamburg, Germany) the cells were harvested and washed with

10 mM Tris Buffer pH 7.4 and stored at −20 °C. The periplasm isolation protocol was adapted from [67] and consisted of the resuspension of the overexpressed cells in 1 mL of 10 mM Tris buffer pH 8.0 containing sucrose (25 % *w*/*v*), EDTA (2 mM) and lysozyme (0.5 mg/mL). After incubation on ice for 20 min, 250 µL of 10 mM Tris buffer pH 8.0 containing sucrose (20% *w*/*v*) and $MgCl_2$ (125 mM) was added. The suspension was centrifuged and the supernatant containing the periplasmic fraction was desalted using a PD10 column (GE, Healthcare, New York, N.Y., USA) to 100 mM potassium phosphate buffer pH 7.4. Afterwards, the solution was shock-frozen with liquid nitrogen and stored at −20 °C for future biocatalysis applications. BSLA was purified mainly from the media. First, proteins were precipitated by adding 50% *v*/*v* saturated ammonium sulphate (2.8 M final concentration) for 5 h at 4 °C. After centrifugation, the solid fraction was dissolved in 100 mL of 100 mM potassium phosphate buffer pH 7.4, filtered through a 0.45 µm filter and loaded into a 5 mL His-Trap previously equilibrated column (GE Healthcare) using a NGC chromatography system (BIORAD, Hercules, C.A., USA). The loaded proteins were washed with equilibration buffer potassium phosphate (100 mM, pH 7.4) containing 500 mM NaCl and 20 mM imidazole. The His-tagged BSLA was eluted with a linear gradient from 0–100% potassium phosphate (100 mM, pH 7.4) containing 500 mM NaCl and 500 mM imidazole. The progress of the purification was monitored at 280 nm. Fractions containing the target protein (as confirmed by SDS-PAGE and activity assay, 50–60% of the gradient) were combined, concentrated and desalted with a PD-10 column (GE Healthcare) to potassium phosphate buffer (100 mM, pH 7.4). The purified enzyme (68–148 µg/L medium) was aliquoted (2.5 U/vial, units determined by tributyrin assay) and freeze dried for 16 h, −80 °C and stored at −20 °C under nitrogen atmosphere.

Protein Sequence of BSLA-His (AKCA5803.1):

MAAEHNPVVMVHGIGGASFNFAGIKSYLVSQGWSRDKLYAVDFWDKTGTNYNNGPVLSR
FVQKVLDETGAKKVDIVAHSMGGANTLYYIKNLDGGNKVANVVTLGGANRLTTGKALPGTDP
NQKILYTSIYSSADMIVMNYLSRLDGARNVQIHGVGHIGLLYSSQVNSLIKEGLNGGGQNTKALEH
HHHHH

Expression and Purification of CALB

An LB-Amp plate (100 µg/mL) was used to freshly grow *E. coli* BL21 (DE3) pET22bCALB from a −80 °C DMSO stock. After incubation at 37 °C for 16 h, a single colony was used to inoculate a 5 mL LB-Amp (100 µg/mL) preculture and grown for 8 h at the same temperature. Large-scale expressions were carried out in 0.5 L of ZYM-5052 media (placed in 2 L shake flask), 2% *v*/*v* of the preculture was used for inoculation. After 17 h expression at 22 °C and 170 rpm, an optical density (600 nm) of approximately 3 was obtained in all cases. Afterwards, cells were spun down, washed with 10 mM potassium phosphate buffer pH 7.4 and stored at −20 °C. ZYM-5052 medium [68]: The main cultures were grown in ZYM-5052 medium containing 50 mL 50xM ($Na_2HPO_4 \cdot 12H_2O$ 448 g/L, KH_2PO_4 170 g/L, NH_4Cl 134 g/L, Na_2SO_4 35.5 g/L), 20 mL 50x5052 (100 g/L α-D-lactose, 250 g/L glycerol and 25 g/L glucose dissolved in ddH_2O) and 2 mL of $MgSO_4$ solution (1 M in ddH_2O) and filled to 1 L with ZY medium (casamino acids 10 g/L—tryptone in this case—yeast extract 5 g/L). Additionally, 0.2 mL of trace element solution was added to the media. The purification of His-tagged CALB was performed from the periplasmic fraction, as described for BSLA in the previous sections.

Protein Sequence of CALB-His (Sequence ID: 4K6G_A):

MALPSGSDPAFSQPKSVLDAGLTCQGASPSSVSKPILLVPGTGTTGPQSFDSNWIPLSTQLGYTPC
WISPPPFMLNDTQVNTEYMVNAITALYAGSGNNKLPVLTWSQGGLVAQWGLTFFPSIRSKVDRLMA
FAPDYKGTVLAGPLDALAVSAPSVWQQTTGSALTTALRNAGGLTQIVPTTNLYSATDEIVQPQVSNS
PLDSSYLFNGKNVQAQAVCGPLFVIDHAGSLTSQFSYVVGRSALRSTTGQARSADYGITDCNPLPAN
DLTPEQKVAAAALLAPAAAAIVAGPKQNCEPDLMPYARPFAVGKRTCSGIVTPAAALEHHHHHH

Bradford Assay

Total protein concentration was determined using Bradford reagent in a microtiter plate (MTP) reader format (96 well plates) [69]. Properly diluted samples were mixed with Bradford reagent (5x), incubated at room temperature (RT) for 5 min and the absorbance measured at 595 nm (in triplicate). The calibration curve was carried out using bovine serum albumin protein, as is standard.

Lipase Activity: Tributyrin Assay

A tributyrin assay for determining lipase activity was performed according to the literature [54]. The assay is based on pH change by acid formation when tributyrin is hydrolyzed by the enzyme. *p*-Nitrophenol was used as pH indicator (colorless at pH 5.5 and yellow at pH 7.5) and the acid concentration was determined by a calibration curve with known amounts of butyric acid (from 0 mM to 40 mM). A negative control was performed by adding buffer instead of an enzyme sample. The substrate consumption (0.8 mM initial concentration) was monitored at 410 nm, 30 °C for 15 min, every 38 s by a microtiter plate reader (in 96-well plates, Synergy 2, BioTek, Winooski, V.T., USA). Plates were shaken for 5 s before every read. The different buffers needed for this assay contained 2.5 mM 3-(*N*-morpholino)propanesulfonic acid (MOPS) (pH 7.2), CHAPS (to dissolve acids) and β-cyclodextrin (to dissolve acids into the solution, to increase the linearity). The activity was determined in U, which is equivalent to μmol acid formed per minute. The assays were done in triplicate. For performing this assay with immobilized enzymes, a larger scale (3 mL) in glass vials with a magnetic stirrer was applied. These were placed on a stirring platform and, for Novozym 435, samples (120 μL) were taken over time and placed in a 96-well plate. If desired, the assay can also be performed with Trioctanoin.

Esterase/Lipase Activity: *p*-Nitrophenol Assay

This protocol was adapted from [53] to an MTP reader equipped for 96 well plates. The enzymatic hydrolysis of *p*-nitrophenyl butyrate with the concomitant formation of *p*-nitrophenol was monitored at 405 nm, 37 °C and recorded for 30 min. For this, a calibration curve of *p*-nitrophenol in potassium phosphate buffer (100 mM, pH 7.4) was prepared (levels from 0-500 μM, 200 μL total volume, in triplicate) and control reactions without enzyme extracts were performed. Lyophilized cell-free extract or pure enzymes were re-dissolved in potassium phosphate buffer (100 mM, pH 7.4) (approximately 20–30 mg/mL) and proper dilutions were added into a preheated potassium phosphate buffer (100 mM, pH 7.4) solution containing 3 mM *p*-nitrophenyl butyrate. The esterase activity measured was corrected by subtracting the activity observed in the controls (no enzyme). By definition, one unit of enzyme (U) is equivalent to 1 μmol of *p*-nitrophenol formed per minute.

Karl Fischer Titration

A Metrohm KF Coulometer Karl Fischer titration setup was used, according to the manufacturer's instructions, to determine the water content in ppm. Samples (100 μL) were taken from the solvent and injected into the system in duplicate. In general, master mixes of toluene after equilibration with $a_W < 0.1$, a_W 0.23 and a_W 0.75 contained 20, 120 and 360 ppm respectively. A deviation of 5–10 ppm per sample was observed.

Equilibration of Solvents/Enzymes to the Desired a_W

All materials, reagents, enzymes and solvents used for the biocatalytic reactions were carefully dried and kept under nitrogen atmosphere with molecular sieves (5 Å) at all times. In all cases, the water content was monitored by Karl Fischer titration and as standard parameter compounds with a water content below 100 ppm were considered dry and suitable for the reaction.

Vapor Phase Method

Oversaturated salt solutions and activated molecular sieves were used to equilibrate the solvents and enzymes needed in the transesterification reaction with a desired water activity [47–50]. In the case of working with dry systems, a master mix was prepared including solvent, substrates (without vinyl acetate) and internal standard, all components previously dried with activated molecular sieves achieving $a_W < 0.1$. Lyophilized BSLA, Novozym 435 and CALB were dried over silica in desiccators under a vacuum at room temperature (20–25 °C) for 24, 48 and 24 h, respectively. In order to achieve higher water activities, the master mix and the enzymes were equilibrated over saturated salt solutions of potassium acetate (KAc) and sodium chloride (NaCl) at 30 °C for 48 h, resulting in a_W 0.23 and a_W 0.75 at 30 °C, respectively [51]. As exceptions, BSLA and free CALB were equilibrated for a shorter period of only 24 h.

Salt Pairs Method

The protocol was adapted from [62]. The enzymes were lyophilized with anhydrous salts (Na_2HPO_4 or NaAc) in a ratio 1:99 (3 mg pure BSLA or CALB enzyme and 297 mg of the respective salt). For the background reaction (no-enzyme), only lyophilized salts were added. An amount of 10 mg of the co-lyophilized enzyme was added under a nitrogen atmosphere to the previously dried reaction components (except vinyl acetate) and a specific amount of water was introduced under the nitrogen atmosphere. The moles of water added to the reaction mixture were calculated in order to generate the couple of hepta- and dihydrated phosphates in the case of Na_2HPO_4 (5 moles of water per mol of salt, $a_W \sim 0.57$) and the couple of tri- and anhydrous acetate in the case of NaAc (1.5 moles of water per mol of salt, $a_W \sim 0.25$) [50]. After overnight equilibration, the last substrate was added (freshly distilled and dry vinyl acetate) to begin the reaction. In the case of the dry system, no water was added ($a_W < 0.1$).

Transesterification Catalyzed by Lipases in Organic Solvents under Fixed Water Activities

The protocol was adapted from [47]. The reaction conditions included substrates 1-propanol, 1-octanol, 2-octanol or benzylacohol (100 mM), vinyl acetate freshly distilled (1 or 5 equiv. in respect to the initial substrate concentration) and decane (ISTD, 500 mM final concentration). A range of 0.5–1 U of purified enzymes, 1 mg of immobilized Novozym 435 were tested as catalysts. Toluene or methyl-*t*-butyl ether were used as media (1 mL total volume in GC airtight vials). Reactions were carried out for 25 h, 30 °C and 1000 rpm (thermoblock Eppendorf, Hamburg, Germany). Negative controls were run for both substrates in absence of enzyme. All reactions were performed in duplicate and monitored over time by gas chromatography.

Analytics: GC and GC-MS

Gas chromatograph (GC) and gas chromatograph-mass spectrometry (GC-MS) methods were adapted from [47]. Samples were injected in a gas chromatograph (GC-2014, Shimadzu, Kyoto, Japan) equipped with a CP Sil 5 column (50 m × 0.53 mm × 1.0 um). Injector and detector temperatures were set to 340 and 360 °C, respectively. The initial column temperature was set to 35 °C for 5 min, followed by an increase of 15 °C/min up to 60 °C for 0.5 min and 15 °C/min up to 160 °C and hold for 2 min. Finally, a burnout was introduced, 30 °C/min up to 325 °C. The retention times for 1-propanol, vinyl acetate, toluene, decane, 1-octanol and 1-octylacetate were 1.69, 1.87, 6.64, 10.52, 11.19 and 12.79 min, respectively. To confirm the product's structure, samples were also injected in a gas chromatograph-mass spectrometer (GC-MS QP2010s, Kyoto, Japan) equipped with a CP Sil 5 (25 m × 0.25 mm × 0.4 μm). The injector, interface and ion source temperatures were set to 315, 250 and 200 °C, respectively. The retention times for vinyl acetate, 1-octanol and 1-octylacetate were 1.78, 11.57 and 12.93 min, respectively.

Amidase Activity Assay: Hydrolysis of Benzyl Chloroacetamide

This protocol was adapted from the literature [36]. The biocatalysis conditions included a total volume 500 μL in a 2 mL Eppendorf tube containing 5 mM benzyl chloroacetamide (stock solution of 500 mM in THF), 100 μg/mL enzyme (as quantified by Bradford assay), in 25 mM potassium phosphate buffer pH 7.0 with 10 % *v/v* THF. The conversion was carried out for 24 h at 37 °C and 500 rpm. Afterwards, the derivatization of 200 μL of reaction mixture was carried out with 50 μL of NBDCl (20 mM in DMSO) for 1 h, at 37 °C and 500 rpm. UV detection at 475 nm was performed.

BSLA Structure

The structure of BSLA was taken from the Protein Data Bank (PDB). The structure downloaded (1R50) was treated by removing all molecules but the protein chain with the software PyMOL. The thus-generated structure was used for visual inspection with PyMOL as well as input for the BioGPS analysis.

4.3. BioGPS Computational Analysis

The BioGPS analysis and projection was taken from the previous published work. The BSLA BioGPS analysis was performed using the BioGPS software provided by Molecular Discovery Ltd. (Borehamwood, Hertfordshire, UK) by projecting the enzyme according to its active site properties in the previously performed analysis. The identification of the BSLA active site and the calculation of its properties has been performed as previously described. Specifically, FLAPsite was used for automatic active site identification.

The active site was mapped using a GRID approach and the resulting computed properties were considered as electrondensity-like fields centered on each atom, which correspond to the so-called pseudo-molecular interaction fields (pseudo-MIFs). Four different properties were mapped: the active site shape (H probe), H-bond donor properties (O probe), H-bond acceptor capabilities (N1 probe), and hydrophobicity (DRY probe). The magnitude of the interaction of the N1 and O probes also includes, implicitly, information about the charge contribution, as these probes already have a partially positive and negative charge, respectively.

The pseudo-MIF points were filtered, by means of a weighted energy-based and space-coverage function, and then used for the generation of quadruplets obtained from all possible combinations of the four pseudo-MIF points. Thus, the BSLA active site was described by a series of quadruplets. Finally, BSLA was projected according to its series of quadruples and scored by the previously performed BioGPS analysis.

5. Conclusions

The longstanding question "what differentiates lipases from esterases?" has led to a list of six parameters that are indicative, but not decisive. Here, we have probed BioGPS and a_W as parameters to distinguish between lipases and esterases, utilizing the minimal serine hydrolase with an α/β fold, BSLA, as a test enzyme. While BioGPS has been used successfully to address similar questions earlier, it was not indicative in this case. The clear assignment of BSLA as either esterase or lipase was a challenging task. The high catalytic activity of BSLA at low a_W clearly demonstrated this serine hydrolase to be a lipase. In future studies, activity at low a_W should, therefore, be utilized to support the differentiation of lipases and esterases.

Author Contributions: Conceptualization, U.H., L.G., P.B. and G.T.; methodology, P.B., V.F. and G.T.; validation, N.v.M., E.A., P.B. and V.F.; formal analysis, P.B., U.H. and L.G.; investigation, N.v.M., E.A., P.B. and V.F.; resources, U.H. and L.G.; data curation, N.v.M., E.A., P.B. and V.F.; writing—original draft preparation, U.H., P.B. and V.F.; writing—review and editing, U.H. and L.G.; visualization, U.H. and L.G.; supervision, P.B., U.H. and L.G.; project administration, U.H. and L.G.; funding acquisition, U.H. and L.G. All authors have read and agreed to the published version of the manuscript.

Acknowledgments: Excellent technical support by the technicians of the BOC group and by Linda Otten is gratefully acknowledged. Bauke Dijkstra and Wim Quax, University of Groningen, the Netherlands kindly provided the *Escherichia coli* (*E. coli*) HB2151 and *E. coli* HB2151 pCANTAB 5E bsla. L.G. is grateful to Molecular Discovery Ltd. for providing software access.

References

1. Bornscheuer, U.T.; Kazlauskas, R. *Hydrolases in Organic Synthesis*; Wiley: Hoboken, NJ, USA, 2005; pp. 61–184.
2. Ben Ali, Y.; Verger, R.; Abousalham, A. Lipases or Esterases: Does It Really Matter? Toward a New Bio-Physico-Chemical Classification. In *Lipases and Phospholipases: Methods and Protocols*; Book Series: Methods in Molecular Biology; Sandoval, G., Ed.; Springer Science+Business Media: New York, NY, USA, 2012; Volume 861, pp. 31–51.
3. Paravidino, M.; Bohm, P.; Gröger, H.; Hanefeld, U. Hydrolysis and Formation of Carboxylic Acid Esters. In *Enzyme Catalysis in Organic Synthesis*; Wiley: Hoboken, NJ, USA, 2012; pp. 249–362.
4. Arpigny, J.L.; Jaeger, K.E. Bacterial lipolytic enzymes: Classification and properties. *Biochem. J.* **1999**, *343*, 177–183. [CrossRef]
5. Jaeger, K.-E.; Kovacic, F. Determination of Lipolytic Enzyme Activities. In *Pseudomonas Methods and Protocols*; Book Series: Methods in Molecular Biology; Filloux, A., Ramos, J.L., Eds.; Springer Science+Business Media: New York, NY, USA, 2014; Volume 1149, pp. 111–134.
6. Holwerda, K.; Verkade, P.E.; De Willigen, A.H.A. Vergleichende Untersuchungen über die Verseifungsgeschwindigkeit einiger Einsäuriger Triglyceride unter Einfluss von Pankreasextrakt. I. Der Einfluss des Verteilungszustandes der Triglyceride auf die Verseifungsgeschwindigkeit. *Rec. Trav. Chim. Pays Bas* **1936**, *55*, 43–57. [CrossRef]
7. Kirk, O.; Christensen, M.W. Lipases from *Candida antarctica*: Unique Biocatalysts from a Unique Origin. *Org. Proc. Res. Develop.* **2002**, *6*, 446–451. [CrossRef]
8. Grochulski, P.; Li, Y.; Schrag, J.D.; Cygler, M. Two conformational states of *Candida rugosa* lipase. *Protein Sci.* **1994**, *3*, 82–91. [CrossRef]
9. Eggert, T.; van Pouderoyen, G.; Pencreac'h, G.; Douchet, I.; Verger, R.; Dijkstra, B.W.; Jaeger, K.-E. Biochemical properties and three-dimensional structures of two extracellular lipolytic enzymes from *Bacillus subtilis*. *Colloids Surf. B Biointerfaces* **2002**, *26*, 37–46. [CrossRef]
10. Adlercreutz, P. Comparison of lipases and glycoside hydrolases as catalysts in synthesis reactions. *Appl. Microbiol. Biotechnol.* **2017**, *101*, 513–519. [CrossRef]
11. Lopes, D.B.; Fraga, L.P.; Fleuri, L.F.; Macedo, G.A. Lipase and esterase—To what extent can this classification be applied accurately? *Ciênc. Tecnol. Aliment. Camp.* **2011**, *31*, 608–613. [CrossRef]
12. Berlemont, R.; Spee, O.; Delsaute, M.; Lara, Y.; Schuldes, J.; Simon, C.; Power, P.; Daniel, R.; Galleni, M. Novel organic solvent-tolerant esterase isolated by metagenomics: Insights into the lipase/esterase classification. *Rev. Argent. Microbiol.* **2013**, *45*, 3–12.
13. Adlercreutz, P. Fundamentals of Biocatalysis in Neat Organic Solvents. In *Organic Synthesis with Enzymes in Non-Aqueous Media*; Carrea, G., Riva, S., Eds.; WILEY-VCH Verlag GmbH & Co. KGaA: Weinheim, Germany, 2008; pp. 3–24.
14. Serdakowski, A.L.; Dordick, J.S. Activating Enzymes for Use in Organic Solvents. In *Organic Synthesis with Enzymes in Non-Aqueous Media*; Carrea, G., Riva, S., Eds.; WILEY-VCH Verlag GmbH & Co. KGaA: Weinheim, Germany, 2008; pp. 47–74.
15. Svensson, I.; Wehtje, E.; Adlercreutz, P.; Mattiasson, B. Effects of water activity on reaction rates and equilibrium positions in enzymatic esterifications. *Biotechnol. Bioeng.* **1994**, *44*, 549–556. [CrossRef]
16. Wehtje, E.; Svensson, I.; Adlercreutz, P.; Mattiasson, B. Continuous control of water activity during biocatalysis in organic media. *Biotechnol. Tech.* **1993**, *7*, 873–878. [CrossRef]
17. Ferrario, V.; Pellis, A.; Cespugli, M.; Guebitz, G.; Gardossi, L. Nature Inspired Solutions for Polymers: Will Cutinase Enzymes Make Polyesters and Polyamides Greener? *Catalysts* **2016**, *6*, 205. [CrossRef]
18. Pellis, A.; Ferrario, V.; Zartl, B.; Brandauer, M.; Gamerith, C.; Acero, E.H.; Ebert, C.; Gardossi, L.; Guebitz, G.M. Enlarging the tools for efficient enzymatic polycondensation: Structural and catalytic features of cutinase 1 from *Thermobifida cellulosilytica*. *Catal. Sci. Technol.* **2016**, *6*, 3430–3442. [CrossRef]
19. Pellis, A.; Ferrario, V.; Cespugli, M.; Corici, L.; Guarneri, A.; Zartl, B.; Acero, E.H.; Ebert, C.; Guebitz, G.; Gardossi, L. Fully renewable polyesters via polycondensation catalyzed by *Thermobifida cellulosilytica* cutinase 1: An integrated approach. *Green Chem.* **2017**, *19*, 490–502. [CrossRef]
20. Valivety, R.H.; Halling, P.J.; Peilow, A.D.; Macrae, A.R. Relationship between water activity and catalytic activity of lipases in organic media. Effects of supports, loading and enzyme preparation. *Eur. J. Biochem.* **1994**, *222*, 461–466. [CrossRef]

21. Valivety, R.H.; Halling, P.J.; Peilow, A.D.; Macrae, A.R. Lipases from different sources vary widely in dependence of catalytic activity on water activity. *Biochim. Biophys. Acta Protein Struct. Mol. Enzym.* **1992**, *1122*, 143–146. [CrossRef]
22. Gupta, R.; Gupta, N.; Rathi, P. Bacterial lipases: An overview of production, purification and biochemical properties. *Appl. Microbiol. Biotechnol.* **2004**, *64*, 763–781. [CrossRef]
23. Eggert, T.; van Pouderoyen, G.; Dijkstra, B.W.; Jaeger, K.-E. Lipolytic enzymes LipA and LipB from *Bacillus subtilis* differ in regulation of gene expression, biochemical properties, and three-dimensional structure. *FEBS Lett.* **2001**, *502*, 89–92. [CrossRef]
24. Boersma, Y.L.; Pijning, T.; Bosma, M.S.; van der Sloot, A.; Godinho, L.F.; Dröge, M.J.; Winter, R.T.; van Pouderoyen, G.; Dijkstra, B.W.; Quax, W.J. Loop Grafting of *Bacillus subtilis* Lipase A: Inversion of Enantioselectivity. *Chem. Biol.* **2008**, *15*, 782–789. [CrossRef]
25. van Pouderoyen, G.; Eggert, T.; Jaeger, K.-E.; Dijkstra, B.W. The crystal structure of *Bacillus subtilis* lipase: A minimal α/β hydrolase fold enzyme. *J. Mol. Biol.* **2001**, *309*, 215–226. [CrossRef]
26. Augustyniak, W.; Brzezinska, A.A.; Pijning, T.; Wienk, H.; Boelens, R.; Dijkstra, B.W.; Reetz, M.T. Biophysical characterization of mutants of *Bacillus subtilis* lipase evolved for thermostability: Factors contributing to increased activity retention. *Protein Sci.* **2012**, *21*, 487–497. [CrossRef]
27. Kamal, Z.; Ahmad, S.; Molugu, T.R.; Vijayalakshmi, A.; Deshmukh, M.V.; Sankaranarayanan, R.; Rao, N.M. In Vitro Evolved Non-Aggregating and Thermostable Lipase: Structural and Thermodynamic Investigation. *J. Mol. Biol.* **2011**, *413*, 726–741. [CrossRef]
28. Ahmad, S.; Kamal, Z.; Sankaranarayanan, R.; Rao, N.M. Thermostable *Bacillus subtilis* Lipases: In Vitro Evolution and Structural Insight. *J. Mol. Biol.* **2008**, *381*, 324–340. [CrossRef]
29. Rajakumara, E.; Acharya, P.; Ahmad, S.; Sankaranaryanan, R.; Rao, N.M. Structural basis for the remarkable stability of *Bacillus subtilis* lipase (Lip A) at low pH. *Biochim. Biophys. Acta Proteins Proteom.* **2008**, *1784*, 302–311. [CrossRef]
30. Frauenkron-Machedjou, V.J.; Fulton, A.; Zhao, J.; Weber, L.; Jaeger, K.E.; Schwaneberg, U.; Zhu, L. Exploring the full natural diversity of single amino acid exchange reveals that 40–60% of BSLA positions improve organic solvents resistance. *Bioresour. Bioprocess.* **2018**, *5*, 2. [CrossRef]
31. Markel, U.; Zhu, L.; Frauenkron-Machedjou, V.J.; Zhao, J.; Bocola, M.; Davari, M.D.; Jaeger, K.-E.; Schwaneberg, U. Are Directed Evolution Approaches Efficient in Exploring Nature's Potential to Stabilize a Lipase in Organic Cosolvents? *Catalysts* **2017**, *7*, 142. [CrossRef]
32. Laane, C.; Boeren, S.; Vos, K.; Veeger, C. Rules for optimization of biocatalysis in organic solvents. *Biotechnol. Bioeng.* **1987**, *30*, 81–87. [CrossRef]
33. Ferrario, V.; Siragusa, L.; Ebert, C.; Baroni, M.; Foscato, M.; Cruciani, G.; Gardossi, L. BioGPS Descriptors for Rational Engineering of Enzyme Promiscuity and Structure Based Bioinformatic Analysis. *PLoS ONE* **2014**, *9*, 109354. [CrossRef]
34. Cross, S.; Baroni, M.; Goracci, L.; Cruciani, G. GRID-Based Three-Dimensional Pharmacophores I: FLAPpharm, a Novel Approach for Pharmacophore Elucidation. *J. Chem. Inf. Model.* **2012**, *52*, 2587–2598. [CrossRef]
35. Veum, L.; Kanerva, L.T.; Halling, P.J.; Maschmeyer, T.; Hanefeld, U. Optimisation of the Enantioselective Synthesis of Cyanohydrin Esters. *Adv. Synth. Catal.* **2005**, *347*, 1015–1021. [CrossRef]
36. Henke, E.; Bornscheuer, U.T. Fluorophoric Assay for the High-Throughput Determination of Amidase Activity. *Anal. Chem.* **2003**, *75*, 255–260. [CrossRef]
37. Mathews, I.; Soltis, M.; Saldajeno, M.; Ganshaw, G.; Sala, R.; Weyler, W.; Cervin, M.A.; Whited, G.; Bott, R. Structure of a Novel Enzyme That Catalyzes Acyl Transfer to Alcohols in Aqueous Conditions. *Biochemistry* **2007**, *46*, 8969–8979. [CrossRef] [PubMed]
38. Mestrom, L.; Claessen, J.G.R.; Hanefeld, U. Enzyme-Catalyzed Synthesis of Esters in Water. *ChemCatChem* **2019**, *11*, 2004–2010. [CrossRef]
39. Land, H.; Hendil-Forssell, P.; Martinelle, M.; Berglund, P. One-pot biocatalytic amine transaminase/acyl transferase cascade for aqueous formation of amides from aldehydes or ketones. *Catal. Sci. Technol.* **2016**, *6*, 2897–2900. [CrossRef]
40. Contente, M.L.; Farris, S.; Tamborini, L.; Molinari, F.; Paradisi, F. Flow-based enzymatic synthesis of melatonin and other high value tryptamine derivatives: A five-minute intensified process. *Green Chem.* **2019**, *21*, 3263–3266. [CrossRef]

41. Hanefeld, U. Reagents for (ir)reversible enzymatic acylations. *Org. Biomol. Chem.* **2003**, *1*, 2405–2415. [CrossRef]
42. Paravidino, M.; Hanefeld, U. Enzymatic acylation: Assessing the greenness of different acyl donors. *Green Chem.* **2011**, *13*, 2651–2657. [CrossRef]
43. Ortiz, C.; Ferreira, M.L.; Barbosa, O.; dos Santos, J.C.S.; Rodrigues, R.C.; Berenguer-Murcia, Á.; Briand, E.L.; Fernandez-Lafuente, R. Novozym 435: The "perfect" lipase immobilized biocatalyst? *Catal. Sci. Technol.* **2019**, *9*, 2380–2420. [CrossRef]
44. Basso, A.; Braiuca, P.; Cantone, S.; Ebert, C.; Linda, P.; Spizzo, P.; Caimi, P.; Hanefeld, U.; Degrassi, G.; Gardossi, L. In Silico Analysis of Enzyme Surface and Glycosylation Effect as a Tool for Efficient Covalent Immobilisation of CalB and PGA on Sepabeads®. *Adv. Synth. Catal.* **2007**, *349*, 877–886. [CrossRef]
45. Ma, L.; Persson, M.; Adlercreutz, P. Water activity dependence of lipase catalysis in organic media explains successful transesterification reactions. *Enzym. Microb. Technol.* **2002**, *31*, 1024–1029. [CrossRef]
46. Paravidino, M.; Sorgedrager, M.J.; Orru, R.V.A.; Hanefeld, U. Activity and Enantioselectivity of the Hydroxynitrile Lyase *Me*HNL in Dry Organic Solvents. *Chem. Eur. J.* **2010**, *16*, 7596–7604. [CrossRef]
47. Secundo, F.; Carrea, G.; Soregaroli, C.; Varinelli, D.; Morrone, R. Activity of different *Candida antarctica* lipase B formulations in organic solvents. *Biotechnol. Bioeng.* **2001**, *73*, 157–163. [CrossRef]
48. Valivety, R.H.; Halling, P.J.; Macrae, A.R. Reaction rate with suspended lipase catalyst shows similar dependence on water activity in different organic solvents. *Biochim. Biophys. Acta Protein Struct. Mol. Enzym.* **1992**, *1118*, 218–222. [CrossRef]
49. Fontes, N.; Partridge, J.; Halling, P.J.; Barreiros, S. Zeolite molecular sieves have dramatic acid-base effects on enzymes in nonaqueous media. *Biotechnol. Bioeng.* **2002**, *77*, 296–305. [CrossRef] [PubMed]
50. Halling, P.J. Salt hydrates for water activity control with biocatalysts in organic media. *Biotechnol. Tech.* **1992**, *6*, 271–276. [CrossRef]
51. Greenspan, L. Humidity fixed points of binary saturated aqueous solutions. *J. Res. Natl. Bur. Stand. Sect. A Phys. Chem.* **1977**, *81*, 89–96. [CrossRef]
52. Johnson, J.R.; Affsprung, H.E.; Christian, S.D. The molecular complexity of water in organic solvents. Part II. *J. Chem. Soc. A* **1966**, 77–78. [CrossRef]
53. Gupta, R.; Rathi, P.; Gupta, N.; Bradoo, S. Lipase assays for conventional and molecular screening: An overview. *Biotechnol. Appl. Biochem.* **2003**, *37*, 63–71. [CrossRef]
54. Mateos-Díaz, E.; Rodríguez, J.A.; de los Ángeles Camacho-Ruiz, M.; Mateos-Díaz, J.C. High-Throughput Screening Method for Lipases/Esterases. In *Lipases and Phospholipases: Methods and Protocols*; Book Series: Methods in Molecular Biology; Sandoval, G., Ed.; Springer Science + Business Media: New York, NY, USA, 2012; Volume 861, pp. 89–100.
55. Híreš, M.; Rapavá, N.; Šimkovič, M.; Varečka, L'.; Berkeš, D.; Kryštofová, S. Development and Optimization of a High-Throughput Screening Assay for Rapid Evaluation of Lipstatin Production by Streptomyces Strains. *Curr. Microbiol.* **2018**, *75*, 580–587. [CrossRef]
56. Nalder, T.D.; Ashton, T.D.; Pfeffer, F.M.; Marshall, S.N.; Barrow, C.J. 4-Hydroxy-N-propyl-1,8-naphthalimide esters: New fluorescence-based assay for analysing lipase and esterase activity. *Biochimie* **2016**, *128–129*, 127–132. [CrossRef]
57. Corici, L.; Ferrario, V.; Pellis, A.; Ebert, C.; Lotteria, S.; Cantone, S.; Voinovich, D.; Gardossi, L. Large scale applications of immobilized enzymes call for sustainable and inexpensive solutions: Rice husk as renewable alternative to fossil-based organic resins. *RSC Adv.* **2016**, *6*, 63256–63270. [CrossRef]
58. Secundo, F.; Carrea, G. Lipase activity and conformation in neat organic solvents. *J. Mol. Catal. B Enzym.* **2002**, *19–20*, 93–102. [CrossRef]
59. Hanefeld, U.; Gardossi, L.; Magner, E. Understanding enzyme immobilisation. *Chem. Soc. Rev.* **2009**, *38*, 453–468. [CrossRef] [PubMed]
60. Stauch, B.; Fisher, S.J.; Cianci, M. Open and closed states of *Candida antarctica* lipase B: Protonation and the mechanism of interfacial activation. *J. Lipid Res.* **2015**, *56*, 2348–2358. [CrossRef] [PubMed]
61. Banik, S.D.; Nordblad, M.; Woodley, J.M.; Peters, G.H. Effect of Water Clustering on the Activity of *Candida antarctica* Lipase B in Organic Medium. *Catalysts* **2017**, *7*, 227. [CrossRef]
62. Ebert, C.; Gardossi, L.; Linda, P. Control of enzyme hydration in penicillin amidase catalysed synthesis of amide bond. *Tetrahedron Lett.* **1996**, *37*, 9377–9380. [CrossRef]

63. Ru, M.T.; Dordick, J.S.; Reimer, J.A.; Clark, D.S. Optimizing the salt-induced activation of enzymes in organic solvents: Effects of lyophilization time and water content. *Biotechnol. Bioeng.* **1999**, *63*, 233–241. [CrossRef]
64. Partridge, J.; Dennison, P.R.; Moore, B.D.; Halling, P.J. Activity and mobility of subtilisin in low water organic media: Hydration is more important than solvent dielectric. *Biochim. Biophys. Acta Protein Struct. Mol. Enzym.* **1998**, *1386*, 79–89. [CrossRef]
65. Valivety, R.H.; Halling, P.J.; Macrae, A.R. *Rhizomucor miehei* lipase remains highly active at water activity below 0.0001. *FEBS Lett.* **1992**, *301*, 258–260. [CrossRef]
66. Syrén, P.-O.; Hult, K. Amidases Have a Hydrogen Bond that Facilitates Nitrogen Inversion, but Esterases Have Not. *ChemCatChem* **2011**, *3*, 853–860. [CrossRef]
67. Larsen, M.W.; Zielinska, D.F.; Martinelle, M.; Hidalgo, A.; Jensen, L.J.; Bornscheuer, U.T.; Hult, K. Suppression of Water as a Nucleophile in *Candida antarctica* Lipase B Catalysis. *ChemBioChem* **2010**, *11*, 796–801. [CrossRef]
68. Studier, F.W. Protein production by auto-induction in high-density shaking cultures. *Protein Expr. Purif.* **2005**, *41*, 207–234. [CrossRef] [PubMed]
69. Bradford, M.M. A rapid and sensitive method for the quantitation of microgram quantities of protein utilizing the principle of protein-dye binding. *Anal. Biochem.* **1976**, *72*, 248–254. [CrossRef]

9

Decoding Essential Amino Acid Residues in the Substrate Groove of a Non-Specific Nuclease from *Pseudomonas syringae*

Lynn Sophie Schwardmann, Sarah Schmitz, Volker Nölle and Skander Elleuche *

Miltenyi Biotec B.V. & Co. KG, Friedrich-Ebert-Straße 68, 51429 Bergisch Gladbach, Germany; l.schwardmann@web.de (L.S.S.); sarahs@miltenyibiotec.de (S.S.); VolkerN@miltenyibiotec.de (V.N.)

* Correspondence: skander.elleuche@miltenyibiotec.de

Abstract: Non-specific nucleases (NSN) are of interest for biotechnological applications, including industrial downstream processing of crude protein extracts or cell-sorting approaches in microfabricated channels. Bacterial nucleases belonging to the superfamily of phospholipase D (PLD) are featured for their ability to catalyze the hydrolysis of nucleic acids in a metal-ion-independent manner. In order to gain a deeper insight into the composition of the substrate groove of a NSN from *Pseudomonas syringae*, semi-rational mutagenesis based on a structure homology model was applied to identify amino acid residues on the protein's surface adjacent to the catalytic region. A collection of 12 mutant enzymes each with a substitution to a positively charged amino acid (arginine or lysine) was produced in recombinant form and biochemically characterized. Mutations in close proximity to the catalytic region (inner ring) either dramatically impaired or completely abolished the enzymatic performance, while amino acid residues located at the border of the substrate groove (outer ring) only had limited or no effects. A K119R substitution mutant displayed a relative turnover rate of 112% compared to the original nuclease. In conclusion, the well-defined outer ring of the substrate groove is a potential target for modulation of the enzymatic performance of NSNs belonging to the PLD superfamily.

Keywords: DNase; kinetic profiles; RNase; semi-rational mutagenesis; substrate specificity

---

1. Introduction

Non-specific nucleases (NSN) are a group of enzymes that hydrolyze deoxyribonucleic acid (DNA) and ribonucleic acid (RNA) in all conformations, including single-stranded and double-stranded or linear and circularized substrates, without sequence specificity [1]. NSNs are ubiquitously distributed among all organisms and are of great potential for versatile biotechnological and clinical applications [2–4].

Enzymes that are highly indiscriminate towards different substrates are generally considered as potential evolutionary starting points for developing novel or more specific catalytic activities [5–7]. Members of the phospholipase D (PLD; Enzyme Commission number (EC) 3.1.4.4) superfamily are known to accept a wide range of ester substrates, including nucleic acids [8–10]. PLDs are mainly represented in eukaryotes and predominately catalyze the hydrolysis of phosphatidylcholine to produce choline and phosphatidic acid [11]. Moreover, PLDs act as important key players in various physiological processes, including cell migration and membrane trafficking [10]. This family of enzymes usually encodes two copies of the conserved $HxK(x)_4D(x)_6GSxN$ motif in one gene.

A structurally related bacterial enzyme (Nuc) has been initially described from the human pathogenic microorganism *Salmonella enterica* subsp. *enterica* serovar *Typhimurium*. Nuc contains a single $HxK(x)_4D(x)_6GSxN$ motif, but forms a homodimer, and is capable of degrading nucleic acids in a non-specific manner [12,13]. This enzyme is among the very few known nucleases that are not dependent on a metal ion in its catalytic region, and is therefore of potential interest for biotechnological applications that take place in buffers supplemented with metal ion chelators, such as ethylenediaminetetraacetic acid (EDTA) or ethylene glycol-bis(β-aminoethyl ether)-N,N,N′,N′-tetraacetic acid (EGTA). The group of metal-ion-independent nucleases mainly consists of two PLD-like, site-specific restriction endonucleases from *Bacillus firmus* and *Bacillus megaterium*, WSV191 from the white spot syndrome virus and GBSV1-NSN from a thermophilic bacteriophage, as well as the restriction glycosylase R.*Pab*I from the hyperthermophilic archeon *Pyrococcus abyssi* [14–18].

So far, only three isozymes of bacterial PLD-like NSNs, beside Nuc from *S. enterica* subsp. *enterica* serovar *Typhimurium*, have been investigated and described in detail with regard to their biochemical and biophysical properties: (1) *Ec*Nuc from *Escherichia coli* has been shown to be applicable during cell lysis and protein purification in EDTA-containing buffers; and (2) two isozymes from the plant pathogenic competitor bacterium *Pantoea agglomerans* were shown to be the result of an ancient gene duplication event followed by diversification [19,20]. These enzymes are completely devoid of catalytic performance towards lipids and exclusively degrade nucleic acids in a non-specific manner.

In this study, another metal-ion-independent NSN (DNase/D157G) from *Pseudomonas syringae* was mutagenized using a semi-rational strategy to gain a deeper insight into the composition of the substrate groove. Homology modeling was applied to identify amino acid residues on the surface of the NSN in the surrounding of the catalytic site, which is buried at the bottom of the putative substrate groove. It has been shown before that positively charged amino acid residues that can interact with the proximal negatively charged phosphate groups in nucleic acids had a stimulating impact on the catalytic activity of human DNase I [21]. Therefore, positively charged amino acids were introduced at positions on the surface of DNase/D157G within the substrate groove. Two regions were defined that were either directly adjacent to the catalytic site (inner ring) or at the border of the substrate groove (outer ring). Substitutions in the inner ring dramatically impaired or completely abolished the catalytic activity, while mutagenesis in the outer ring had no or little effect. DNase variant K119R/D157G displayed increased activity, the temperature optimum of variant S143R/D157G was shifted from 50 °C to 40 °C, and N95K/D157G and S143K/D157G exhibited an increased tolerance towards 50 mM of EDTA.

2. Results

2.1. Identification of Amino Acid Residues within the Substrate Groove of a NSN from *Pseudomonas* sp.

Phylogenetic analyses revealed a highly conserved NSN within the genomes of *Pseudomonas* species. These enzymes are related to Nuc from *S. enterica* subsp. *enterica* serovar *Typhimurium* (~57% identity in 159 amino acids overlap) and are highly active at neutral pH, in the presence of salt concentrations up to 250 mM, and in a temperature range between 4 and 50 °C (Supplementary Materials Figure S1). It has been shown that amino acid residues D157, E157, and G157 occur naturally in homologues from the genus *Pseudomonas*. A comparison of the catalytic activities in our laboratory revealed that an enzyme variant containing G157 is superior over a variant with a negatively charged amino acid at position 157 (unpublished results). Therefore, the natural amino acid sequence from a NSN of *P. syringae* containing a single amino acid substitution at position 157 (DNase/D157G) was used in our study as the starting variant for semi-rational mutagenesis to generate double-mutants.

A homology model of the enzyme DNase/D157G was produced based on the crystal structure from *S. enterica* subsp. *enterica* serovar *Typhimurium* (Figure 1). The enzyme is modelled as a hypothetical homodimer, with the catalytic site buried within a putative substrate groove at the dimeric interface. The catalytic site is composed of amino acids H122, K124, G136, S137, and N139, that are part of the $HxK(x)_4D(x)_6GSxN$ motif, while D129 has been proposed to be of structural relevance [12]. Fifteen different amino acid residues were identified that are present on the protein surface within the substrate groove. These amino acid residues were either assigned to be part of an inner ring that is directly adjacent to the catalytic site or to an outer ring that surrounded the inner ring amino acid residues. The following amino acids were identified as being located on the protein surface close to the catalytic site: Y63, S64, T66, I120, and S141, while P68, H91, G92 D94, N95, A97, A101, K119, A142, and S143 are defined as being part of the outer ring (Figure A1 Appendix A).

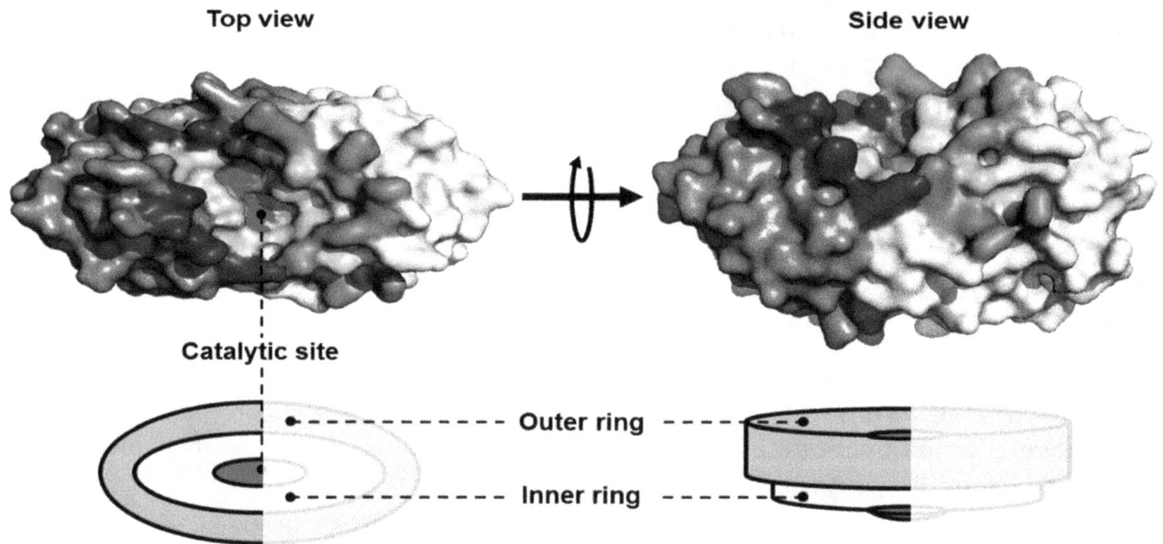

Figure 1. Homology model of non-specific nucleases (NSNs) from *Pseudomonas syringae* in top and side views. Low resolution model of DNase/D157G using Nuc (PDB: 1BYS_A) as a template. PyMOL was used to highlight a hypothetical dimeric structure with dark and light grey monomers. The conserved residues of the HxK(x)$_4$D(x)$_6$GSxN motif are given in dark and light red. Naturally occurring amino acids of the outer and inner rings are highlighted in dark and light blue, while substituted positively charged amino acid residues are indicated in dark and light green. Cartoon illustrations at the bottom are used to simplify the orientation of amino acid residues of the inner and outer ring within the potential substrate groove surrounding the catalytic site.

Surface-presented amino acid residues were exchanged for lysine or asparagine to improve substrate binding and modulate the enzymatic performance or to identify amino acids essential for catalytic activity. Histidine was not considered due to its bulkiness, aggravating the risk for interference with the structure of the protein. Potential steric hindrance was determined by in silico mutagenesis, and the following substitutions were selected: Y63K, S64K, T66R, P68R, H91R, D94K, N95K, K119R, I120K, S141K, A142R, and S143R. Due to the fact that K119 is the only positively charged amino acid within the substrate groove, this lysine was replaced with asparagine. Amino acid residues G92, A97, and A101 were not mutagenized due to expected clashes with adjacent amino acids.

2.2. *Production and Purification of Recombinant Nuclease Mutants*

Recombinant DNase/D157G double-mutants were produced in *Escherichia coli* Veggie BL21 (DE3), except for mutant P68R/D157G, because this expression strain could not be transformed with the corresponding expression plasmid. Therefore, *Escherichia coli* Veggie BL21 (DE3) pLysS was used for the production of this mutant. All recombinant mutant enzymes were produced in soluble form and could be purified using a two-step approach combining affinity and ion exchange chromatography. Purification strategy was optimized using the Äkta purifier (Figure 2). The purification level of all recombinant nucleases was visualized with sodium dodecyl sulfate polyacrylamide gel electrophoresis (SDS-PAGE) and Western blotting analyses using an anti-HIS antibody.

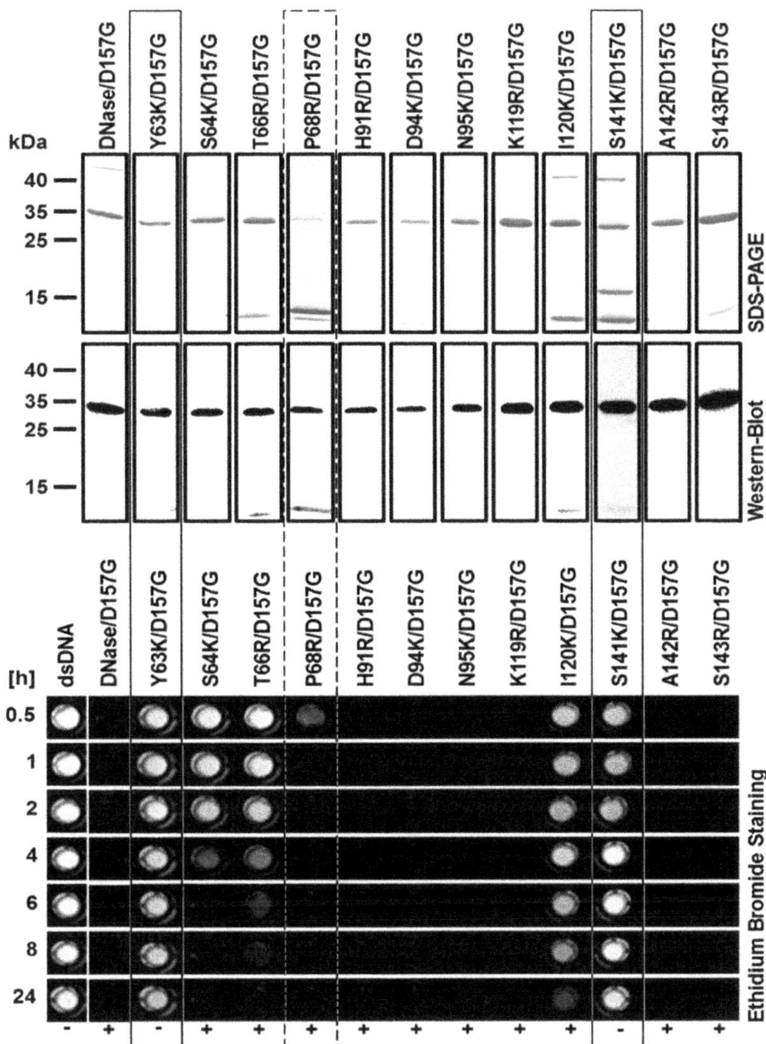

Figure 2. Biochemical and enzymatical analysis of purified nuclease mutants. SDS-PAGE results showed the purity of HIS-tagged nucleases (29 kDa) after a two-step purification approach composed of affinity and ion exchange chromatography (upper row). Western blotting analyses confirmed the presence of HIS-tagged nucleases (middle row). Qualitative activity assays used ethidium bromide staining in 96-well plates (lower row): white colored dots indicate the presence of ethidium bromide intercalating into DNA and grey dots indicate incomplete degradation, while a complete loss of fluorescence is due to complete DNA degradation. Solid line boxes—inactive variants Y63K/D157G and S141K/D157G were purified in non-optimized gravity flow experiments. Dashed line box—variant P68K/D157G was produced in expression strain *E. coli* Veggie BL21(DE3) pLysS. Note: "+" indicates activity and "−" indicates inactivity. SDS-PAGE results, including all purification steps, are shown in Figure A2 Appendix A.

A qualitative assay using ethidium bromide to visualize non-degraded, double-stranded DNA (dsDNA) was applied to demonstrate that mutants Y63K/D157G and S141K/D157G were inactive, while mutant S64K/D157G completely degraded dsDNA after 6 h, mutant T66R/D157G after 24 h, and mutant I120K/D157G partly hydrolyzes DNA after 24 h. These five mutants contain amino acid substitutions within the inner ring of the substrate groove. The remaining mutants as well as the positive control DNase/D157G completely hydrolyzed dsDNA after 30 min (Figure 2).

To ensure that the correct mutants were purified and characterized, the molecular masses of the purified enzymes were determined by mass spectrometry in addition to sequence verification of generated plasmids. Measured molecular weights are in accordance with predicted masses (Table 1). Variant P68R/D157G could not be properly identified due to impurities.

Table 1. Molecular masses of nuclease variants.

Variant	Theoretical MW (Da) [1]	Measured MW (Da)
DNase/D157G	28,900.30	28,900.61
Y63K/D157G	28,780.19	28,780.30
S64K/D157G	28,856.29	28,856.19
T66R/D157G	28,870.28	28,871.01
P68R/D157G	28,874.27	n.d. [2]
H91R/D157G	28,834.24	28,834.07
D94K/D157G	28,828.28	28,828.56
N95K/D157G	28,829.26	28,829.40
K119R/D157G	28,843.21	28,842.25
I120K/D157G	28,830.21	28,829.26
S141K/D157G	28,856.29	28,856.10
A142R/D157G	28,900.30	28,900.58
S143R/D157G	28,884.30	28,884.27

[1] Theoretical molecular weights (MW) of monomers were calculated using *Compute pI/MW*. [2] n.d.—not detectable.

2.3. Biochemical Properties of Nucleases

In vitro activity assays using dsDNA at 25 °C without the addition of ethidium bromide were conducted to confirm the qualitative plate assays. The reaction was stopped after 1 h of incubation and samples were loaded onto an agarose gel to investigate the level of hydrolysis of sheared dsDNA (Figure 3). As expected, inactive mutant enzymes Y63K/D157G and S141K/D157G were not capable of degrading dsDNA, while recombinant enzymes S64K/D157G, T66R/D157G, and I120K/D157G, respectively, exhibited reduced activity levels compared to the initial nuclease variant DNase/D157G and the remaining mutants.

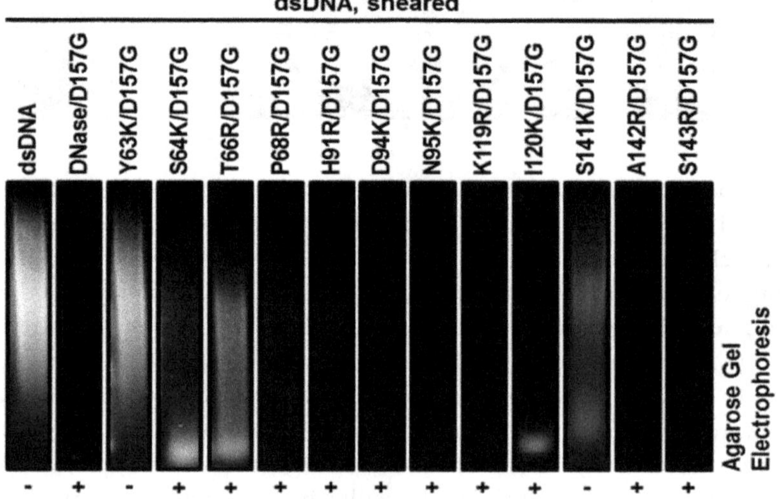

Figure 3. Digestion of sheared dsDNA by nuclease mutants. Sheared dsDNA (UltraPure™ Salmon Sperm DNA Solution) exhibited an average size of ≤ 2000 bps. Negative control containing substrate without enzymes is indicated as "dsDNA". The positive control DNase/D157G and mutants S64K/D157G, H91R/D157G, D94K/D157G, N95K/D157G, K119R/D157G, I120K/D157G, A142R/D157G, and S143R/D157G were able to completely hydrolyze sheared dsDNA. S64K/D157G, T66R/D157G, and I120K/D157G partially degraded sheared dsDNA within 1 h at 25 °C. Y63K/D157G and S141K/D157G did not exhibit activity towards sheared dsDNA. Note: "+" indicates activity and "−" indicates inactivity, without any quantification of the activity level.

After 1 h of incubation, mutant enzyme T66R/D157G only initiated the hydrolysis of dsDNA with some low molecular weight fragments visible at the bottom of the agarose gel. Therefore, the catalytic activity of this mutant was investigated with regard to its degradation velocity. Identical concentrations of mutant enzyme T66R/D157G were incubated with substrate for 30 min, 1 h, 2, h, 4 h, and 24 h, respectively. Reactions were stopped and the samples were loaded onto an agarose gel, revealing that the substrate becomes slowly degraded and is still not completely digested after 4 h of incubation (Figure 4).

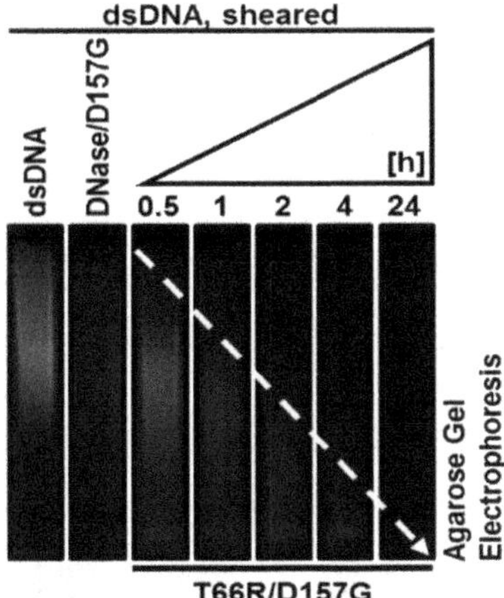

Figure 4. Degradation of sheared dsDNA by mutant T66R/D157G over time. Sheared dsDNA (UltraPureTM Salmon Sperm DNA Solution) exhibited an average size of ≤ 2000 bps. Reaction was stopped after indicated times (between 0.5 and 24 h). White dashed arrow indicates level of nucleic acid molecular weights. The dsDNA is completely hydrolyzed after 24 h of incubation at 25 °C.

To investigate the substrate promiscuity of active recombinant nuclease mutants, the enzymatic hydrolysis was studied towards the following substrates: unsheared dsDNA, single-stranded genomic DNA (ssDNA), circularized DNA, and RNA from bacteriophage MS2. Mutant enzymes with amino acid substitutions in the outer ring of the substrate groove that were active towards sheared dsDNA (Figure 3) also degraded unsheared dsDNA, ssDNA, circularized DNA, and RNA (Figure 5, Figure A3).

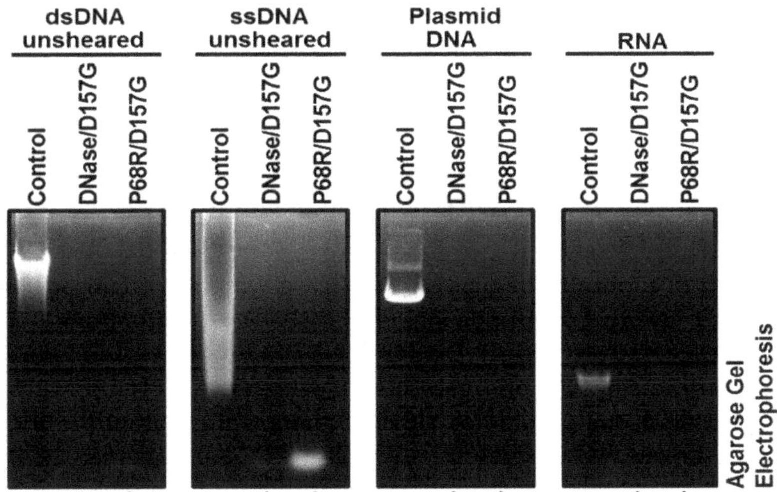

Figure 5. Substrate specificity of mutants with amino acid substitutions in the outer ring. Positive control DNase/D157G and mutant P68R/D157G were incubated for 1 h at 25 °C with different types of DNA and RNA. Further outer ring mutants H91R/D157G, D94K/D157G, N95K/D157G, K119R/D157G, A142R/D157G, and S143R/D157G were also able to degrade all types of nucleic acids (Figure A3 Appendix A). "Control" indicates negative controls containing substrate but no enzyme in the reaction mixture.

Nuclease mutant enzyme Y63K/D157G was also inactive towards unsheared dsDNA, ssDNA, and circularized DNA, while S141K/D157G showed some activity on all substrates except sheared dsDNA (Figures 3 and 6). However, mutant Y63K/D157G exhibited some activity towards RNA. In good agreement with previous results, mutant enzyme T66R/D157G exhibited reduced activity compared to control DNase/D157G and active mutant enzymes when incubated with both DNA and RNA (Figure 6). In contrast to sheared dsDNA, mutant S64K/D157G completely digested unsheared dsDNA, ssDNA, and circularized DNA, and partially digested RNA, while I120K/D157G was also active on every substrate, but only completely degraded RNA within 1 h at 25 °C (Figure 6).

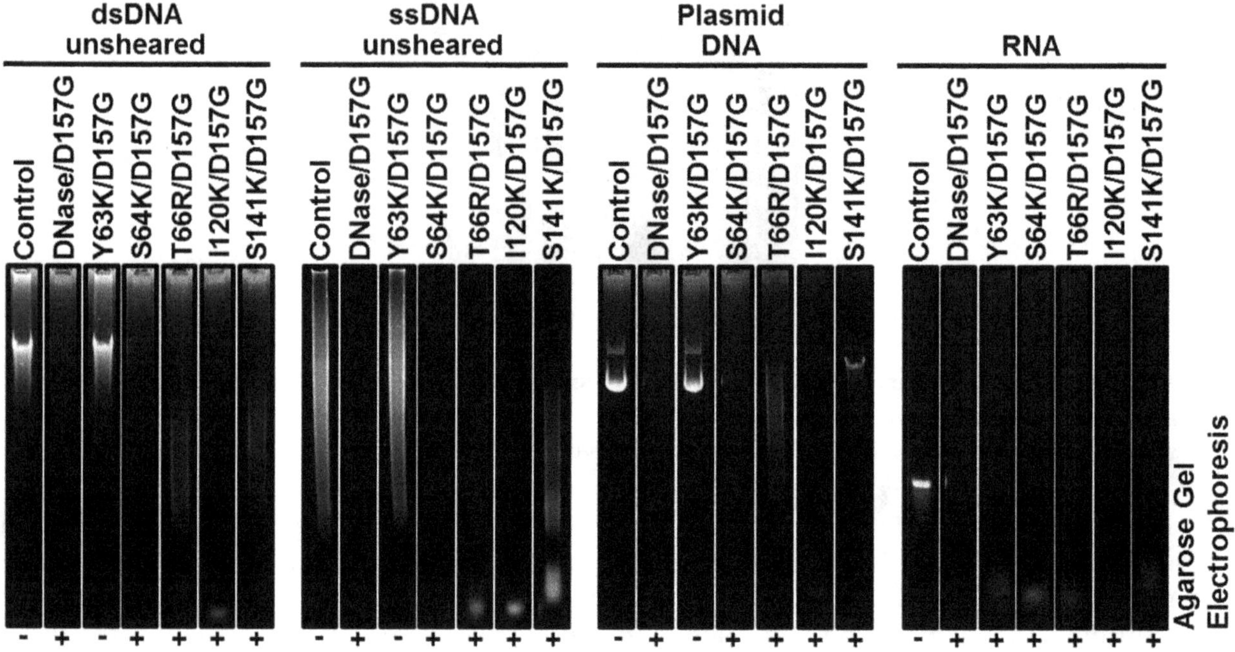

Figure 6. Substrate specificity of mutants with amino acid substitutions in the inner ring. Mutant Y63K/D157G was not able to degrade any type of nucleic acid. Mutants S64K/D157G, T66R/D157G, and I120K/D157G partially degraded all types of nucleic acids within 1 h at 25 °C, while mutant S141K/D157G exhibited low levels of degradation activity. "Control" indicates negative controls containing substrate but no enzyme in the reaction mixture.

Furthermore, pH and temperature optima of active mutants were determined. Every mutant as well as the initial variant DNase/D157G displayed optimal activity at pH 7.0. The temperature optima were around 50 °C for the initial variant DNase/D157G and all active mutants, except for mutants S64K/D157G and S143R/D157G, which showed optimal activities at 60–70 °C and 40 °C, respectively.

2.4. Enzyme Kinetics

Michaelis–Menten kinetics using sheared dsDNA confirmed the result of the quality activity assays: mutant enzymes Y63K/D157G and S141K/D157G did not exhibited any activity in these assays towards sheared dsDNA and the catalytic performances of mutant enzymes S64K/D157G, T66R/D157G, and I120K/D157G were lower compared to the initial nuclease variant DNase/D157G and remaining active mutant enzymes at 25 °C. Relative turnover rates were below 10% compared with DNase/D157G and the catalytic efficiency (k_{cat}/K_M) was below 0.2 s^{-1} μM^{-1} (Table 2). The turnover number of mutant enzyme K119R/D157G (1034 s^{-1}) exclusively surpassed the catalytic activity of DNase/D157G (924 s^{-1}), but the latter mutant enzyme also exhibited the lowest substrate affinity of all mutants (K_M = 428 μM).

Table 2. Kinetic characteristics of nuclease mutants [1].

Variant	K_M (μM)	k_{cat} (s^{-1})	k_{cat}/K_M (s^{-1} μM^{-1})	Relative Turnover Rate (%)
DNase/D157G	357	924	2.58	100
Y63K/D157G	n.d. [2]	n.d.	n.d.	-
S64K/D157G	115	8.0	0.07	0.9
T66R/D157G	335	62	0.19	6.7
P68R/D157G	358	106	0.3	11
H91R/D157G	284	290	1.02	31
D94K/D157G	224	170	0.76	18
N95K/D157G	273	455	1.66	49
K119R/D157G	428	1034	2.42	112
I120K/D157G	115	3.0	0.03	0.3
S141K/D157G	n.d.	n.d.	n.d.	-
A142R/D157G	352	707	2.01	77
S143R/D157G	83	153	1.84	17

[1] All experiments were done at 25 °C. [2] Not detectable, below the detection limit.

2.5. EDTA Tolerance of Nuclease Mutants

As metal-ion-independent proteins, NSNs of the PLD-like family are known to be tolerant in the presence of chelating agents, such as EDTA. DNase/D157G displayed > 80% relative activity at a concentration of 20 mM EDTA and still preserved > 60% residual activity in the presence of 50 mM EDTA. Active positively charged mutant enzymes were also tested in the presence of EDTA in concentrations between 1 mM and 50 mM, without any abnormalities, except that mutant enzymes P68R/D157G, N95K/D157G, I120K/D157G, and S143K/D157G were almost not affected by any concentration of EDTA (> 80% residual activity at concentrations up to 50 mM), while mutant enzymes T66R/D157G and K119R/D157G only showed residual activities of 25% and 34% in the presence of 50 mm EDTA, respectively.

3. Discussion

Commercially available NSNs are of high potential for the elimination of nucleic acids during protein downstream processing, to reduce the viscosity, or for prevention of cell clumping in cell sorting approaches [1,20,22]. The prototype non-specific nuclease is the metal-ion-dependent NSN from *Serratia marcescens* that is commercially sold under the trademark "Benzonase® Nuclease" (Merck KGaA, Darmstadt, Germany). This enzyme has been investigated in detail with regard to protein maturation, secretion, catalytic mechanism, and biotechnological applications [2]. Further NSNs, especially metal-ion-independent enzymes, have only been rarely investigated to date [13,16,18–20,23].

In this study, a semi-rational approach was used to select suitable mutation sites within the substrate groove of an NSN from *Pseudomonas syringae*, which were substituted by site-directed mutagenesis against positively charged amino acid residues to modulate the affinity for negatively charged substrates. Critical amino acid residues for the enzymatic performance of different nucleases were routinely identified by site-directed mutagenesis in previous studies [21,24–26].

Under natural conditions, it has been hypothesized that either positive mutations are first installed in a protein, while negative and neutral mutations are accumulated over time (Neo-Darwinian hypothesis), or neutral mutations pave the way for a flexible evolution and positive or negative mutations are installed as a response to certain conditions (competing hypothesis) [27]. By comparing homology models of proteins with singular amino acid substitutions, it is difficult to determine which of the specific mutations evokes an advantageous, neutral, or even deleterious effect on the protein performance, with certain effects on secondary and tertiary structures often being totally unpredictable [28]. Therefore, in silico mutagenesis was exclusively used to predict steric hindrance between introduced charged amino acid residues and amino acids of the catalytic site, or the predicted substrate groove followed by experimental testing of produced mutants.

Fifteen amino acid residues on the protein surface near the catalytic site were identified in a NSN that is highly conserved within the genus *Pseudomonas*. In silico mutagenesis revealed that twelve amino acids could be substituted against either arginine or lysine, without any steric effects in the molecular model. Eleven mutants could be produced in recombinant form in *E. coli* Veggie BL21 (DE3). However, transformation of *E. coli* Veggie BL21 (DE3), with an expression plasmid encoding for mutant P68R/D157G, did not result in any clones, but the recombinant enzyme could be produced in expression strain *E. coli* Veggie BL21 (DE3) pLysS with a very low yield. The additional plasmid pLysS encodes for T7 lysozyme to lower the background expression level of genes under the control of the T7 promoter. Therefore, it can be hypothesized that background expression of the gene-encoding mutant P68R/D157G in pLysS-less expression strains leads to lethality of the host strain. It is worth mentioning that in another experiment, the same results were monitored when proline at position 68 was replaced with negatively charged aspartate in our control. Nevertheless, mutant P68R/D157G only exhibited a turnover rate of 11% compared with the original variant DNase/D157G. However, the reduced activity may also be dependent on protein impurities that were still present after application of a two-step purification approach.

Artificially increasing the positive charge of the putative substrate groove in the NSN did not accelerate the catalytic performance of the enzyme. In another study, the activity of human DNase I also dropped with the addition of basic amino acids compared with the wild-type enzyme. However, suboptimal conditions for the wild-type enzyme, such as increased salt concentrations, accelerated the performance of the mutated enzyme variants [21]. Eleven out of twelve mutants in our portfolio displayed reduced activity at optimal conditions, while K119R/D157G was the only variant exhibiting an increased turnover number compared with DNase/D157G. However, this enzyme was the sole exception with a basic amino acid replaced by another basic amino acid. Two mutations within the inner ring of the substrate groove, namely Y63K and S141K, completely abolished the ability to hydrolyze sheared dsDNA, while mutant S141K/D157G was still capable of partially hydrolyzing unsheared dsDNA and circular plasmid DNA. Both mutants were also active towards RNA, but it is important to note that although all buffers were prepared under sterile conditions, the possibility that the observed activity is an artefact due to contamination with RNase cannot be excluded. Interestingly, asparagine at position 95 within the outer ring of the substrate groove was slightly impaired with regard to k_{cat}. A comparable effect has also been observed in another mutant, in which asparagine was replaced by serine. The latter amino acid occurs naturally in the homologous NSN in some members of the genus *Pseudomonas*.

Mutants were also probed for stability and activity effects in the presence of chelating agents. A slight performance improvement was detected for mutants N95K/D157G and S143K/D157G with regard to the tolerance of high concentrations of EDTA, which were not affected by concentrations up to 50 mM. Metal-ion-dependent DNases are usually completely inhibited by low concentrations of EDTA (1–5 mM), and even some metalloproteins that tolerate EDTA are dramatically impaired by concentrations of 50 mM [29,30]. However, the related NSNs from *Escherichia coli* and *Pantoea agglomerans* were already inhibited at concentrations above 20 mM of EDTA, which is similar to the results obtained with mutant enzymes T66R/D157G and K119R/D157G [19,20]. Nevertheless, these data are in line with the crystal structure of Nuc from *S. enterica* subsp. *enterica* serovar *Typhimurium*, demonstrating that NSNs of the superfamily of PLD proteins are not metal-ion-dependent [12].

The optimal growth temperature of *Pseudomonas syringae* is at 28 °C, but the highest activity of the nuclease and its mutants was determined to take place between 40 and 70 °C [31], These results are in line with previous observation of enzymes derived from psychro-, meso-, and thermophiles that displayed a temperature optimum, which is above their preferred growth temperatures [19,20,29,32,33].

Detailed enzyme characterizations are always a prerequisite for understanding the functionality of enzymes and to enable the modulation of their catalytic performances [34]. It has been shown that the enzyme family of NSNs from the genus *Pseudomonas* is a promising model protein for modifying the catalytic performance with regard to turnover number, temperature optimum, or EDTA tolerance by single amino acid substitutions. Due to their enzymatic properties, PLD-like NSNs from bacteria are of great potential for versatile biotechnological applications, and for this reason the discovery of novel enzymes, the optimization of available candidates, and the development of further applications are all highly needed [4,20,35–37]. Finally, the discovery of novel candidates and the extensive characterization in combination with straight-forward protein engineering techniques will lead to the production of more tailor-made enzymes for specific biotechnological applications.

4. Materials and Methods

4.1. Strain and Culture Conditions

Escherichia coli strains Veggie BL21 (DE3) and Veggie BL21(DE3) pLysS (both from Merck KGaA, Darmstadt, Germany) were used for gene expression and protein production. *E. coli* strain NEB® 5-alpha (New England Biolabs, Frankfurt/Main, Germany) was used for plasmid propagation and maintenance.

4.2. Computational Sequence Analysis and Structure Modelling

Protein sequence data of a non-specific nuclease (WP_050543862.1) from the gram-negative, ice-nucleating bacterium *Pseudomonas syringae* was identified and biochemically characterized in our laboratory (unpublished results). The naturally occurring amino acid substitution D157G in related homologous sequences was shown to be beneficial for catalytic activity of the enzyme compared to the wild-type polypeptide sequence. The three-dimensional model of the bacterial nuclease DNase/D157G was generated by the SWISS-MODEL online server, using the crystal structure of a homologous nuclease (PDB ID:1BYS_A) from *Salmonella enterica* subsp. *enterica* serovar *Typhimurium* as a template. The structure model was analyzed and visualized using the PyMOL software package (PyMOL Molecular Graphics System, Version 2.0 Schrödinger, LLC, New York, NY, USA). Amino acid residues that are located on the surface of the predicted substrate groove were identified and chosen to be substituted against positively charged amino acids (lysine or asparagine). Fifteen amino acid residues were identified to be orientated towards the protein surface close to the catalytic region. These amino acids were assigned to two groups: (1) inner ring (closely located to the catalytic site): Y63, S64, T66, I120, S141, and S143; (2) outer ring (distantly located to the catalytic region): P68, H91, G92, D94, N95, A97, A101, K119, and A142 (Figure A1 Appendix A). In silico mutagenesis was performed to discriminate between lysine and asparagine residue, replacing amino acid residues that are located on the surface of the predicted substrate groove. The substitution with the preferred amino acid residue, either asparagine or lysine, did not result in any steric clashes with adjacent amino acids in the protein for 12 out of 15 amino acids. Therefore, amino acid residues G92, A97, and A101 were excluded from site-directed mutagenesis.

4.3. Cloning of Nuclease Variants

Genes-encoding nuclease variants with amino acid substitutions were codon-optimized for expression in *E. coli* and synthesized by ATUM (Newark, CA, USA). Flanking *Nco*I and *Aat*II restriction sites were used for unidirectional ligation into linearized vector pET24d(+) (Merck KGaA, Darmstadt, Germany), equipped with a double HIS tag. Sequence verification of inserted genes was done by Eurofins Genomics (Ebersberg, Germany).

4.4. Gene Expression and Protein Purification

Expression of genes in *E. coli* Veggie BL21 (DE3) was performed as described earlier [20]. In brief, nuclease variants were produced with *E. coli* Veggie BL21(DE3) in 1 L cultures in 2 L Erlenmeyer shaking flasks. It was not possible to transform *E. coli* Veggie BL21 (DE3) with a plasmid coding for the P68K/D157G mutant in our control. Therefore, *E. coli* BL21 (DE3) pLys was used as an alternative expression host for this NSN variant. Cells were grown under constant shaking (250 rpm) at 37 °C until an optical density at 600 nm (OD_{600}) of 0.6–0.8 was reached. Afterwards, gene expression was induced by the addition of 0.4 mM isopropyl β-d-1-thiogalactopyranoside (IPTG). Cells were harvested 4 h post-induction by centrifugation for 15 min at 4 °C and 2880× *g*.

Cells were disrupted by high-pressure homogenization (constant cell disruption systems, Constant Systems Limited, Northants, UK) at 5 °C and 1250 bar. Pelleted cells were dissolved in lysis buffer (50 mM $NaPO_4$, pH 7.3): 1 g per 5 mL with a minimal volume of 20 mL. Crude protein extract was incubated for 30 min at 37 °C to enable digestion of nucleic acids by the recombinantly expressed nuclease. Afterwards, the sample was centrifuged at 4000× *g* for 30 min at 4 °C. HIS-tagged fusion enzymes were purified in a two-step approach using a combination of affinity (AC) and ion-exchange chromatography (IEX). Initially, gravity flow experiments using Ni sepharose 6 Fast Flow and SP Sepharose Fast Flow cation exchange chromatography resins (both GE Healthcare, Munich, Germany) were done to test activity of partly purified mutant enzymes. Afterwards, the Äkta purifier (GE Healthcare, Munich, Germany) was used to optimize the purification strategy. Supernatant from cell disruption was loaded onto a HisTrap FF Crude histidine-tagged protein purification column (GE Healthcare, Munich, Germany) equilibrated with 50 mM $NaPO_4$, 50 mM NaCl, 5 mM imidazole, pH 7.3. The loaded column was connected to an Äkta protein purifier system and washed with 10 column volumes of 50 mM $NaPO_4$, 50 mM NaCl, 50 mM imidazole, pH 7.3, prior to the elution with 50 mM $NaPO_4$, 50 mM NaCl, 500 mM imidazole, pH 7.3. For the purification by IEX, the eluate from the Ni sepharose affinity purification was diluted 1:4 with IEX running buffer (25 mM $NaPO_4$, pH 6.0) to reach a conductivity of 7–8 mS/cm. A HiTrap-SP FF column (GE Healthcare, Munich, Germany) was equilibrated with 10 column volumes of IEX running buffer, before the sample was loaded and washed with 10 column volumes of 50 mM sodium phosphate, 300 mM NaCl, pH 6.0, prior to the elution with 50 mM sodium phosphate, 700 mM NaCl, pH 6.0. All steps were performed at a flow velocity of 1 mL/min. Finally, a PD-10 desalting column (GE Healthcare, Munich, Germany) was used to replace the IEX elution buffer with storage buffer (50 mM $NaPO_4$, 25 mM NaCl, pH 7.3). SDS-PAGE (ProGel Tris Glycin 4–20%, Anamed Elektrophorese GmbH, Groß-Bierberau/Rodau, Germany) in combination with a Western blot using a nitrocellulose blotting membrane (GE Healthcare, Munich, Germany) was used for visualization of recombinant nuclease mutants. An anti-HIS horseradish peroxidase (HRP) antibody (Miltenyi Biotec B.V. & Co. KG, Bergisch Gladbach, Germany) and the Immobilon™ Western HRP substrate (Merck, Darmstadt, Germany) were used to detect HIS-tagged proteins.

Enzyme concentrations were measured using a (micro-) Bradford approach in 96-well plate format. A plate reader (EMax, Molecular Devices, San Jose, CA, USA) was used to determine absorbances at 590 nm, which were evaluated with the software SoftmaxPro V5 (Molecular Devices, San Jose, CA, USA). The identities of the nuclease mutants based on their molecular masses were verified with the micrOTOF-Q II Benchtop Mass Spectrometer (Bruker, Billerica, MA, USA).

4.5. Enzyme Activity Assays

Enzyme activity was determined both qualitatively and quantitatively based on the depolymerization of nucleic acids. All activity assays were done at pH 7, which has been determined to be the optimal pH. Due to the limited stability of nuclease mutants at elevated temperatures, all assays were done at 25 °C. No thermal effect on the stability was observed for any mutant at this temperature. (1) Qualitative ethidium bromide staining: this assay was adopted from [4]. Recombinant enzyme was incubated with 5 µg sheared double-stranded genomic DNA (dsDNA), namely UltraPure™ Salmon Sperm DNA Solution (Thermo Fisher Scientific, Darmstadt, Germany). This method is based on quenching of fluorescence of ethidium bromide intercalated into DNA. Repeated fluorescence recordings were taken using a VWR® imager (VWR international, Radnor, PA, USA) over a duration of up to 24 h. (2) Qualitative visualization by agarose gel electrophoresis: this assay was adopted from [20]. Substrate specificity was tested with sheared dsDNA, unsheared dsDNA, namely deoxyribonucleic acid from calf thymus (Sigma-Aldrich, St. Louis, MO, USA), single-stranded genomic DNA (ssDNA) from calf thymus (Sigma-Aldrich, St. Louis, USA), RNA from bacteriophage MS2 (Sigma-Aldrich, St. Louis, USA), and circularized plasmid DNA. Qualitative levels of activity are exclusively interpreted as "+" (active enzyme) and "−" (inactive enzyme) in Figure 3, Figure 5, and Figure 6, and do not allow any quantification of activities. (3) Quantitative measurements were done in Corning® 96-well UV-transparent plates (Merck KGaA, Darmstadt, Germany) using the Victor™ X4 Multilabel Plate Reader (PerkinElmer, Rodgau, Germany), as described previously [20]. Standard activity assays were conducted in 50 mM sodium phosphate buffer at pH 7.3. For the generation of pH profiles, reactions were performed in 50 mM sodium acetate buffer at pH 5 and 6, in 50 mM sodium phosphate buffer at pH 6, 7, and 8, and in Tris/HCl buffer at pH 7, 8, and 9. Temperature profiles were conducted in the range of 10 to 90 °C. EDTA tolerance was tested at the following concentrations: 0, 1, 2, 5, 10, 20, and 50 mM. (4) Kinetic parameters were determined with V_{max} and K_M obtained from Michaelis–Menten technique by non-linear regression, as described previously [20,38]. It has been speculated that the initial reaction rate of high molecular weight substrates is reduced at high substrate concentrations due to the extension of the lag phase [39]. Therefore, maximum reaction rates were evaluated for each substrate concentration to determine defined kinetic parameters. All experiments were done in triplicate. The error level was below 10%.

5. Conclusions

The composition of the substrate groove was investigated by a combination of structural modelling, multiple sequence alignment, and site-directed mutagenesis. Amino acid residues that are in close proximity to the catalytic site (inner ring of the substrate groove) are of tremendous importance for proper activity of NSN, while amino acids at the border of the substrate groove (outer ring) are promising targets for modulation of the enzymatic properties with regard to turnover number, EDTA tolerance, and temperature preference.

6. Patents

A patent application describing the utilization of non-specific nucleases from the genus *Pseudomonas* and their application potential in cell-sorting approaches has been submitted by Miltenyi Biotec B.V. & Co. KG.

Author Contributions: Conceptualization, S.E.; methodology, L.S.S. and S.S.; validation, L.S.S., S.S., V.N., and S.E.; resources, V.N.; visualization, L.S.S. and S.E.; writing—original draft preparation, S.E.; writing—review and editing, S.E. and V.N.; All authors approved the final manuscript.

Acknowledgments: The authors thank Jens Hellmer (Miltenyi Biotec B.V. & Co. KG) for mass spectrometry analyses and Marek Wieczorek (Miltenyi Biotec B.V. & Co. KG) for discussion.

Appendix A

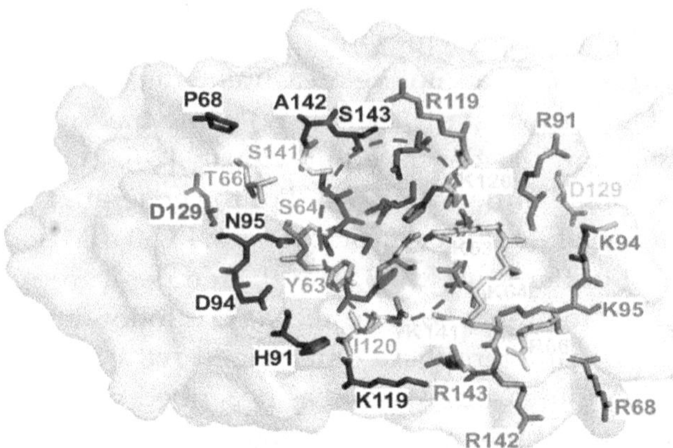

Figure A1. Homology model of NSN from *Pseudomonas syringae* in top view. Amino acid residues of the conserved $HxK(x)_4D(x)_6GSxN$ motif are given as sticks in dark and light red in the respective monomer. Numbering of amino acids that are part of the catalytic site was omitted for clarity (encircled in red, dashed line), except for amino acid residue D129, which is part of the $HxK(x)_4D(x)_6GSxN$ motif, but not part of the catalytic site. Naturally occurring amino acids of the outer and inner rings are also given as sticks and highlighted in dark and light blue, while substituted positively charged amino acid residues are indicated in dark and light green.

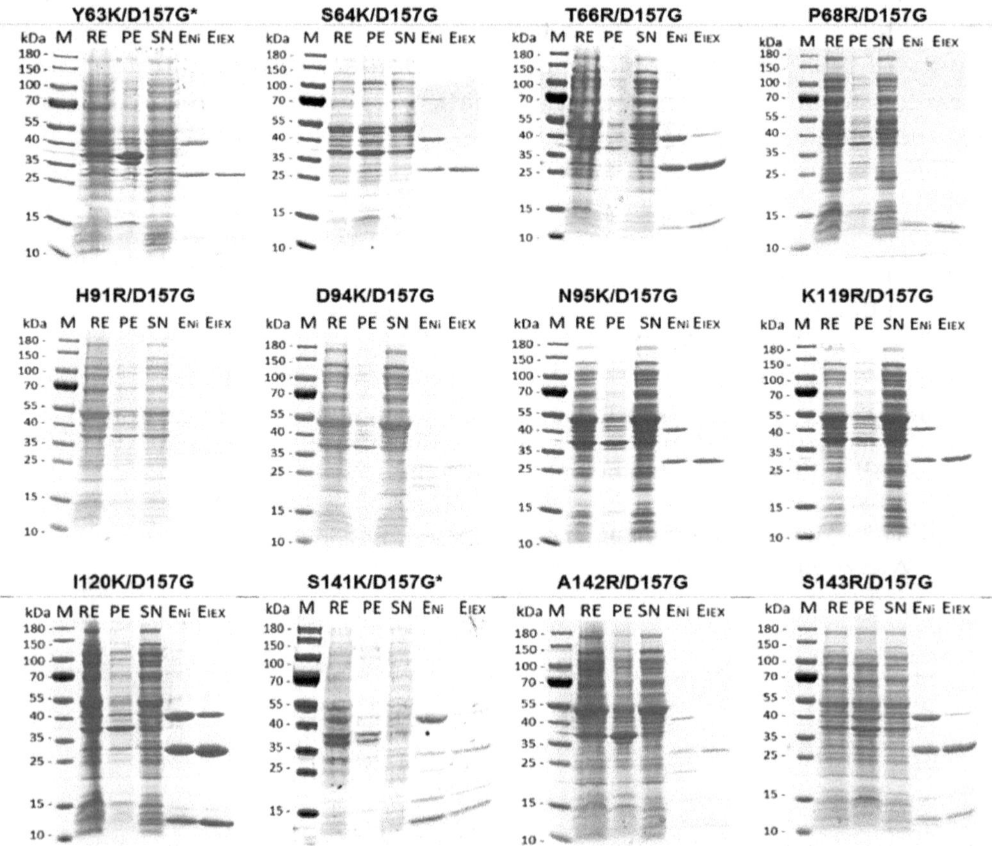

Figure A2. SDS-PAGE results illustrating the purification of all mutant enzymes. M—protein marker, RE—crude extract, PE—pellet, SN—supernatant, E_{Ni}—elution fraction Ni-agarose, E_{IEX}—elution fraction ion exchange chromatography. Asterisks (*) indicate that inactive mutants Y63K/D157G and S141K/D157G were purified using non-optimized gravity flow purification approaches, resulting in lower purities.

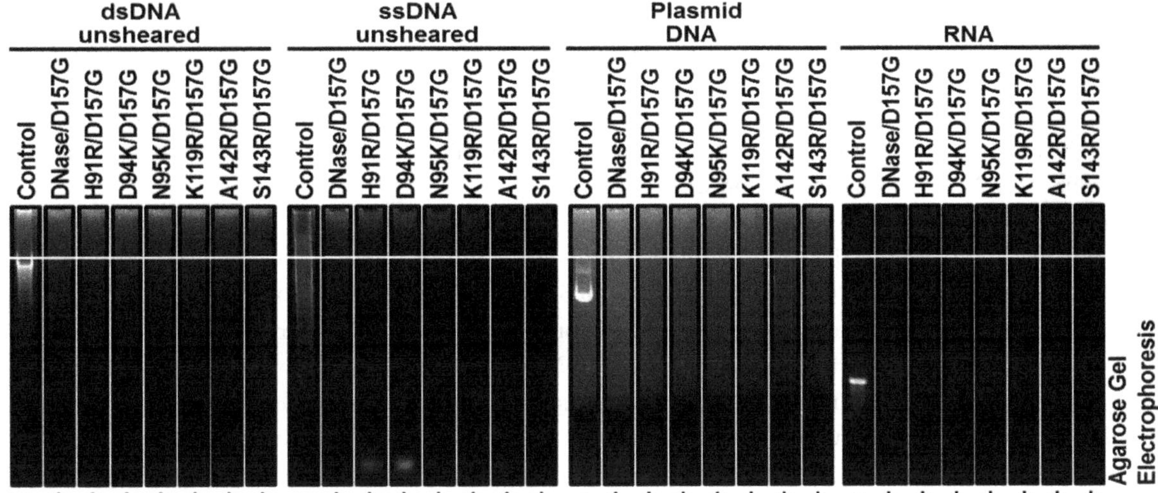

Figure A3. Substrate specificity of mutants with amino acid substitutions in the outer ring. All incubations were done for 1 h at 25 °C. "Control" indicates negative controls containing substrate but no enzyme in the reaction mixture.

References

1. Rangarajan, E.S.; Shankar, V. Sugar non-specific endonucleases. *FEMS Microbiol. Rev.* **2001**, *25*, 583–613. [CrossRef]
2. Benedik, M.J.; Strych, U. *Serratia marcescens* and its extracellular nuclease. *FEMS Microbiol. Lett.* **1998**, *165*, 1–13. [CrossRef] [PubMed]
3. Dang, G.; Cui, Y.; Wang, L.; Li, T.; Cui, Z.; Song, N.; Chen, L.; Pang, H.; Liu, S. Extracellular sphingomyelinase Rv0888 of *Mycobacterium tuberculosis* contributes to pathological lung injury of *Mycobacterium smegmatis* in mice via inducing formation of neutrophil extracellular traps. *Front. Immunol.* **2018**, *9*, 677. [CrossRef] [PubMed]
4. Vafina, G.; Zainutdinova, E.; Bulatov, E.; Filimonova, M.N. Endonuclease from gram-negative Bacteria *Serratia marcescens* is as effective as Pulmozyme in the hydrolysis of DNA in sputum. *Front. Pharmacol.* **2018**, *9*, 114. [CrossRef] [PubMed]
5. Khersonsky, O.; Tawfik, D.S. Enzyme promiscuity: A mechanistic and evolutionary perspective. *Annu. Rev. Biochem.* **2010**, *79*, 471–505. [CrossRef] [PubMed]
6. Yan, X.; Wang, J.; Sun, Y.; Zhu, J.; Wu, S. Facilitating the evolution of esterase activity from a promiscuous enzyme (Mhg) with catalytic functions of amide hydrolysis and carboxylic acid perhydrolysis by engineering the substrate entrance tunnel. *Appl. Environ. Microbiol.* **2016**, *82*, 6748–6756. [CrossRef]
7. Rao, S.J.; Shukla, E.; Bhatia, V.; Lohiya, B.; Gaikwad, S.M.; Kar, A.; Pal, J.K. The *Leishmania donovani* IMPACT-like protein possesses non-specific nuclease activity. *Int. J. Biol. Macromol.* **2018**, *119*, 962–973. [CrossRef] [PubMed]
8. Ponting, C.P.; Kerr, I.D. A novel family of phospholipase D homologues that includes phospholipid synthases and putative endonucleases: Identification of duplicated repeats and potential active site residues. *Protein Sci. A Publ. Protein Soc.* **1996**, *5*, 914–922. [CrossRef]
9. Rudolph, A.E.; Stuckey, J.A.; Zhao, Y.; Matthews, H.R.; Patton, W.A.; Moss, J.; Dixon, J.E. Expression, characterization, and mutagenesis of the *Yersinia pestis* murine toxin, a phospholipase D superfamily member. *J. Biol. Chem.* **1999**, *274*, 11824–11831. [CrossRef]
10. Nelson, R.K.; Frohman, M.A. Physiological and pathophysiological roles for phospholipase D. *J. Lipid Res.* **2015**, *56*, 2229–2237. [CrossRef]
11. Liscovitch, M.; Czarny, M.; Fiucci, G.; Tang, X. Phospholipase D: Molecular and cell biology of a novel gene family. *Biochem. J.* **2000**, *345*, 401–415. [CrossRef] [PubMed]
12. Stuckey, J.A.; Dixon, J.E. Crystal structure of a phospholipase D family member. *Nat. Struct. Biol.* **1999**, *6*, 278–284. [CrossRef] [PubMed]

13. Zhao, Y.; Stuckey, J.A.; Lohse, D.L.; Dixon, J.E. Expression, characterization, and crystallization of a member of the novel phospholipase D family of phosphodiesterases. *Protein Sci.* **1997**, *6*, 2655–2658. [CrossRef] [PubMed]
14. Bao, Y.; Higgins, L.; Zhang, P.; Chan, S.H.; Laget, S.; Sweeney, S.; Lunnen, K.; Xu, S.Y. Expression and purification of *Bmr*I restriction endonuclease and its N-terminal cleavage domain variants. *Protein Expr. Purif.* **2008**, *58*, 42–52. [CrossRef] [PubMed]
15. Grazulis, S.; Manakova, E.; Roessle, M.; Bochtler, M.; Tamulaitiene, G.; Huber, R.; Siksnys, V. Structure of the metal-independent restriction enzyme *Bfi*I reveals fusion of a specific DNA-binding domain with a nonspecific nuclease. *Proc. Natl. Acad. Sci. USA* **2005**, *102*, 15797–15802. [CrossRef] [PubMed]
16. Song, Q.; Zhang, X. Characterization of a novel non-specific nuclease from thermophilic bacteriophage GBSV1. *BMC Biotechnol.* **2008**, *8*, 43. [CrossRef] [PubMed]
17. Wang, D.; Miyazono, K.I.; Tanokura, M. Tetrameric structure of the restriction DNA glycosylase R.*Pab*I in complex with nonspecific double-stranded DNA. *Sci. Rep.* **2016**, *6*, 35197. [CrossRef]
18. Li, L.; Lin, S.; Yanga, F. Functional identification of the non-specific nuclease from white spot syndrome virus. *Virology* **2005**, *337*, 399–406. [CrossRef]
19. Schmitz, S.; Börner, P.; Nölle, V.; Elleuche, S. Comparative analysis of two non-specific nucleases of the phospholipase D family from the plant pathogen competitor bacterium *Pantoea Agglomerans*. *Appl. Microbiol. Biotechnol.* **2019**, *103*, 2635–2648. [CrossRef]
20. Schmitz, S.; Nölle, V.; Elleuche, S. A non-specific nucleolytic enzyme and its application potential in EDTA-containing buffer solutions. *Biotechnol. Lett.* **2019**, *41*, 129–136. [CrossRef]
21. Pan, C.Q.; Lazarus, R.A. Hyperactivity of human DNase I variants. Dependence on the number of positively charged residues and concentration, length, and environment of DNA. *J. Biol. Chem.* **1998**, *273*, 11701–11708. [CrossRef] [PubMed]
22. Miltenyi, S.; Hübel, T.; Nölle, V. Process for Sorting Cells by Microfabricated Components Using a Nuclease. US Patant 10,018,541 B2, 10 July 2018.
23. Belkebir, A.; Azeddoug, H. Characterization of *Lla*KI, a new metal ion-independent restriction endonuclease from *Lactococcus lactis* KLDS4. *ISRN Biochem.* **2012**, *2012*, 287230. [CrossRef] [PubMed]
24. Friedhoff, P.; Kolmes, B.; Gimadutdinow, O.; Wende, W.; Krause, K.L.; Pingoud, A. Analysis of the mechanism of the *Serratia* nuclease using site-directed mutagenesis. *Nucleic Acids Res.* **1996**, *24*, 2632–2639. [CrossRef] [PubMed]
25. Zhang, Y.; Li, Z.H.; Zheng, W.; Tang, Z.X.; Shi, L.E. Enzyme activity and thermostability of a non-specific nuclease from *Yersinia enterocolitica supsp. palearctica* by site-directed mutagenesis. *Electron. J. Biotechnol.* **2016**, *24*, 32–37. [CrossRef]
26. Franke, I.; Meiss, G.; Pingoud, A. On the advantage of being a dimer, a case study using the dimeric *Serratia* nuclease and the monomeric nuclease from *Anabaena* sp. strain PCC 7120. *J. Biol. Chem.* **1999**, *274*, 825–832. [CrossRef] [PubMed]
27. Bommarius, A.S.; Paye, M.F. Stabilizing biocatalysts. *Chem. Soc. Rev.* **2013**, *42*, 6534–6565. [CrossRef] [PubMed]
28. Schaefer, C.; Rost, B. Predict impact of single amino acid change upon protein structure. *BMC Genom.* **2012**, *13*, S4. [CrossRef]
29. Elleuche, S.; Fodor, K.; Klippel, B.; von der Heyde, A.; Wilmanns, M.; Antranikian, G. Structural and biochemical characterisation of a NAD(+)-dependent alcohol dehydrogenase from *Oenococcus oeni* as a new model molecule for industrial biotechnology applications. *Appl. Microbiol. Biotechnol.* **2013**, *97*, 8963–8975. [CrossRef]
30. Marcal, D.; Rego, A.T.; Fogg, M.J.; Wilson, K.S.; Carrondo, M.A.; Enguita, F.J. Crystallization and preliminary X-ray characterization of 1,3-propanediol dehydrogenase from the human pathogen *Klebsiella pneumoniae*. *Acta Crystallogr. Sect. F Struct. Biol. Cryst. Commun.* **2007**, *63*, 249–251. [CrossRef]
31. Xin, X.F.; Kvitko, B.; He, S.Y. *Pseudomonas syringae*: What it takes to be a pathogen. *Nat. Rev. Microbiol.* **2018**, *16*, 316–328. [CrossRef]
32. Daniel, R.M.; Danson, M.J.; Eisenthal, R.; Lee, C.K.; Peterson, M.E. The effect of temperature on enzyme activity: New insights and their implications. *Extremophiles* **2008**, *12*, 51–59. [CrossRef] [PubMed]

33. Feller, G.; Narinx, E.; Arpigny, J.L.; Zekhnini, Z.; Swings, J.; Gerday, C. Temperature dependence of growth, enzyme secretion and activity of psychrophilic Antarctic bacteria. *Appl. Microbiol. Biotechnol.* **1994**, *41*, 477–479. [CrossRef]
34. Erb, T.J. Back to the future: Why we need enzymology to build a synthetic metabolism of the future. *Beilstein J. Org. Chem.* **2019**, *15*, 551–557. [CrossRef] [PubMed]
35. Fang, X.J.; Tang, Z.X.; Li, Z.H.; Zhang, Z.L.; Shi, L.E. Production of a new non-specific nuclease from *Yersinia enterocolitica subsp. palearctica*: Optimization of induction conditions using response surface methodology. *Biotechnol. Biotechnol. Equip.* **2014**, *28*, 559–566. [CrossRef] [PubMed]
36. Maciejewska, N.; Walkusz, R.; Olszewski, M.; Szymanska, A. New nuclease from extremely psychrophilic microorganism *Psychromonas ingrahamii* 37: Identification and characterization. *Mol. Biotechnol.* **2019**, *61*, 122–133. [CrossRef] [PubMed]
37. Anisimova, V.E.; Shcheglov, A.S.; Bogdanova, E.A.; Rebrikov, D.V.; Nekrasov, A.N.; Barsova, E.V.; Shagin, D.A.; Lukyanov, S.A. Is crab duplex-specific nuclease a member of the *Serratia* family of non-specific nucleases? *Gene* **2008**, *418*, 41–48. [CrossRef] [PubMed]
38. MacLellan, S.R.; Forsberg, C.W. Properties of the major non-specific endonuclease from the strict anaerobe *Fibrobacter succinogenes* and evidence for disulfide bond formation in vivo. *Microbiology* **2001**, *147*, 315–323. [CrossRef]
39. Friedhoff, P.; Meiss, G.; Kolmes, B.; Pieper, U.; Gimadutdinow, O.; Urbanke, C.; Pingoud, A. Kinetic analysis of the cleavage of natural and synthetic substrates by the *Serratia* nuclease. *Eur. J. Biochem.* **1996**, *241*, 572–580. [CrossRef]

10

Screening and Comparative Characterization of Microorganisms from Iranian Soil Samples Showing ω-Transaminase Activity toward a Plethora of Substrates

Najme Gord Noshahri [1], Jamshid Fooladi [1,*], Christoph Syldatk [2], Ulrike Engel [2], Majid M. Heravi [3], Mohammad Zare Mehrjerdi [4] and Jens Rudat [2,*]

1 Department of Biotechnology, Faculty of Biology Science, Alzahra University, 1993893973 Tehran, Iran
2 BLT_II: Technical Biology, Karlsruhe Institute of Technology (KIT), Fritz-Haber-Weg 4, 76131 Karlsruhe, Germany; christoph.syldatk@kit.edu (C.S.); ulrike.engel@kit.edu (U.E.)
3 Department of Chemistry, School of Sciences, Alzahra University, 1993891176 Tehran, Iran; mmheravi@alzahra.ac.ir
4 Higher Education Complex of Shirvan, 9468194477 Shirvan, Iran; mzarem@um.ac.ir
* Correspondence: jfooladi@alzahra.ac.ir (J.F.); jens.rudat@kit.edu (J.R.)

Abstract: In this study, soil microorganisms from Iran were screened for ω-transaminase (ω-TA) activity based on growth on minimal media containing (*rac*)-α-methylbenzylamine (rac-α-MBA) as a sole nitrogen source. Then, for the selection of strains with high enzyme activity, a colorimetric o-xylylendiamine assay was conducted. The most promising strains were identified by 16S rDNA sequencing. Five microorganisms showing high ω-TA activity were subjected to determine optimal conditions for ω-TA activity, including pH, temperature, co-solvent, and the specificity of the ω-TA toward different amine donors and acceptors. Among the five screened microorganisms, *Bacillus halotolerans* turned out to be the most promising strain: Its cell-free extract showed a highly versatile amino donor spectrum toward aliphatic, aromatic chiral amines and a broad range of pH activity. Transaminase activity also exhibited excellent solvent tolerance, with maximum turnover in the presence of 30% (*v/v*) DMSO.

Keywords: ω-transaminase; α-methylbenzylamine; biocatalysis; chiral amine; biotransformation

1. Introduction

Chiral amines as building blocks are prevalent in pharmaceuticals and chemical industries [1]. For instance, chiral amines are present in roughly 40% of FDA approved pharmaceuticals [2,3]. Furthermore, chiral β-amino acids (β-aas) have attracted significant attention recently [4]. They are critical building blocks in highly important medicines and natural compounds such as maraviroc (HIV remedy) [5,6], taxol (cytostatic drug against breast and ovarian cancer), jaspamide (insecticidal and antifungal agent), theopalauamide (antifungal compound), and dolastatin (antitumor agent) [7]. β-aas are also employed in the production of peptidomimetics that are stable against degradation by proteolytic enzymes [4,8].

The significant challenge in industry is to use optically pure amines and amino acids from cost-effective production [1,9]. For the mass production of such compounds, nowadays biocatalysts are frequently used to comply with green chemistry as well be economically feasible [1,5]. Among various enzymatic methods for producing chiral amine compounds, ω-transaminases (ω-TAs) are promising enzymes to synthesize chiral amines. They show high substrate spectra, enantioselectivity, and no requirement to regenerate external cofactors [10–13]. They belong to pyridoxal-5′–phosphate-(PLP)-dependent enzymes that transfer an amino group from an amino donor to amino accepters such as prochiral ketones or aldehydes [14,15]. In addition, ω-TAs can be used for providing optically enriched chiral amino acids via the kinetic resolution of racemic amines or asymmetric synthesis from prochiral ketones [12].

A restricted substrate scope for large molecules [16], as well as rather low enzyme stability under extreme reaction conditions associated with pH, temperature [17], and organic solvents [5,13] have been major hurdles of using ω-TAs. Thus, these limitations should be circumvented by additional screening approaches to fill the toolbox of industrially applicable enzymes.

Basically, there are different approaches to discover new ω-TAs: metagenomic screenings [18,19], site-directed mutagenesis of well-established enzymes [20], and screening new biocatalysts by identifying novel microorganisms from as-yet unexplored screening sites [21–23].

Results of several studies indicated that Iranian landscapes contain highly diverse microbial storages due to considerable variety in geographical features, including mountain range peaks with volcano activity, sandy deserts (Dasht-e Lut is one of the hottest points of the Earth), rivers [24], forest areas, hot springs, glaciers, and frequent snowing. Iran also has two coastlines to its north and south. Besides the extremely versatile environments in Iran, industries such as petroleum (oil well) and petrochemical units releasing chemical substances cause influences on the microorganism populations of the above-mentioned country. With such a variety of environments, Iran's soil can be considered a rich source for screening microorganisms, which are able to produce special extremophile biocatalysts [25–28].

The purpose of this study is to identify novel ω-TA from Iranian soil microorganisms to be examined as a catalyst in the synthesis of enantiopure β-amino acid (β-aa) and other chiral amino compounds. Relevant properties of this novel ω-TA, such as the effect of temperature, pH, and solvent and substrate spectra were also investigated.

2. Results and Discussion

2.1. Microorganisms Screening and Identification

In this study, 42 strains were isolated by enrichment culture against (*rac*)-α-methylbenzylamine (rac-MBA) as a sole nitrogen source. Induction and amount of ω-TA is affected by nitrogen source in media. In complex media such as Luria-Bertani (LB) medium, some of the microorganisms produce enzymes with lower activity [29]. Therefore, minimal media (MIM) containing (*rac*)-α-MBA as a model amine compound [22,23] was used in the experiment.

They were subsequently validated through *o*-xylylenediamine (*o*-XDA) assay, with the color changing from yellow to black after 4 h. *o*-XDA after deamination undergoes spontaneous cyclization, tautomerization, and irreversible polymerization to form a black product [1]. Five strains showing ω-TA activity were found to be promising. Four of these strains were identified by 16S rDNA sequencing to be members of the genus *Bacillus* close to *Bacillus halotolerans* (BaH) (99.93% similarity) and *Bacillus endophyticus* (BaE) (99.76% similarity). Two subspecies belong to *Bacillus subtilis* with 99.92% and 99.93% similarity to *stercoris* (BaS) and *inaquosorum* (BaI), respectively. The fifth bacterium is associated to *Rhizobium radiobacter* (RhZ) with 99.77% similarity (Table 1). All of the strains except BaE were isolated from first method of screening.

Table 1. Results of 16SrDNA sequencing.

Isolate	Closest Relative in GenBank	% Similarity/Sequence Length (bp)
BaH	*Bacillus halotolerans* (LPVF01000003)	99.9/1388
BaE	*Bacillus endophyticus* (AF295302)	99.7/1273
BaS	*Bacillus subtilis* subsp. *stercoris* (JHCA01000027)	99/1385
BaI	*Bacillus subtilis* subsp. *inaquosorum* (AMXN01000021)	99.9/1406
RhZ	*Rhizobium radiobacter* (AJ389904)	99.4/1338

Number in brackets is GenBank accession number.

2.2. Enzymatic Properties and Substrate Specificity

Appropriate temperature, pH, and organic solvents play important roles in ω-TA activity [5,30]. To survey such effects, the optimum reaction condition against these three factors as investigated in the screened bacteria. Several studies have been performed by applying cell and crude extract to convert different substrates [2,18,21,29,31–33], so instead of using high-priced purified protein, we applied cell-free extracts to optimize reaction parameters and to investigate substrate scope.

The relative values of ω-TA activity were calculated based on producing acetophenone as a product of *(S)*-MBA deamination. The highest activity was considered as 100%.

2.2.1. Effect of Temperature on ω-TA Activity

The cell-free extracts were incubated at defined temperatures (25–65 °C) for one hour. The crude extracts of BaH, RhZ, and BaS shared similar expected mesophilic temperature profiles between 30 and 65 °C, with highest activity at 35 °C, which is reported for other ω-TAs [21,34]. They exhibited reasonable activity of around 70% up to 45 °C. Among all of them, 40–50% of activity remained at 50 °C, which decreased to approximately 10–20% at temperatures above 60 °C (Figure 1a).

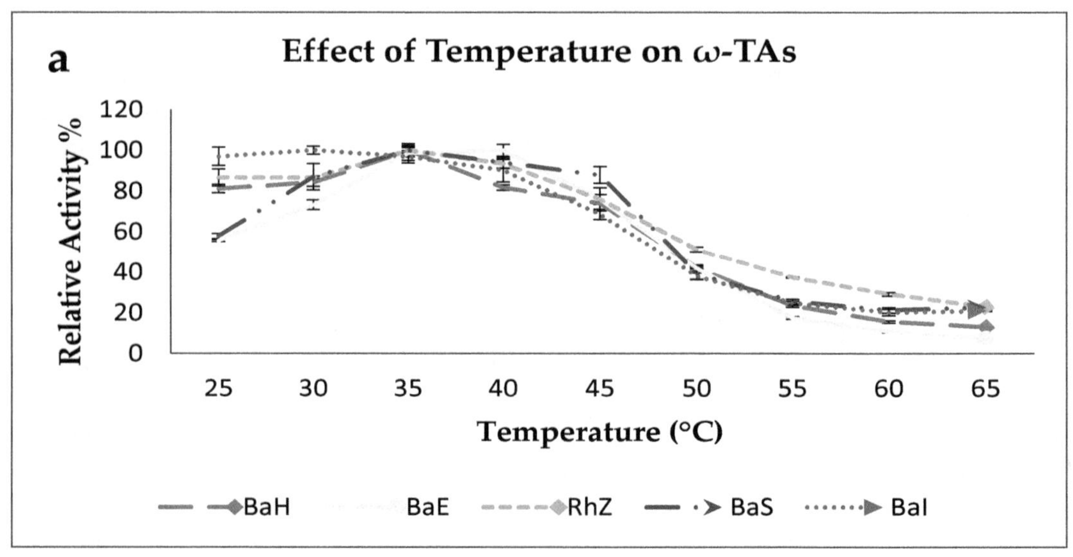

Figure 1. *Cont.*

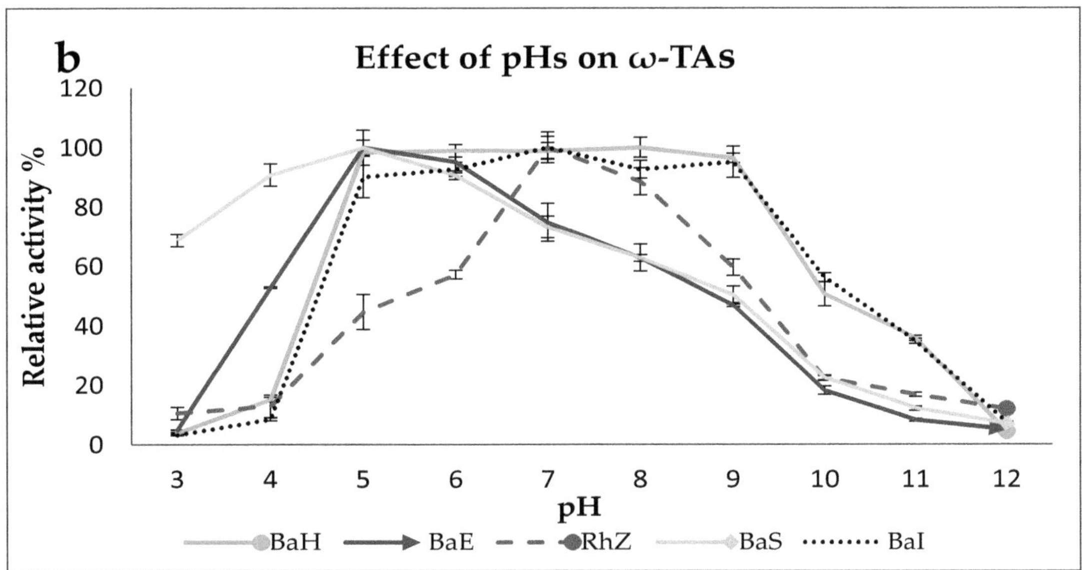

Figure 1. The effect of temperature (**a**) and pH (**b**) on the amination of pyruvate and (*S*)-MBA as amino donor using 1 mg/mL of cell-free extracts of screened strains: *Bacillus halotolerans* (BaH), *Bacillus endophyticus* (BaE), *Rhizobium radiobacter* (RhZ), *Bacillus subtilis* subsp. *stercoris* (BaS) and *Bacillus subtilis* subsp. *inaquosorum* (BaI).Reaction conditions are given in the method section. Formation of acetophenone was detected by HPLC at 254 nm. The value of 100% corresponded to the highest activity. Error bars represent ± standard deviation.

For ω-TABaE and ω-TABaI, the optimum reaction temperature was 40 °C and 30 °C, respectively. No significant change of activity was observed at 35 °C. Thus, for surveying the effect of pH and solvent on enzyme activity on all crude extracts, the temperature was maintained at 35 °C. Additionally, ω-TABaI showed approximately the same activity at temperatures between 20 and 30 °C (Table S1).

2.2.2. Effect of pH on ω-TA Activity

In contrast to other ω-TAs that usually show their maximum activity at slightly alkaline pH [12,21,31], ω-TABaI and ω-TABaH exhibited an almost constant activity in the pH range of 5 to 9. Surprisingly, for ω-TABaS and ω-TABaE, the highest enzyme activity was observed under acidic conditions (pH 5). Notably, 70% of ω-TABaS activity remained at pH 3, Furthermore, ω-TARhz preferred neutral conditions (Figure 1b). No enzyme activity was observed at pH 12. In the literature data, there is no report concerning ω-TA activity in acidic conditions. ω-TA activity at different pHs was mostly tested in pH range from 6 to 11 [4,17,18].

2.2.3. Effect of Organic Solvents on ω-TA Activity

Organic solvents are often added to ω-TA reactions to increase substrate solubility. However, this mostly causes negative effects on enzyme activity [10]. Therefore, the examination of the tolerance of the applied enzyme to different solvents is highly required [5]. To analyze whether solvents affect enzyme activity, four different solvents with a 0%–20% with the exception of DMSO which increased up to 40% *v/v* concentration were tested. The ω-TA activity in aqueous buffer without addition of organic solvents was set as 100%. The results were expressed as relative activity in %.

Dimethyl sulfoxide (DMSO) is a common solvent that was used in concentrations ranging from 5%–20% *v/v* in different projects [5]. DMSO as co-solvent led to the highest activity, followed by methanol, N,N-Dimethylformamide (DMF), and isopropanol. In some cases (BaI and Rhz), the reactions without any co-solvent showed better activity than methanol. It is remarkable to mention that ω-TABaH activity was enhanced threefold with a concentration of DMSO up to 30%. The results again suggest that DMSO is the most suitable organic solvent for catalytic activation of ω-TAs. However, methanol in some cases also increased enzyme activity (Figure 2). Methanol was mentioned as a second co-solvent after DMSO for enhancing ω-TA activity in Pawer et al.'s experiment [2].

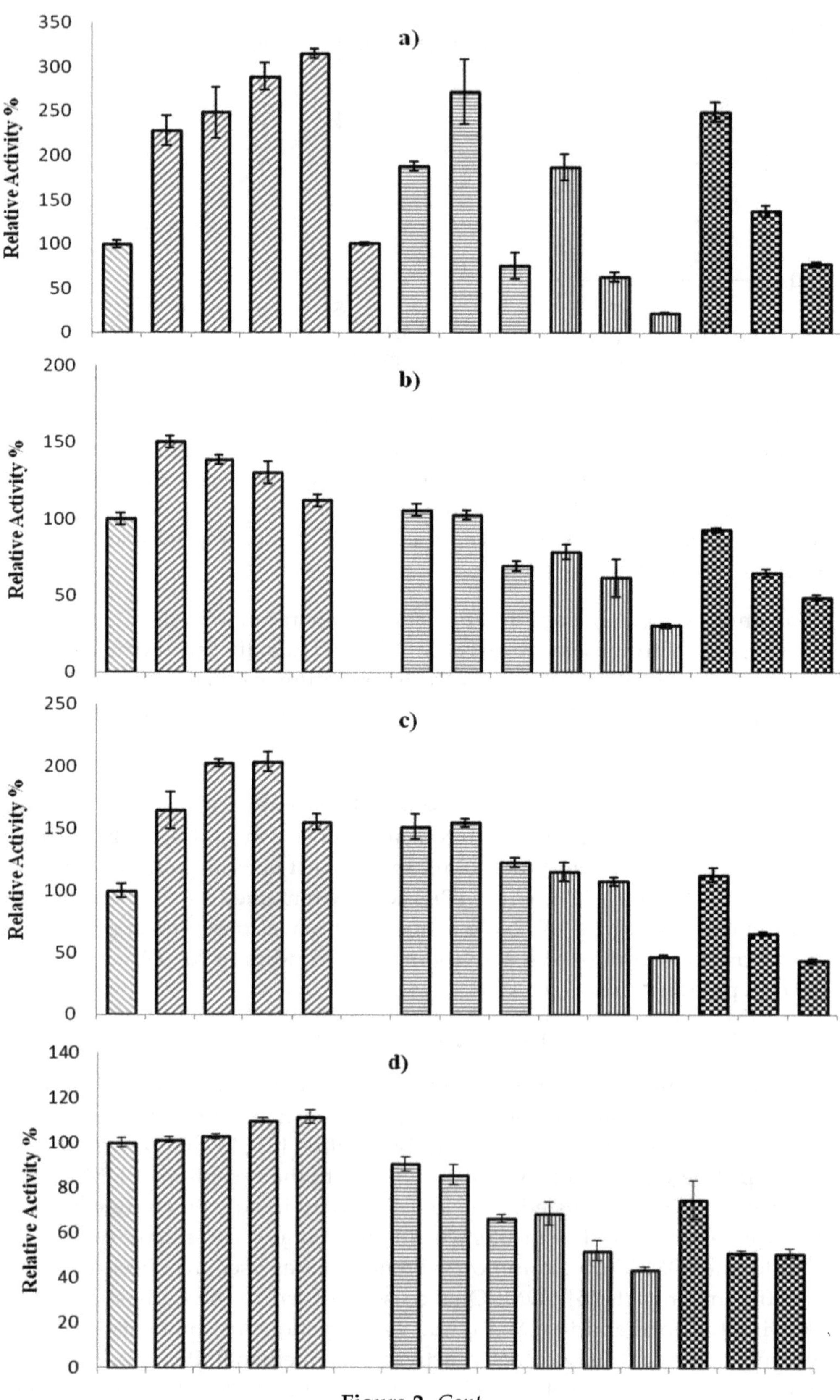

Figure 2. *Cont.*

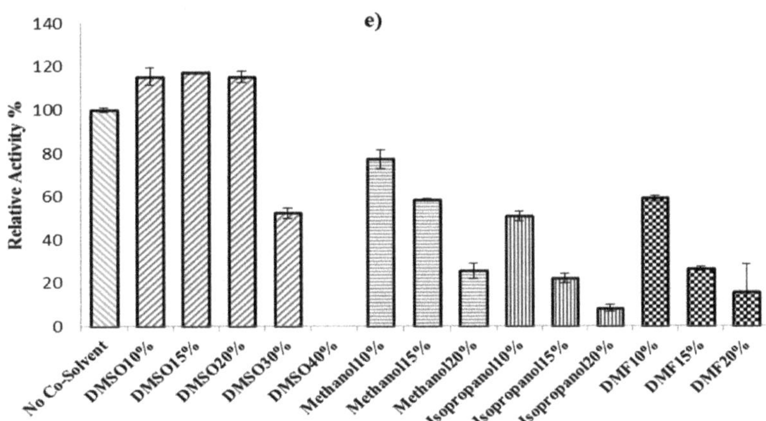

Figure 2. Effect of co-solvent on ω-TA activity of screened strains. (**a**) BaH; (**b**) BaE; (**c**) BaS; (**d**) RhZ; (**e**) BaI. The value of 100% corresponds to enzyme activity in the no co-solvent condition. 1 mg/mL cell-free extract was added to react under the following conditions: (*S*)-MBA (10 mM), pyruvate (10 mM), PLP (0.1 mM), and co-solvent (0–30% *v/v*) in Britton–Robinson buffer (pH 7) at 35 °C. Formation of acetophenone was detected by HPLC at 254 nm. Error bars represent ± standard deviation.

2.2.4. Substrate Specificity and Enantioselectivity

The substrate spectrum of all strains was investigated in order to find the most suitable amino donor and acceptor.

Regarding amino donors, 28 substrates, including aliphatic and aromatic amino compounds were reacted to pyruvate as an amino acceptor to investigate amino donor specificity, and alanine generation was monitored by the HPLC technique. Racemates as well as the pure, *(S)* and *(R)* enantiomers (if available) of amino donors were chosen to explore the enantioselectivity of ω-TA. The results of the substrate screening are illustrated in Table 2. HPLC analysis proved that all enzymes show predominant formation of an *(S)*-configuration. This is in accordance with most previously reported publications in the literature indicating that *(R)*-stereoselectivity are rather rare in wild-type microorganisms [35]. All cell-free extracts showed high activity against *(S)*- α-methylbenzylamine (*S*-MBA) **1s**.

BaH showed considerable activity toward a wide range of substrates, including *(S)*, *(R)*, and racemates of aromatic and aliphatic amino donors. Although this cell-free extract also showed activity toward some tested *(R)*-enantiomers (**1r**, **2r**, **4r**, **5r**, **7r**, **8r**, **13r**), (S)-selectivity appeared to be not so strict in this strain. Notably, it is difficult to conclude that BaH showed (*R*)-selectivity since in this study, crude extract was applied. There is a possibility that (*R*) converted to (*S*)-enantiomer via enzymatic racemization [36], and then the latter was subjected to ω-TA activity. On the other hand, (*R*)-preference towards **13r** in compare to **13s** leads to another hypothesis, in which both R and S transaminases are present in crude extract.

While the cell-free extract of BaH revealed varying enantiopreference toward all tested amino donors to different degrees, the substrate specificity was found to resemble what has previously been reported by Mathew et al. [5,30], as a preference to react with aromatic β-amino acids is clearly visible [5]. *(S)*-MBA **1s** is converted with the highest activity, which therefore was set as 100%. Although a few ω-TAs are described with activity toward aromatic β-amino acids (56), it is remarkable that BaH uses β-phenylalanine (**13**, **13s**, **13r**) and (*rac*)-β-homophenylalanine (**14**) as an amino donor; both of these were inert for crude extraction from other bacteria in this experiment, which implies the presence of a broad binding pocket in the active site of ω-TABaH. ω-TA activity against **14** has been reported by Mathew et al. for ω-TABG, with 50% relative activity in comparison with **13** **[30]**. A recent study by Buß et al. explained the lack of ω-TA activity against **14** (by an ω-TA that was highly active against **13**) by sterice hindrance caused by the additional carbon atom located between the amino group and phenyl group [36]. Several substituted derivatives of β-phenylalanine are converted as well.

Table 2. Amino donor screened assays were carried out in Britton–Robinson buffer, pH 7 with amino donor (10 mM), pyruvate (10 mM), PLP (0.1 mM), and 15% (*v/v*) DMSO. The reaction was initiated by adding crude extract to a final concentration of 1 mg protein per ml. Reactions were performed at 35 °C, 600 rpm for 20 h. Relative activity was measured by HPLC detection of alanine formation at 338 nm. All reactions were conducted in triplicate, and each cell-free extract was separately analyzed for alanine content as a control.

		Amino Donor	Relative Activity%				
			BaH	**BaE**	**BaS**	**BaI**	**RhZ**
1	*(rac)*	α-methylbenzylamine	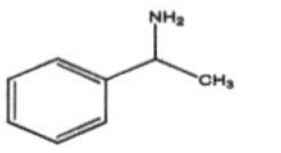 53	48	45	60	36
1s	*(S)*		100	100	100	100	100
1r	*(R)*		23	0	0	6	0
2	*(rac)*	3-amino-3-(4-chlorophenyl) propionic acid	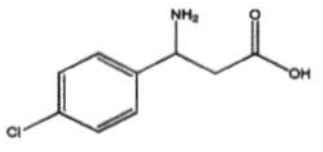 19	0	0	0	0
2s	*(S)*		20	0	0	0	0
2r	*(R)*		16	0	0	0	0
3	*(rac)*	3-amino-3-(3,4-dichloro-phenyl)propionic acid	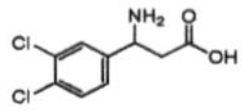 20	0	0	0	0
4	*(rac)*	3-amino-3-(4-nitrophenyl)propionic acid	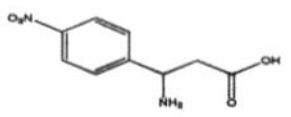 51	0	0	0	0
4r	*(R)*		19	0	0	0	0
5r	*(R)*	(R)3-amino-3-(2-nitro-phenyl)-propionic acid	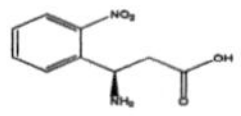 16	0	0	0	0
6	*(rac)*	3-amino-3-(4-methoxyphenyl) propionic acid	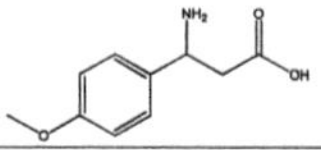 21	0	0	0	0
6s	*(S)*		17	0	0	0	0
7s	*(S)*	3-amino-3-(4-hydroxy-phenyl)propionic acid	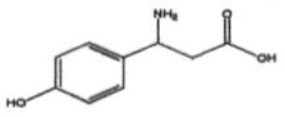 18	0	0	0	0
7r	*(R)*		15	0	0	0	0
8	*(rac)*	3-amino-3-(4-bromophenyl)-propionic acid	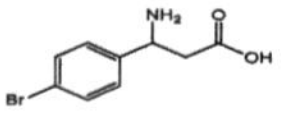 21	0	0	0	0
8r	*(R)*		20	0	0	0	0
9	(rac)	3-amino-3-(4-fluorophenyl)propionic acid	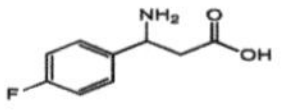 43	0	0	0	0
9s	(S)		16	0	0	0	0
10	*(rac)*	Methyl(-3-amino-3-phenyl propanoate hydrochloride)	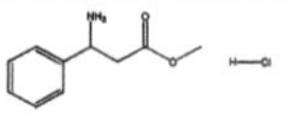 11	0	0	0	0
11	*(rac)*	3-amino-3-(4-isopropylphenyl) propionic acid	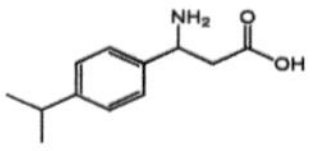 20	0	0	0	0
12	*(rac)*	1-aminocyclopropane-1-carboxylic acid	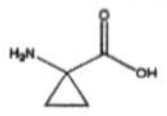 17	17	0	30	0
13	*(rac)*	β-phenylalanine	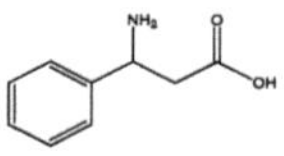 17	0	0	0	0
13s	*(S)*		16	0	0	0	0
13r	*(R)*		22	0	0	0	0
14	*(rac)*	β-Homophenylalanine	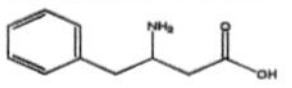 16	0	0	0	0
15		Isopropylamine	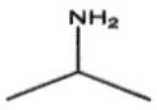 20	0	0	21	0
16		Sec-butylamine	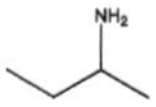 25	7	0	27	0
17s	*(S)*	3-amino butyric acid	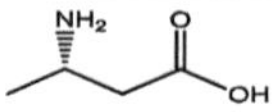 91	59	54	20	135

The activity against *(S)*-MBA was set as 100. An amount under 1% was considered as zero activity.

BaE presented only (*S*)-selectivity in relation to tested substrates. Similar to BaH, the strain's crude extract exhibited activity toward **1s**, almost double that of **1**. BaE and BaH share the same activity toward **12**, while BaI showed almost double the activity in comparison with those. The relative activity of ω-TA of BaI toward (*rac*)-MBA **1** exceeded 60%, which might be related to the presence of racemase in the crude extract of bacteria [37,38].

Crude extracts of RhZ and BaS displayed (*S*)-selectivity toward the tested amino donors with similar preference profiles. RhZ showed more activity toward **17s** as an aliphatic substrate than MBA **1s,** which was not observed for other screened strains.

Amino acceptors play a critical role in ω-TA activity [30]. Eight ketones were reacted with (*S*)-MBA to explore the amino acceptor specificities by HPLC detection of acetophenone production (Table 3). Aliphatic ketones such as sodium pyruvate **7b** and α-ketoglutaric acid **8b** showed high conversion, whereas most aromatic ketones were not accepted by the investigated enzymes.

Table 3. Amino acceptor screened assays were carried out in Britton–Robinson buffer, pH 7 with (*S*)-MBA (10 mM), amino acceptor (10 mM), PLP (0.1 mM), and 15% (*v/v*) DMSO. The reaction was started by the addition of crude extract to a final concentration of 1 mg/mL in the following conditions: 35 °C, 600 rpm for 20 h. Relative activity was measured by HPLC detection of acetophenone formation at 254 nm. All reactions were conducted in triplicate, and each cell-free extract was separately analyzed as a control.

	Amino Acceptors		Relative Activity%				
			BaH	**BaE**	**BaS**	**BaI**	**RhZ**
1b	Ethyl benzoylacetate	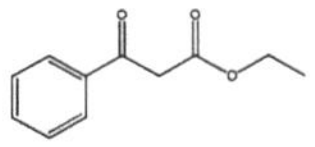	40.95	0	20.84	14.87	0
2b	Ethyl 3-(4-methoxyphenyl)-3-oxopropanoate	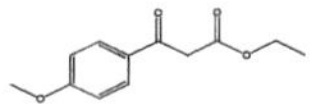	0	0	0	60.48	0
3b	Sodium 3-oxo-3-phenylpropanoate	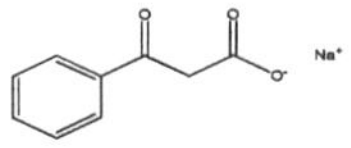	0	0	0	0	0
4b	Ethyl(4-flurobenzoyl)acetate	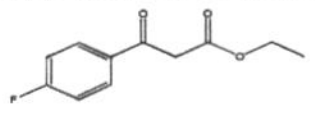	0	0	0	0	0
5b	Ethyl(4-chlorobenzoyl)acetate	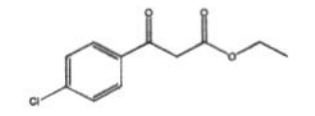	0	0	0	0	0
6b	Ethyl 4-nitrobenzoylacetate	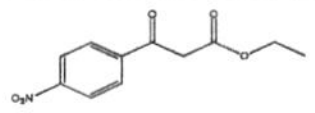	0	0	0	0	0
7b	Pyruvate	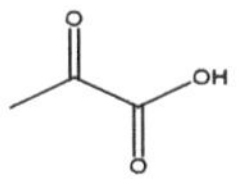	100	100	43.73	100	36.39
8b	α-ketoglutarate	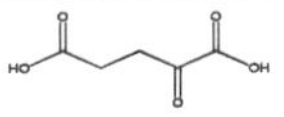	17.09	7.31	100	0	100

In each column, the highest activity was defined as 100%. Activity less than 1% was considered zero.

Cell-free extract of BaH, BaS, and BaI exhibited some activity toward bulkier substrates, including **1b** or **2b**. It is supposed that β-amino acid esters were produced by using these acceptors, but this needs to be investigated in further experiments. BaH, BaS, and BaI showed activity toward **1b** as amino acceptor. BaI was the only extract to show activity toward acceptor **2b**.

The substrate scope of ω-TA is determined by the active site, which consists of a large and a small binding pocket. The former can accommodate bulky residues, including aryl groups, whereas the latter is loaded by a small group such as a methyl substituent [14]. So most ω-TAs have a limitation in accepting bulky substrates without applying protein engineering. Voss et al. reported that after double mutation on ω-TA 3FCR from *Ruegeria* sp. TM1040, this enzyme gained activity toward bulkier substrates such as phenylpropylamine and phenylbutylamine [16]. Strain BaH showed a good ability to convert bulky substrates and showed different substrate spectra with known ω-TAs [15,35]. Purification and further investigation of this enzyme is intended for further studies.

3. Materials and Methods

3.1. Chemicals

All solvents and chemicals used in this study were purchased from Sigma-Aldrich (St Louis, MO, USA), Carl Roth GmbH (Karlsruhe, Germany), and Peptech (Burlington, MA, USA).

3.2. Screening and Identification

3.2.1. Enrichment of Microorganisms on (rac)-α-MBA as Sole Nitrogen Source

Due to the role of transaminase in nitrogen metabolism [29], areas enriched with nitrogen compounds were chosen as potential sites for screening novel ω-transaminase. Soil samples (2 g) from a petrochemical site, oil well, and an agriculture field in Iran were separately suspended in 200 mL sterile minimal medium (MIM) in 500 mL Erlenmeyer flasks containing 10 mM (*rac*)-α-MBA, 100 mM glycerol, 1 g/L $MgSO_4 .7H_2O$, 0.02 mg/L H_3BO_3, 0.2 mM $CaCl_2$, 0.1 mg/L $MnSO_4$. $4H_2O$, 0.1 mg/L $CuSO_4 .5H_2O$, 0.1 mg/L $NiSO_4$. $6H_2O$, 2.0 mg/L $NaMoO_4$, 0.05 mg/L $CoCl_2$, 0.1 mg/L $ZnCl_2$, 4 mg/L $FeSO_4$. $7H_2O$, and potassium phosphate buffer (50 mM, pH 7.0) [39]. $FeSO_4$ and (*rac*)-α-MBA were separately sterile filtrated and added to the medium. Autoclaving causes precipitation of ferric hydrate [40]. Each sample was prepared in duplicate. Enrichment went through two strategies:

In the first method, Erlenmeyer flasks were incubated for 24 h at 180 rpm at 35 °C (depending on the original temperature of the soil samples). After 24 h of incubation, a 100 μL sample was transferred to 20 mL of fresh minimal medium and incubated for 24 h under the same conditions as described before. This serial dilution was repeated 3 times. After that, the culture was diluted with PBS buffer (10: 90 μL) and poured to MIM agar plates (mineral medium, 10 mM (*rac*)-α-MBA with 1.5% agar) and incubated at 35 °C. As soon as colonies were visible, single colonies were streaked on new MIM agar plates to obtain uniform colonies under the same temperature.

In the second method, Erlenmeyer flasks were transferred to a dark place, without shaking, at 35 °C. During incubation, 1 mL of concentrated MIM (5X), 0.5% *v/v* glycerol, and 10 mM (*rac*)-α-MBA were added to the flasks every week. After 3 months, the culture broth was centrifuged (1000 *g*, 2 min) and 100 μL of supernatant was spread on MIM agar plates. The colonies were isolated after 3 days of incubation at 35 °C. This was continued until the isolation of single colonies was achieved.

3.2.2. Selection of the Most Promising Strains

To identify whether isolated strains use (*rac*)-α-MBA as a sole nitrogen source or atmospheric nitrogen, *o-Xylylenediamine* was applied as a smart amino donor. Each colony was cultivated in a shake flask containing 20 mL MIM medium with 12 mM (*rac*)-α-MBA for three days at 35 °C at 120 rpm. The cells were harvested by centrifugation at 6000× *g* for 10 min at 4 °C and resuspended in HEPES buffer (50 mM, pH 7.5). The cell concentration was adjusted to around 20 mg/mL by measuring dry cell mass based on Buß et al. [36]. The cells were frozen at −80 °C and thawed at room temperature for cell disruption.

Reaction was started by adding 50 μL of whole cells to 150 μL reaction solution containing 7.5 mM *o-Xylylenediamine,* 5 mM pyruvate, 1 mM PLP, and 10% DMSO in HEPES buffer (50 mM, pH 7.5) [41] in a 96-well plate. Each reaction was conducted in triplicate. The plate was incubated overnight at 35 °C, 150 rpm.

3.2.3. Identification of Bacteria

The positive strains were grown separately overnight in LB broth at 35 °C, 150 rpm. DNA was extracted by ZR soil Microbe DNA Kit^{TM}. PCR was done with a Q5 high-fidelity PCR kit (NEB,

Germany). Universal Primers (27F and 1492R) were applied to amplify 1.5 kb 16S rDNA fragments. PCR was conducted following the manufacturer's instructions for Q5 polymerase. Gene sequencing was performed by Eurofins Company (Ebersberg, Germany). The resulting 16S rRNA gene sequences were compared with available gene sequences in the EZ bioCloud database [42].

3.3. *Enzymatic Properties and Substrate Specificity*

3.3.1. Preparation of Crude Extract

The inoculum (3 mL) was prepared by picking a single colony and cultivating it for three days in MIM with 12 mM (*rac*)-α-MBA. The medium (400 mL) was inoculated by the addition of 1% (*v/v*) of the latter culture into 1 L shaking flasks containing MIM with 0.5 g/L yeast extract and 12 mM (*rac*)-α-MBA overnight at 150 rpm and 35 °C. Cells were harvested before reaching stationary phase by centrifugation at 6000× *g* for 10 min at 4 °C. After washing the pellet with sodium phosphate buffer (50 mM, pH 7), cells were resuspended in 10 mL lysis buffer (sodium phosphate (50 mM, pH 7), 0.1 μM PLP, 100μg/mL lysozyme]. Afterwards, they were incubated for 1 h at room temperature, followed by sonification (3 cycles: 30 s pulse, 20 s pause; 60% amplitude) on ice. Subsequently, the cell debris was removed by centrifugation (20,000 rpm, 4 °C, 15 min) in a JA-30–50 rotor (Coulter–Beckman centrifuge). The protein concentration of crude extract was determined by using the Roti®-Quant universal kit (Carl Roth, Karlsruhe, Germany) following the manufacturer's instructions. Eventually, they were mixed with 15% glycerol and preserved in −80 °C for further analysis.

3.3.2. Effect of Temperature, pH, and Co-Solvent

The effects of temperature and pH on enzyme activity were examined at different temperatures (25–65 °C at pH7) and various pHs (3–12 at 35 °C) by using Britton–Robinson buffer (0.04 M H_3BO_3, 0.04 M H_3PO_4, 0.04 M CH_3COOH) containing an amino donor (10 mM), amino acceptor (10 mM), PLP (0.1 mM), and DMSO (10%). The total reaction volume was set to 0.25 mL. The reactions were initiated by adding crude extract of enzyme (~1 mg/mL).

To study the effect of solvent on ω-TA activity, reactions were carried out applying various organic solvents with 0–20% (*v/v*) (for DMSO 0–40%). (rac)-α-MBA (10 mM) was used as amino donor and pyruvate (10 mM) as amino acceptor in the presence of PLP (0.1 mM) and cell-free extract (~1 mg/mL). The reactions were carried out in Britton–Robinson buffer (pH 7) at 35 °C as described above.

Each reaction was conducted in triplicate and incubated (ThermoMixer, Eppendorf) at 600 rpm for 1 h. As a control, the reaction was conducted without adding enzyme. Furthermore, the amount of acetophenone in crude extract of enzyme was also evaluated. Reactions were stopped by heating to 95 °C for 5 min. After centrifugation, the supernatant was analyzed by HPLC to detect produced acetophenone according to Section 3.4. Results are shown as relative activity.

3.3.3. Substrate Specificity and Enantioselectivity

The reactions were carried out according to Section 3.3.2. Various amino donors (10 mM) listed in Table 1 were tested using sodium pyruvate (10 mM) as amino acceptor in the Britton–Robinson buffer (pH 7) and 15% (*v/v*) DMSO. The reaction was carried out at 35 °C, 600 rpm, for 20 h (ThermoMixer, Eppendorf). The produced alanine was analyzed by HPLC according to Section 3.4.

Different amino acceptors (10 mM) were tested with (S)-α-MBA as amino donor (10 mM) in Britton–Robinson buffer, pH 7, at 35 °C, 600 rpm, for 20 h. The production of acetophenone was analyzed by HPLC according to Section 3.4.

Each reaction was conducted in triplicate, and the average of three independent reactions was used for evaluation. Enzyme inactivation was applied by heating to 95 °C for 5 min. The supernatant was used for analysis after centrifugation. In addition, the amounts of alanine and acetophenone were measured in enzyme crude extracts as blanks.

3.4. HPLC Analytics

All samples were analyzed by Agilent 1100 series HPLC system (Santa Clara, CA, USA). For the analysis of alanine, derivatization by using ortho-phthalaldehyde was carried out according to Brucher et al. [43] and Buß et al. [44], with an automated precolumn derivatization. A reversed-phase C_{18} column (150 × 4.6 mm HyperClone 5 μm ODS, Phenomenex, Germany) with isocratic elution with 35% (*v/v*) methanol and 65% (*v/v*) sodium phosphate buffer (40 mM, pH 6.5) at a flow rate of 0.8 mL min^{-1} and detection at 338 nm at 25 °C was used.

The acetophenone concentration in the samples was determined chromatographically by isocratic elution with acetonitrile/water (50/50, *v/v*) at flow rate of 0.6 mL min^{-1} with UV detection at 254 nm [45], at 25 °C applying a C_{18} Hypersil-keystone column (250 × 4.6 mm 5 μ Hypersil). The injection volume was adjusted to 1 μL.

4. Conclusions

Our motivation for screening novel microorganisms exhibiting transaminase activity resulted from several challenges for the industrial application of ω-TAs, e.g., lacking activity at acidic pH, high temperature, unnatural substrates, (*R*)-configured molecules, and elevated concentrations of organic solvents (needed due to low substrate solubility in aqueous systems). We succeeded in finding promising new strains from a high variety of Iranian soil samples by enrichment culture using (*rac*)-α-methylbenzylamine (α-MBA) as a sole nitrogen source.

In particular, a *Bacillus halotolerans* (BaH) strain was isolated from a petroleum refinery, exhibiting ω-TA activity using a broad spectrum of amino donors over a pH range of 5–9 at elevated concentrations of DMSO and other organic solvents up to 30% (*v/v*). Two other strains isolated from an agricultural field (*Bacillus endophyticus*, BaE) and an oilfield (*Bacillus subtilis*, BaS) showed the highest ω-TA activity against α-MBA at pH 5 with 70% remaining activity at pH 3 (BaS), whereas ω-TAs are usually described to prefer slightly alkaline conditions. The enzymes dedicated to these extraordinary activities will be purified and subjected to in-depth studies for application-technical characterization.

Author Contributions: Conceptualization, J.F. and J.R., methodology, N.G.N., M.Z.M., and U.E., validation, J.R., M.M.H., and U.E., formal analysis, N.G.N., investigation, N.G.N., data curation, N.G.N., writing—original draft preparation, N.G.N., writing—review and editing, N.G.N., J.F., J.R., U.E., M.M.H., M.Z.M., and C.S., visualization, M.Z.M., supervision, J.F., J.R., and C.S., project administration, J.F.

Acknowledgments: This study was financially supported by grant No: 970201 of the Biotechnology Development Council of the Islamic Republic of Iran. The publication of this article was funded by Karlsruhe Institute of Technology (Helmholtz).

References

1. Gomm, A.; Lewis, W.; Green, A.P.; O'Reilly, E. A New Generation of Smart Amine Donors for Transaminase-Mediated Biotransformations. *Chem. A Eur. J.* **2016**, *22*, 12692–12695. [CrossRef]
2. Pawar, S.V.; Hallam, S.J.; Yadav, V.G. Metagenomic Discovery of a Novel Transaminase for Valorization of Monoaromatic Compounds. *RSC Adv.* **2018**, *8*, 22490–22497. [CrossRef]
3. Aleku, G.A.; France, S.P.; Man, H.; Mangas-Sanchez, J.; Montgomery, S.L.; Sharma, M.; Leipold, F.; Hussain, S.; Grogan, G.; Turner, N.J. A Reductive Aminase from Aspergillus Oryzae. *Nat. Chem.* **2017**, *9*, 961. [CrossRef]
4. Kim, G.H.; Jeon, H.; Khobragade, T.P.; Patil, M.D.; Sung, S.; Yoon, S.; Won, Y.; Choi, I.S.; Yun, H. Enzymatic Synthesis of Sitagliptin Intermediate Using a Novel ω-Transaminase. *Enzym. Microb. Technol.* **2019**, *120*, 52–60. [CrossRef]

5. Mathew, S.; Nadarajan, S.P.; Chung, T.; Park, H.H.; Yun, H. Biochemical Characterization of Thermostable ω-Transaminase from Sphaerobacter Thermophilus and its Application for Producing Aromatic β-and γ-Amino Acids. *Enzym. Microb. Technol.* **2016**, *87*, 52–60. [CrossRef]
6. Haycock-Lewandowski, S.J.; Wilder, A.; Åhman, J. Development of a Bulk Enabling Route to Maraviroc (UK-427,857), a CCR-5 Receptor Antagonist. *Org. Process. Res. Dev.* **2008**, *12*, 1094–1103. [CrossRef]
7. Slomka, C.; Zhong, S.; Fellinger, A.; Engel, U.; Syldatk, C.; Bräse, S.; Rudat, J. Chemical Synthesis and Enzymatic, Stereoselective Hydrolysis of a Functionalized Dihydropyrimidine for the Synthesis of β-Amino Acids. *AMB Express* **2015**, *5*, 85. [CrossRef]
8. Weiner, B.; Szymański, W.; Janssen, D.B.; Minnaard, A.J.; Feringa, B.L. Recent Advances in the Catalytic Asymmetric Synthesis of β-Amino Acids. *Chem. Soc. Rev.* **2010**, *39*, 1656–1691. [CrossRef]
9. Turner, N.J.; Truppo, M.D. Biocatalytic Routes to Nonracemic Chiral Amines. In *Chiral Amine Synthesis: Methods, Developments and Applications*; Nugent, T.C., Ed.; Wiley-VCH: Weinheim, Germany, 2010; pp. 431–459.
10. Mathew, S.; Jeong, S.S.; Chung, T.; Lee, S.H.; Yun, H. Asymmetric Synthesis of Aaromatic β-Amino Acids Using ω-Transaminase: Optimizing the Lipase Concentration to Obtain Thermodynamically Unstable β-Keto Acids. *Biotechnol. J.* **2016**, *11*, 185–190. [CrossRef]
11. Malik, M.S.; Park, E.S.; Shin, J.S. Features and Technical Applications of ω-Transaminases. *Appl. Microbiol. Biotechnol.* **2012**, *94*, 1163–1171. [CrossRef]
12. Jiang, J.; Chen, X.; Feng, J.; Wu, Q.; Zhu, D. Substrate Profile of an ω-Transaminase from Burkholderia Vietnamiensis and its Potential for the Production of Optically Pure Amines and Unnatural Amino Acids. *J. Mol. Catal. B Enzym.* **2014**, *100*, 32–39. [CrossRef]
13. Cerioli, L.; Planchestainer, M.; Cassidy, J.; Tessaro, D.; Paradisi, F. Characterization of a Novel Amine Transaminase from Halomonas Elongata. *J. Mol. Catal. B Enzym.* **2015**, *120*, 141–150. [CrossRef]
14. Voss, M.; Das, D.; Genz, M.; Kumar, A.; Kulkarni, N.; Kustosz, J.; Kumar, P.; Bornscheuer, U.T.; Höhne, M. In Silico Based Engineering Approach to Improve Transaminases for the Conversion of Bulky Substrates. *ACS Catal.* **2018**, *8*, 11524–11533. [CrossRef]
15. Rudat, J.; Brucher, B.R.; Syldatk, C. Transaminases for the Synthesis of Enantiopure Beta-Amino Acids. *AMB Express* **2012**, *2*, 11. [CrossRef]
16. Pavlidis, I.V.; Weiß, M.S.; Genz, M.; Spurr, P.; Hanlon, S.P.; Wirz, B.; Iding, H.; Bornscheuer, U.T. Identification of (S)-Selective Transaminases for the Asymmetric Synthesis of Bulky Chiral Amines. *Nat. Chem.* **2016**, *8*, 1076–1082. [CrossRef]
17. Kelly, S.A.; Magill, D.J.; Megaw, J.; Skvortsov, T.; Allers, T.; McGrath, J.W.; Allen, C.C.; Moody, T.S.; Gilmore, B.F. Characterisation of a Solvent-Tolerant Haloarchaeal (R)-Selective Transaminase Isolated from a Triassic Period Salt Mine. *Appl. Microbiol. Biotechnol.* **2019**, *103*, 5727–5737. [CrossRef]
18. Leipold, L.; Dobrijevic, D.; Jeffries, J.W.; Bawn, M.; Moody, T.S.; Ward, J.M.; Hailes, H.C. The Identification and Use of Robust Transaminases from a Domestic Drain Metagenome. *Green Chem.* **2019**, *21*, 75–86. [CrossRef]
19. Ferrandi, E.E.; Previdi, A.; Bassanini, I.; Riva, S.; Peng, X.; Monti, D. Novel Thermostable Amine Transferases from Hot Spring Metagenomes. *Appl. Microbiol. Biotechnol.* **2017**, *101*, 4963–4979. [CrossRef]
20. Genz, M.; Vickers, C.; van den Bergh, T.; Joosten, H.-J.; Dörr, M.; Höhne, M.; Bornscheuer, U. Alteration of the Donor/Acceptor Spectrum of the (S)-Amine Transaminase from Vibrio Fluvialis. *Int. J. Mol. Sci.* **2015**, *16*, 26953–26963. [CrossRef]
21. Kelly, S.A.; Megaw, J.; Caswell, J.; Scott, C.J.; Allen, C.C.; Moody, T.S.; Gilmore, B.F. Isolation and Characterisation of a Halotolerant ω-Transaminase from a Triassic Period Salt Mine and Its Application to Biocatalysis. *ChemistrySelect* **2017**, *2*, 9783–9791. [CrossRef]
22. Shin, J.S.; Kim, B.G. Kinetic Resolution of α-Methylbenzylamine with o-Transaminase Screened from Soil Microorganisms: Application of a Biphasic System to Overcome Product Inhibition. *Biotechnol. Bioeng.* **1997**, *55*, 348–358. [CrossRef]
23. Pavkov-Keller, T.; Strohmeier, G.A.; Diepold, M.; Peeters, W.; Smeets, N.; Schürmann, M.; Gruber, K.; Schwab, H.; Steiner, K. Discovery and Structural Characterisation of New Fold Type IV-Transaminases Exemplify the Diversity of This Eenzyme Fold. *Sci. Rep.* **2016**, *6*, 38183. [CrossRef] [PubMed]
24. Yazdi, A.; Emami, M.H.; Shafiee, S.M. Dasht-E Lut in Iran, the Most Complete Collection of Beautiful Geomorphological Phenomena of Desert. *Open J. Geol.* **2014**, *4*, 249–261. [CrossRef]
25. Ataee, N.; Fooladi, J.; Namaei, M.H.; Rezadoost, H.; Mirzajani, F. Biocatalysts Screening of Papaver Bracteatum Flora for Thebaine Transformation to Codeine and Morphine. *Biocatal. Agric. Biotechnol.* **2017**, *9*, 127–133. [CrossRef]

26. Ghasemi, Y.; Rasoul-Amini, S.; Ebrahiminezhad, A.; Kazemi, A.; Shahbazi, M.; Talebnia, N. Screening and Isolation of Extracellular Protease Producing Bacteria from the Maharloo Salt Lake. *Iran. J. Pharm. Sci.* **2011**, *7*, 175–180.
27. Alghabpoor, S.S.; Panosyan, H.; Trchounian, A.; Popov, Y. Purification and Characterization of a Novel Thermostable and Acid Stable α-Amylase from Bacillus Sp. Iranian S1. *Int. J. Eng. Trans. B Appl.* **2013**, *26*, 815–820. [CrossRef]
28. Shirsalimian, M.; Amoozegar, M.; Sepahy, A.A.; Kalantar, S.; Dabbagh, R. Isolation of Extremely Halophilic Archaea from a Saline River in the Lut Desert of Iran, Moderately Resistant to Desiccation and Gamma Radiation. *Microbiology* **2017**, *86*, 403–411. [CrossRef]
29. Shin, J.S.; Kim, B.G. Comparison of the ω-Transaminases from Different Microorganisms and Application to Production of Chiral Amines. *Biosci. Biotechnol. Biochem.* **2001**, *65*, 1782–1788. [CrossRef]
30. Mathew, S.; Bea, H.; Nadarajan, S.P.; Chung, T.; Yun, H. Production of Chiral β-Amino Acids Using ω-Transaminase from Burkholderia Graminis. *J. Biotechnol.* **2015**, *196*, 1–8. [CrossRef]
31. Guo, F.; Berglund, P. Transaminase Biocatalysis: Optimization and Application. *Green Chem.* **2017**, *19*, 333–360. [CrossRef]
32. Dreßen, A.; Hilberath, T.; Mackfeld, U.; Billmeier, A.; Rudat, J.; Pohl, M. Phenylalanine Ammonia Lyase from Arabidopsis Thaliana (AtPAL2): A Potent MIO-Enzyme for the Synthesis of Non-Canonical Aromatic Alpha-Amino Acids: Part I: Comparative Characterization to the Enzymes from Petroselinum Crispum (PcPAL1) and Rhodosporidium Toruloides (RtPAL). *J. Biotechnol.* **2017**, *258*, 148–157.
33. Schätzle, S.; Höhne, M.; Robins, K.; Bornscheuer, U.T. Conductometric Method for the Rapid Characterization of the Substrate Specificity of Amine-Transaminases. *Anal. Chem.* **2010**, *82*, 2082–2086. [CrossRef]
34. Schätzle, S.; Höhne, M.; Redestad, E.; Robins, K.; Bornscheuer, U.T. Rapid and Sensitive Kinetic Assay for Characterization of ω-Transaminases. *Anal. Chem.* **2009**, *81*, 8244–8248. [CrossRef] [PubMed]
35. Koszelewski, D.; Tauber, K.; Faber, K.; Kroutil, W. ω-Transaminases for the Synthesis of Non-Racemic α-Chiral Primary Amines. *Trends Biotechnol.* **2010**, *28*, 324–332. [CrossRef] [PubMed]
36. Buß, O.; Dold, S.M.; Obermeier, P.; Litty, D.; Muller, D.; Grüninger, J.; Rudat, J. Enantiomer Discrimination in β-Phenylalanine Degradation by a Newly Isolated Paraburkholderia Strain BS115 and Type Strain PsJN. *AMB Express* **2018**, *8*, 149. [CrossRef] [PubMed]
37. Conti, E.; Stachelhaus, T.; Marahiel, M.A.; Brick, P. Structural Basis for the Activation of Phenylalanine in the Non-Ribosomal Biosynthesis of Gramicidin S. *EMBO J.* **1997**, *16*, 4174–4183. [CrossRef] [PubMed]
38. Yamada, M.; Kurahashi, K. Further Purification and Properties of Adenosine Triphosphate-Dependent Phenylalanine Racemase of Bacillus Brevis Nagano. *J. Biochem.* **1969**, *66*, 529–540. [CrossRef]
39. Yun, H.; Lim, S.; Cho, B.K.; Kim, B.G. ω-Amino Acid: Pyruvate Transaminase from Alcaligenes Denitrificans Y2k-2: A New Catalyst for Kinetic Resolution of β-Amino Acids and Amines. *Appl. Environ. Microbiol.* **2004**, *70*, 2529–2534. [CrossRef]
40. Temple, K.L.; Colmer, A.R. The Autotrophic Oxidation of Iron by a New Bacterium: Thiobacillus Ferrooxidans. *J. Bacteriol.* **1951**, *62*, 605.
41. Buß, O.; Voss, M.; Delavault, A.; Gorenflo, P.; Syldatk, C.; Bornscheuer, U.; Rudat, J. β-Phenylalanine Ester Synthesis from Stable β-Keto Ester Substrate Using Engineered ω-Transaminases. *Molecules* **2018**, *23*, 1211. [CrossRef]
42. Yoon, S.H.; Ha, S.M.; Kwon, S.; Lim, J.; Kim, Y.; Seo, H.; Chun, J. Introducing EzBioCloud: A taxonomically united database of 16S rRNA gene sequences and whole-genome assemblies. *Int. J. Syst. Evol. Microbiol.* **2017**, *67*, 1613. [PubMed]
43. Brucher, B.; Rudat, J.; Syldatk, C.; Vielhauer, O. Enantioseparation of Aromatic β^3-Amino Acid by Precolumn Derivatization with o-Phthaldialdehyde and N-Isobutyryl-l-Cysteine. *Chromatographia* **2010**, *71*, 1063–1067. [CrossRef]
44. Buß, O.; Muller, D.; Jager, S.; Rudat, J.; Rabe, K.S. Improvement in the Thermostability of a β-Amino Acid Converting ω-Transaminase by Using FoldX. *ChemBioChem* **2018**, *19*, 379–387. [CrossRef] [PubMed]
45. Gao, S.; Su, Y.; Zhao, L.; Li, G.; Zheng, G. Characterization of a (R)-Selective Amine Transaminase from Fusarium Oxysporum. *Process Biochem.* **2017**, *63*, 130–136. [CrossRef]

11

Challenges and Opportunities in Identifying and Characterising Keratinases for Value-Added Peptide Production

Juan Pinheiro De Oliveira Martinez [1], Guiqin Cai [1], Matthias Nachtschatt [1], Laura Navone [1], Zhanying Zhang [1], Karen Robins [1,2] and Robert Speight [1,*]

1 Science and Engineering Faculty, Queensland University of Technology, Brisbane, QLD 4000, Australia; j5.martinez@hdr.qut.edu.au (J.P.D.O.M.); guiqin.cai@qut.edu.au (G.C.); m.nachtschatt@qut.edu.au (M.N.); laura.navone@qut.edu.au (L.N.); jan.zhang@qut.edu.au (Z.Z.); sustainbiotech@iinet.net.au (K.R.)

2 Sustain Biotech, Sydney, NSW 2224, Australia

* Correspondence: robert.speight@qut.edu.au

Abstract: Keratins are important structural proteins produced by mammals, birds and reptiles. Keratins usually act as a protective barrier or a mechanical support. Millions of tonnes of keratin wastes and low value co-products are generated every year in the poultry, meat processing, leather and wool industries. Keratinases are proteases able to breakdown keratin providing a unique opportunity of hydrolysing keratin materials like mammalian hair, wool and feathers under mild conditions. These mild conditions ameliorate the problem of unwanted amino acid modification that usually occurs with thermochemical alternatives. Keratinase hydrolysis addresses the waste problem by producing valuable peptide mixes. Identifying keratinases is an inherent problem associated with the search for new enzymes due to the challenge of predicting protease substrate specificity. Here, we present a comprehensive review of twenty sequenced peptidases with keratinolytic activity from the serine protease and metalloprotease families. The review compares their biochemical activities and highlights the difficulties associated with the interpretation of these data. Potential applications of keratinases and keratin hydrolysates generated with these enzymes are also discussed. The review concludes with a critical discussion of the need for standardized assays and increased number of sequenced keratinases, which would allow a meaningful comparison of the biochemical traits, phylogeny and keratinase sequences. This deeper understanding would facilitate the search of the vast peptidase family sequence space for novel keratinases with industrial potential.

Keywords: keratinase; serine protease; metalloprotease; peptidase; keratin hydrolysis; keratin waste; valorisation; bioactive peptides

---

1. Introduction

Millions of tonnes of waste keratin are produced every year in the poultry, meat processing, leather and wool textile industries. The global poultry meat processing industry alone produces 40×10^6 tonnes of waste feathers annually [1]. With the transition away from the fossil fuel-centric economy to a sustainable circular economy, the valorisation of keratin materials addresses the waste problem and facilitates the integration of waste keratin into new value chains to enable a circular economy.

Traditionally, keratin waste has been sent to landfill or rendering, or used as fertilizer, feather meal or incinerated [2,3]. There is, however, an opportunity for livestock industries to produce higher value products from waste keratin. There are multiple thermochemical methods available to prepare hydrolysed keratin for various value-adding opportunities [4]. However, the use of peptidases with keratinolytic activity for keratin hydrolysis protects the integrity of the keratin amino acids in most cases and allows control over the peptide size in the hydrolysate that is not readily achievable with other methods [5]. This degree of control allows the production of bespoke medical biomaterials, smart biocomposites, protein feed supplements with enhanced nutritional and bioactive properties as well as personal care products with enhanced functional and bioactive properties.

Identifying peptidases with keratinolytic activity is an inherent problem associated with the search for new enzymes. Keratinase activity however appears to be dependent on the accessibility of the keratin substrate to the enzyme [6,7]. Thermochemical or biochemical treatment of the keratin, with emphasis on the reduction of the disulphide bond and disruption of other important bonds involved in the structural stability of keratin like isopeptide, hydrogen and glycolytic bonds [6,8–10], appears to be the prerequisite for enzymatic hydrolysis. Sulphitolysis, which involves reduction of the disulphide bond in keratin, often acts synergistically with keratinases in nature [6,7]. Although destabilization of the keratin structure is a prerequisite for keratin hydrolysis, not all peptidases can hydrolyse keratin. Peptidases like trypsin, papain and pepsin cannot hydrolyse keratin as efficiently as peptidases with keratinolytic activity, even if the reduction of disulphide bond has already occurred [11]. The elucidation of the unique characteristics of peptidases with keratinolytic activity that differentiate them from the other peptidases, would be an important breakthrough in the search for new and robust keratinases for the valorisation of keratin waste.

This paper reviews twenty sequenced peptidases with keratinolytic activity from the serine protease and metalloprotease families by comparing their biochemical characteristics and will highlight the difficulties associated with the interpretation of these data.

2. Keratin: A Complex and Strong Structure

Keratins are important structural proteins produced by vertebrate epithelia that have various physiological function. Keratins can act as a protective barrier to water, against infection or cushion tissue from mechanical impact. The two main types of keratins proteins are α-keratin and β-keratins. These two types are further divided into acidic or basic, soft or hard, and have different molecular weights [4,9,12,13]. The following section describes the complexity of the keratin structure, which provides insight into the resistance of keratin to hydrolysis. This review will concentrate on hard α-keratin and β-keratin, γ-keratins and the keratin-associated proteins, which are common to mammalian hair, bristles, wool, hooves, horns and feathers.

α-Keratin has an α-helix structure, which is stabilized by hydrogen bonding and the presence of multiple cysteines forming disulphide bridges. α-Keratin is characterized by a lower sulphur content compared to other keratins and a molecular mass of 60–80 kDa [4]. Hard α-keratin is the major protein of mammalian fibres, nails, hooves and horns. In contrast, hard β-keratins are characteristic of the hard, cornified epidermis of reptiles and birds, e.g., feathers, claws and scales, and have a twisted β-sheet-like structure. They also form the major component of the fibre cuticle. The β-keratin pleated sheets consist of β-strands, which are laterally packed and can have a parallel or antiparallel orientation. The β-sheets are held together by hydrogen bonds and the planar nature of the peptide bond, which results in the stable pleated β-sheet [13]. β-Keratins have a molecular mass of 10–22 kDa. A third type of keratin, γ-keratin, is a globular protein with a high sulphur content and a molecular weight of about 15 kDa. This keratin, along with keratin-associated proteins, form the matrix between the microfibrils and microfibrils of the fibre cortex of mammalian fibres and stabilize the structure of the cortex via extensive disulphide bridge formation.

The complex structural organization of all mammalian fibres is very similar [8]. The hair fibre consists of an outermost cuticle layer, which is composed of overlapping flattened scale-like cells that form a protective sheath around the cortex [8]. The major protein of the fibre cuticle is β-keratin [4]. The cortex is composed of hard α-keratin intermediate filaments embedded in a sulphur-rich matrix. These filaments surround the medulla when present, as is the case for coarser fibres. The cell membrane complex binds the cuticle and cortical cells.

The cuticle layer is laminated and consists of the following layers—the cuticle filament-associated surface membrane, the cystine-rich exocuticular *a*-layer, the lower exocuticle and the endocuticle, which contains only low levels of sulphur-containing amino acids and constitutes the inner lining of the cuticle [8]. The outermost layer of the cuticle provides a hydrophobic barrier, which protects the fibre surface from water and chemical compounds. This cuticle filament-associated surface membrane is 2–7 nm thick and composed of highly cross-linked proteins and lipids. The major fatty acid of the cuticle surface lipids found in human and animal hair is 18-methyleicosanoic acid [14]. It is covalently linked to the protein matrix below by a thioester linkage and the protein matrix is cross-linked by isopeptide bonds [15]. An isopeptide bond results from the transglutaminase-catalysed formation of an amide bond between the amino acid side chains of the amino acid residues in the keratin protein, for example, lysine and glutamine [9].

The cortical cells are assembled as keratin intermediate filaments and have a diameter of 7–8 nm in all mammalian fibres [8]. These intermediate filaments form ordered aggregates or microfibrils and macrofibrils depending on species and function (Figure 1). The hard α-keratin intermediate filaments are assembled from tetramers, a pair of laterally aligned and antiparallel dimeric molecules. On average, keratin intermediate filaments contain eight tetramers. In the case of wool, the cortex region is composed of an orthocortex and paracortex with different intermediate filament/matrix packing. The proportion of ortho- and paracortex in the wool fibre determines the degree of crimping [13].

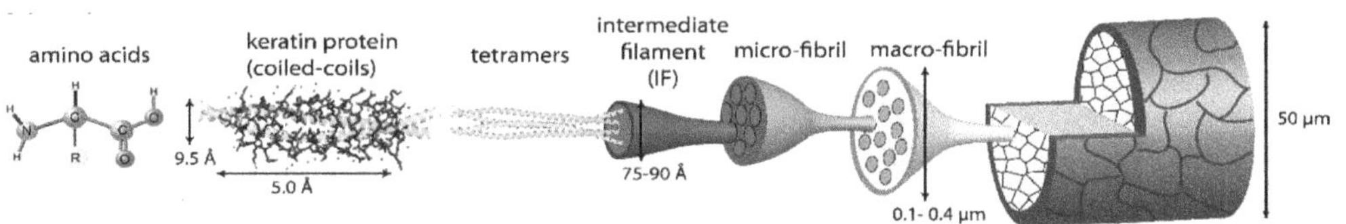

Figure 1. Structure of keratin. Adapted from work in [12] under the Creative Commons Attribution 4.0 International license (https://creativecommons.org/licenses/by/4.0/deed.en).

Keratin peptide heterodimers are formed when a type I (acidic) polypeptide chain and a type II (basic) polypeptide chain align in parallel. Each polypeptide chain is composed of a central α-helical region (about 46 nm in length) with non-helical head and tail domains [13]. The head and tail domains are rich in cysteine, glycine and tyrosine amino acids. Disulphide and isopeptide bonds are formed with other keratin intermediate filaments, cysteine-rich matrix proteins and keratin associated proteins, which stabilize the fibre [8–10]. The disulphide bonds along with the N-acetyl glucosamine-glycosylated serine and threonine in the head and tail domains also stabilize the heterodimers [6].

3. Thermochemical Methods of Keratin Degradation

There are multiple thermochemical methods available to prepare hydrolysed keratin for various value-adding opportunities, with specific processes chosen depending on the end-use [4]. Thermochemical methods include solubilization of keratin in organic solvents, ionic liquids or by hydrothermal methods; oxidation or reduction of the disulphide bridges; disruption of the hydrogen bonds with compounds like urea; and acid or base hydrolysis.

The composition of the final hydrolysate will depend on the method used to hydrolyse the keratin. Some of the thermochemical processes result in a hydrolysate containing a highly diverse mix of keratin-derived peptides and free amino acids and others are more specific. However, in most cases, the amino acid composition is modified. The processes and hydrolysate products will be described in more detail in the following section.

After solubilization of keratin with solvents like with *N,N*-dimethylformamide or dimethyl sulfoxide, precipitation is required with acetone and drying to produce a powder of keratin [16]. The major drawback of this method is the use of large quantities of solvents, which need to be recycled or incinerated. Solubilization can also be achieved with ionic liquids. Xie et al. used the ionic liquid, 1-butyl-3-methylimidazolium chloride for the solubilization of wool keratin, which disrupted the hydrogen bonds in the keratin macromolecules [17]. The keratin peptides were precipitated from the resulting hydrolysate with methanol. Ionic liquids are more expensive than traditional solvents and extraction of the keratins from the ionic liquid can be difficult.

Hydrothermal treatment is usually carried out at temperatures of 80–140 °C and steam pressures of 10–15 psi. Acid or base can be added to speed up the process of solubilization [18]. Under conditions of high temperature and pressure, the thermally unstable amino acids, glutamine and asparagine are degraded [19]. If base is added to this process then lysine, methionine and tryptophan are also destroyed [20,21]. Modified amino acids, lysinoalanine and lanthionine are also formed from lysine and cystine, respectively. Heating of proteins leads to a degree of racemization of the free and bound L-amino acids [22–24].

Reduction with reducing agents like thioglycolate [4], dithiothreitol [25], 2-mercaptoethanol [26], sodium sulphite [27], bisulphites [28] or cysteine [29] combined with high concentrations of compounds like urea, thiourea or surfactants, which disrupt the hydrogen bonds stabilising the keratin structure, results in the production of kerateine [30]. Kerateine contains cysteine thiol and cysteine sulfonate in place of the disulphide bonds. Kerateine is less soluble in water and can be re-cross-linked if exposed to an oxidant [4].

The microstructure of wool keratin after treatment for 4 h at 65 °C with 2-mercaptoethanol, EDTA, high concentrations of urea and pH 9 was investigated by Cardamone [31]. Analysis of the hydrolysate revealed a defined mixture of microfibrillar and intermediate filaments. This mixture of subunits was suitable for producing self-assembling biomaterials.

Oxidation of keratin by oxidants like peracetic acid [32] or peroxycarboximidic acid [33] leads to the formation of keratose. Keratose contains sulfonic acid groups and cysteic acid instead of the disulphide bonds [4]. These keratoses are hydroscopic, water soluble and the disulphide bridges cannot spontaneously re-form under oxidative conditions. Keratoses are not as stable as kerateines.

Oxidative sulphitolysis has been patented and commercialized to produce three functional keratin protein and peptide products. These products are based on S-sulphonated keratin intermediate filaments, S-sulphonated keratin high-sulphur proteins and keratin peptides [34]. The process aims at maintaining the structural integrity of the keratin proteins. The cystine groups in the wool keratin are converted to S-sulfocysteine using sodium sulphite or sodium metabisulfite and then oxidized with cupraammonium hydroxide. The intermediate filaments and peptides can undergo crosslinking by reductive desulfonation of the cysteines in the filaments and peptides and subsequent reformation of the intermolecular disulphide bonds.

One of the disadvantages of alkaline hydrolysis of keratins is the modification or degradation of amino acids (Table 1). Alkaline hydrolysis of keratins at higher temperatures results in the degradation of the thermally unstable amino acids, asparagine, glutamine, arginine, serine, threonine and cysteine [5]. Lysinoalanine and 8-aminoalanine are formed under alkaline conditions [35,36]. Another modification that occurs is the racemization of free or bound L-amino acids to the D-enantiomers [23,37,38]. Free amino acids racemize ten times slower than bound amino acids [24]. Following, for example, prolonged treatment of wool keratin at 70 °C and pH 9–11, lanthionyl residues [31] and dehydroalanine [39] are formed from cystine. Cystine and hydroxy amino acids were destroyed if the alkaline treatment was performed in the presence of reducing agents [40].

Table 1. Amino acid modification during alkaline treatment.

Amino Acid	Degradation Products	Reference
Asparagine	Aspartate, ammonia	[41]
Glutamine	Glutamate, ammonia	[41]
Arginine	Ornithine, citrulline, 3-aminopiperidin-2-one	[42]
Serine	Glycine, alanine, oxalic acid, lactic acid, ammonia	[43]
Threonine	Glycine, alanine, α-aminobutyric acid, ammonia	[44]
Cysteine	Pyruvic acid, sodium sulfide, ammonia	[45]
* Cystine, lysine, arginine	Lanthionine, lysinoalanine, ornithinalanine; * dehydroalanine [39]	[35,36]
L-amino acids	D-amino acids	[37]

Note: * Dehydroalanine is probably formed from the cleavage of the C-S bond in cystine.

Acid hydrolysis of keratins leads to the loss of some amino acids like serine, threonine, tyrosine and cystine and the conversion of asparagine, glutamine, methionine and tryptophan into other compounds ([5,19,46] Table 2). Polypeptides, resulting from the acid hydrolysis of keratin, have a more amorphous structure than alkaline hydrolysates, because most of the hydrogen bonds are broken during this process [47]. A typical acid hydrolysis of keratin uses hydrochloric acid [48,49] or sulphuric acid [50] at high temperatures.

Zhang et al. showed that acid hydrolysis was not as effective as other treatments mentioned above [49]. Wool keratin was hydrolysed with 4M hydrochloric acid at 95 °C for 24 h, resulting in 33% solubilization of the wool keratin. Increasing the treatment time had no effect on the yield, suggesting that there is a recalcitrant portion of the keratin resistant to acid hydrolysis.

Thermochemical methods offer cheap and versatile processes for hydrolysing keratin for a variety of applications. However, the use of harsh chemicals and conditions, the lack of ability to control the processes in most cases and the often unfavourable modification of the amino acids or peptides present environmental problems and peptide mixes that would be unsuitable for some applications. Using enzymes working under mild conditions to catalyse the hydrolysis offers a favourable alternative.

Table 2. Amino acid modification during acid treatment.

Amino Acid	Degradation Products	Reference
Asparagine	Aspartate, ammonia	[19,41]
Glutamine	Glutamate, ammonia	[19,41]
Methionine	Methionine sulfoxide	[19]
Tryptophan	Oxindolylalanine, dioxindolylalanine	[46]

4. Microbial Degradation of Keratin

The first peptidases with keratinolytic activity were found in *Bacillus* sp. and *Streptomyces* sp. and belong to the serine peptidase family [51]. The ability to degrade keratin is widespread and has been identified in bacteria and fungi [4,52]. Keratin-degrading microorganisms have been isolated from many sources like skin, feathers, hair, nails, soil, geothermal hot stream and wastewater, which is reflected in the optimum pH and temperature of the keratinase activity of these microorganisms. The pH optimums of keratinases range from pH 5.5 for the fungal keratinase from *Trichophyton mentagrophytes* [53] to pH 12.5 for the keratinase from *Brevibacillus* sp. AS-S10-11 [54]. Although, temperature optimums vary from 30 °C for the keratinase from *Brevibacterium luteolum* [55] to 100 °C for the keratinase from *Fervidobacterium islandicum* AW-1 [56].

Publications from 2018 and 2019 report the isolation of diverse species of bacteria like *Streptomyces* sp. [57], *Aeromonas hydrophila* FB3 [58], *Pseudomonas putida* KT2440 [59] and *Serratia marcescens* EGD-HP20 [60,61] with keratinolytic activity. However, the number of *Bacillus* strains with keratinolytic activity prevailed over any other genus of bacteria [62–86]. Valorisation of waste feathers [5,65,66,87] and the replacement of the traditional, highly polluting hide dehairing step used in the leather industry with a more environmentally friendly enzymatic step using keratinases [2,55,79,88] were the dominant themes of these papers.

Despite the interest in the enzymatic hydrolysis of keratin, mechanisms of keratin degradation in microorganisms are not fully understood. There is evidence that microbial degradation of keratin proceeds via a consortium of enzymes (Figure 2 [6,7,89]).

Disruption of the keratin structure is an essential step in the breakdown of keratin by keratinases. Various mechanisms have been suggested for fungal systems. Disulphide bond reductases and the intracellular cysteine dioxygenase can break the structure-stabilizing disulphide bridges in keratin [6,7,90]. Cysteine dioxygenase in conjunction with aspartate aminotransferase produces the reducing agent, sulphite, from cysteine, which is secreted into the surroundings and contributes to the chemical reduction of the disulphide bond. The reduction of the disulphide bonds aids access of the endoproteases (serine protease family), exoproteases (metalloprotease family) and oligopeptidase (metalloprotease family) to the keratin fibres or feathers. It has also been found that the membrane-bound redox system of the cell can cleave the disulphide bonds in keratin. The mechanical pressure exerted by fungal mycelia penetrating the keratin structure can also contribute to the disruption of this structure, facilitating access of the keratinase to the substrate. In nature, these mechanisms act synergistically with keratinases and speed up the degradation of keratin. Auxiliary proteins, like lytic polysaccharide monooxygenases (LPMOs), have been found associated with keratin degradation [6]. It is thought that they contribute to α- and β-keratin degradation. Until now LPMOs were thought to be associated with cellulose, chitin, hemicellulose and starch degradation only. It is possible that these enzymes hydrolyse the glycolytic bond between N-acetylglucosamine and serine and threonine in the head and tail region of the intermediate filaments, which contributes to the destabilization of the keratin structure.

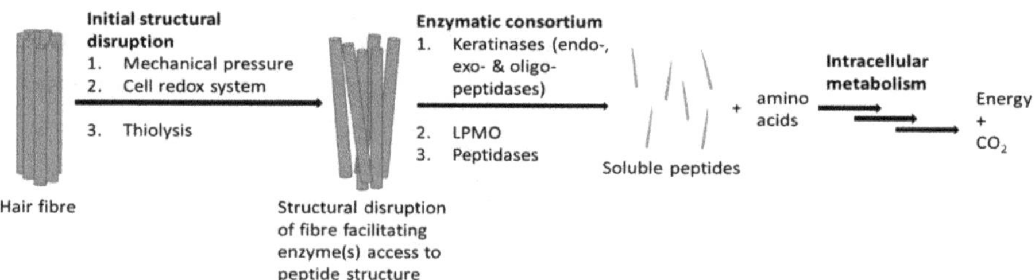

Figure 2. Possible mechanisms for microbial degradation of keratin (LPMO = lytic polysaccharide monooxygenase).

However, examples of peptidases with keratinolytic activity that do not need the assistance of disulphide reducing enzymes or agents have also been reported. Pillai et al. isolated a serine protease from *Bacillus subtilis* P13 with reductase and keratinase activities [91]. The isolated enzyme was able to decompose feathers and dehair hides.

He et al. analysed the enzyme consortium involved in the hydrolysis of feathers by a specific strain of *Bacillus subtilis* and identified four of the enzymes involved in keratin hydrolysis [74]: a serine protease with keratinase and disulphide bond-reducing activity; a peptidase T; a γ-glutamyltransferase, which generates a free cysteinyl group from glutathionine; and a cystathionine γ-synthase, which catalyses the production of L-cystathionine from homoserine ester and cysteine. The L-cystathionine is further converted to methionine and ammonia is released.

5. Characterisation and Comparison of Keratinases from S1, S8 and M4 Peptidase Families

Many articles characterising organisms capable of degrading keratin and their possible industrial applications have been published. Yet, there are few articles that report enzyme sequences and investigate the molecular and biochemical characteristics of the enzymes produced by these organisms [92,93]. The first paper that explored the molecular aspects of a keratinase produced by *Bacillus licheniformis* was published by Lin et al., 1995. Since then, more than 40 keratinases have been sequenced. To date, peptidases with keratinolytic activity from six different peptidase families have been identified: S1, S8, M4, M5, M14 and M28. Most of the characterized keratinases are produced by *Bacilli* and are members of the S8 serine peptidase family. There are currently over 127,000 peptidase sequences from the S1 (70919), S8 (38270), M4 (6403), M5 (145), M14 (11202) and M28 (904) families deposited on the MEROPS peptidases database. These 127,000 peptidase sequences represent an enormous unmined potential for the discovery of new peptidases with keratinolytic activity if the requisite properties of a peptidase with keratinolytic activity can be identified.

The S1 family sequences, when pairwise aligned, show a minimum value of 27.27% and a maximum of 97.22% identity, with an average of 61.48% for the four available sequences. The S8 family has a minimum of 13.69% and a maximum of 99.72% identity, with an average of 63.29% for the 13 available sequences, and the M4 family has 25.56% identity between the two available sequences. Although many of the characterized enzymes have been produced by the native unmodified organism [94–98], several examples involve heterologous expression. Different organisms have been used for recombinant production, including yeast such as *Komagataella Pastoris* (*Pichia Pastoris*) [99] and bacteria such as *Escherichia coli* [100–110] and *Streptomyces lividans* [111]. Including the pre-pro-domains with the catalytic domain in heterologous systems have been shown to maintain enzyme activity and secretion [99,102,107,110] and inclusion of C-domains, when present, is important for substrate binding and recognition [105]. Replacing the native signal peptide for the *E. coli* signal peptide when expressing in *E. coli* has also led to higher levels of expression [101].

In this section the biochemical data of twenty sequenced peptidases with keratinolytic activity from the S1 and S8 peptidase families (serine proteases) and the M4 peptidase family (metalloprotease) are compared (Table 3). Difficulties associated with the interpretation of these data are also highlighted. The selection is based on the availability of sequence and biochemical data. The M5, M14 and M28 peptidase families were excluded because each family had only one biochemically characterized example with full sequence data available.

Table 3. Keratinolytic microorganisms and their keratinases from the S1, S8 and M4 keratinases selected for this study.

Organism	Strain	Keratinase Name	Accession No. [1]	Reference
S1A Peptidases				
Actinomadura viridilutea	DZ50	KERDZ	KU550701	[94]
Actinomadura keratinilytica	Cpt29	KERAK-29	ASU91959	[95]
Streptomyces fradiae	Var. k11	SFP2	AJ784940	[99]
Nocardiopsis sp.	TOA-1	NAPase	AY151208	[111]
S8A Peptidases				
Bacillus circulans	DZ100	SAPDZ	AGN91700	[100]
Bacillus licheniformis	RPk	KerRP	EU502844	[96]
Stenotrophomonas maltophilia	BBE11-1	KerSMD	KC814180	[101]
Stenotrophomonas maltophilia	BBE11-1	KerSMF	KC763971	[101]
Bacillus pumilus	A1	KerA1	ACM47735	[97]
Bacillus pumilus	CBS	SAPB	CAO03040	[102]
Bacillus pumilus	KS12	rK_{27}	HM219183	[103]
Bacillus tequilensis	Q7	KerQ7	AKN20219	[104]
Bacillus cereus	DCUW	Vpr	ACC94305	[105,112]
Bacillus altitudinis	RBDV1	KBALT	APZ77034	[63]
Thermoactinomyces sp.	YT06	YT06 Protease	WP_037995056	[98]
Thermoactinomyces sp.	CDF	Protease C2	ADD51544	[106]
Meiothermus taiwanensis	WR-220	rMtaKer	5WSL	[107]
Brevibacillus sp.	WF146	WF146 Protease	AAQ82911	[108]
M4 Peptidases				
Geobacillus stearothermophilus	AD-11	RecGEOker	KJ783444	[109]
Pseudomonas aeruginosa	KS-1	KerP	HM452163	[110]

Note: [1] NCBI GenBank nucleotide accession number.

5.1. *S1, S8 and M4 Peptidase Families*

The S1 family is the largest family of serine proteases. The active site of S1 peptidases contains the catalytic triad, His, Asp and Ser. All enzymes characterized in this family are endopeptidases. The four peptidases in Table 3 belong to the S1A family represented by chymotrypsin as the type-example. The hydrophobic amino acid at the P1 site determines the specificity of these peptidases [113,114].

The S8 family is currently the second largest serine protease family and the most widely characterized to date [114,115]. Most of the keratinases are found in the subfamily S8A including the 14 keratinases in Table 3. They are represented by subtilisin as the type-example. Their active site contains the catalytic triad of Asp, His and Ser. In general, these enzymes are endopeptidases [116], active between neutral and moderately alkaline pH and many are thermostable [117]. Most enzymes in this family are not specific, usually cleaving after a hydrophobic residue in the peptide substrate [114,117]. S1 and S8 families are examples of convergent evolution as they catalyse the same reaction but have no sequence homology. Two calcium-binding sites contribute to thermal stability in many members of these families [114,117].

Two keratinases in Table 3 belong to the M4 family. They are characterized by a catalytic zinc ion tetrahedrally coordinated in the active site by a histidine and glutamate present in a HEXXH motif, another glutamate residue and water [118]. Most members of this family are endopeptidases and active at neutral pH. The preferred cleavage site occurs at a hydrophobic residue followed by leucine, phenylalanine, isoleucine or valine. These peptidases are stabilized by Ca^{2+} [119].

Independent of their families, keratinases usually cleave aromatic and hydrophobic amino acid residues at the P1 position. Keratins are composed of 50 to 60% aromatic and hydrophobic residues, which could partially explain the keratinase specificity for keratin [120–122]. Most of these peptidases are stabilized by divalent cations like Ca^{2+} and are extracellular [119,123,124].

5.2. *Problems Associated with Keratinase Assays*

There are several issues with the current methods used to characterize keratinases. The assays are not standardized in the literature in terms of reaction conditions and substrates. The most common method used to measure keratinase activity is a colorimetric assay that uses the commercially available derivative of wool, keratin azure [125] or azokeratin (sulfanilic acid-azokeratin [126]). However, batch variability and the fact that the chromogenic agents are only bound to the outer portion of the substrate compromises reproducibility. Quantification of the soluble peptides generated by hydrolysis of keratin has also been used to determine the effectiveness of keratinases on keratin substrates. Common quantification methods used are Bradford [95,127], Lowry [128,129] or measurement of absorption at 280 nm [106,111] (see Table 4). Each of these methods have several limitations. The Coomassie Blue dye used in the Bradford method preferentially reacts with arginine and lysine in the peptide mix and alkaline pH and detergents interfere with the reaction. The Folin–Ciocalteu dye used in the Lowry method oxidizes the aromatic amino acids residues in the protein and is affected by reducing agents. Only tyrosine, tryptophan and cysteine absorb at 280 nm and other compounds like DNA in the solution can interfere with the measurement [130]. The simplest and probably most accurate method for quantifying keratinase activity is the measurement of weight loss when the insoluble keratin substrates like mammalian hair fibres, feathers or wool are solubilised through hydrolysis [96,127].

Table 4 describes a variety of substrates that have been used to assay keratinase activity in the literature. The substrates that were used include keratin azure (wool), keratin powder, soluble keratin, keratin (undefined), feathers and feather meal powder. It was not possible to ascertain the source and integrity of most of these keratin substrates from the papers. The pretreatment of these substrates is also an important aspect in determining the keratinase activity. Keratin powder and solubilized keratin were generally obtained from commercial sources; however, the sources and preparation were not described. Pretreatments like autoclaving and milling [103,107], or treating with solvents at high temperatures [106], are known methods for keratin powder preparation from the literature. In the case of the rK_{27} keratinase, the feather powder used in the assay was autoclaved and dried at 60 °C [103]. These preparation methods, as already described in Section 3, would compromise the keratin structure. The keratinases, KerRP [96], Ker-A1 [97] and SAPB [102] were assayed on keratins of unknown source. In the WF146 protease assay, the feather substrate was washed with ethanol prior to use in the assay, which would likely remove the protective lipid layer [108].

Co-treatment can also affect the integrity of the keratin substrates during enzymatic hydrolysis [125,131,132]. Except for KerQ7 [104], all assays in Table 4 were carried under alkaline conditions between pH 8 and pH 12.5 and temperatures ranging from 50 to 80 °C. These conditions would most likely contribute to the weakening the keratin structure. Keratinase assays with SAPDZ [100], KerQ7 [104], KERDZ [94], and KERAK-29 [95] were supplemented with the divalent cations Ca^{2+} or $Mn2^{+}$. Divalent cations are known to stabilize serine proteases [114,117]. Other keratinase studies added reducing agents, like β-mercaptoethanol (protease C2 [106], WF146 protease [108]) or dithiothreitol (SFP2 [99]) to the assay mixture. Reducing agents are known to break the disulphide bond leading to disruption of the keratin.

Table 4. Keratinase pH and temperature optimums of the selected S1, S8 and M4 keratinases with associated assay conditions. See text for further details of the assays.

Protein	* pH	* Temp (°C)	Conditions	PT	CT
			S1A Peptidases		
KERDZ	11	80	10 g/L keratin azure, 50 mM bicarbonate-NaOH buffer, pH 11 mixed 1:1 with the enzyme, 30 min, 80 °C, 200 rpm (Abs_{595nm}).	-	2 mM $CaCl_2$
KERAK-29	10	70	1 mL of 10 g/L keratin azure, 100 mM Glycine-NaOH buffer mixed 1:1 with the enzyme, pH 10, 20 min, 70 °C (Abs_{595nm}).	-	5 mM $MnSO_4$
SFP2	10	60	5 mg keratin azure, 50 mM Tris-HCl, pH 8.5, 1 h, 37 °C (Abs_{595nm}).	-	10 mM DTT
NAPase	12.5	60	60 mg wool keratin powder, Glycine-NaOH, pH 10 or 50 mM KCl-NaOH, pH 12.5, 30 °C, 2 h (Abs_{280nm}).	Not specified	-
			S8A Peptidases		
SAPDZ	12.5	85	10 g/L keratin azure, 100 mM KClNaOH, 250 rpm, 20 min incubation, 85 °C (Abs_{595nm}).	-	5 mM $CaCl_2$
KerRP	9 (11)	60 (65–70)	0.8% w/v keratin diluted 1:1 in enzyme, 1 h incubation, 60 °C (Abs_{280nm}).	Not specified	-
KerSMD	8	60	1% w/v soluble keratin, 50 mM Gly-NaOH, 20 min, 50 °C (Folin–Ciocalteu).	Not specified	-
KerSMF	9	60	1% w/v soluble keratin, 50 mM Gly-NaOH, 20 min, 50 °C (Folin–Ciocalteu).	Not specified	-
KerA1	9 (10)	60 (65)	0.8% w/v keratin diluted 1:1 in enzyme solution, 1h, 50 °C (Abs_{280nm}).	Not specified	-
SAPB	10.6	65	1% keratin w/v, on 100 mM glycine-NaOH Buffer, pH 10.6, 30 min, 55 °C. 2 mM $CaCl_2$ supplemented.	Not specified	-
rK_{27}	9	70	20 mg feather powder, Gly-NaOH 50 mM, 1 h (Abs_{280nm}).	Washed with Triton X-100 (1%), rinsed with water, autoclaved, dried in an oven at 60 °C for 1 h, milled then sieved with 2 mm pore size.	-
KerQ7	7	30	10 g/L keratin azure, 50 mM HEPES buffer, 30 min, 200 rpm (Abs_{595nm}).	-	1 mM $CaCl_2$
Vpr	8.5	50	2% w/v chopped feather keratin, 50 °C, 15 min, pH 7.5.	-	-
KBALT	8	85	5 mg keratin azure, 0.8 mL buffer, 15 min incubation, pH 6 to 12, 25 to 95 °C (Abs_{595nm}).	-	-
YT06 protease	8–9	65	1% soluble keratin, 50 mM Gly-NaOH, pH 9, 20 min (Folin–Ciocalteu).	Not specified	-
Protease C2	11	60–80	5% keratin powder, 50 mM Tris-HCl pH 8, 60 min 60 °C (Abs_{280nm}).	100 °C incubation in DMSO for 2 h. Protein precipitated with acetone 2:1 v/v [133]	0.5% β-ME
rMtaKer	10	65	1% feather powder on 50 mM HEPES, pH8.0, 25–95 °C. Supplemented with 10 mM $CaCl_2$, 150 mM NaCl (Ninhydrin).	Chicken feathers rinsed, air-dried, ground by ball mill.	-
WF146 protease	-	80	10 mg of feathers, 50 °C or 80 °C, 1 ml Tris-HCl 50 mM buffer, pH 8.0, 10 mM $CaCl_2$, multiple time points from 0 to 20 h (Abs_{280nm}).	70 Ethanol wash, rinse water, dry, cut 2–3 mm long	1% β-ME
			M4 Peptidases		
RecGEOker	9	60	4 mg keratin azure, 50 mM Tris-HCl, pH 7.8, 1 h (Wool-Folin–Ciocalteu; Abs_{595nm}).	-	-
KerP	9	50	20 mg chicken feathers, Glycine-NaOH buffer, pH 10, 60 °C, 60 min (Abs_{280nm})	-	-

Note: Source organism, accession numbers and references can be found in Table 3; * in some cases pH and temperature optimums were determined on both casein and keratin substrates. The casein optimums are in brackets; Temp = temperature; PT = pretreatment; CT = co-treatment; β-ME = β-mercaptoethanol; DTT = dithiothreitol. Quantification methods, where available, are in brackets after the assay condition description.

The challenges with the keratinase assays discussed above highlight the need for standardized assays and substrates used to test keratinases and the challenges faced in attempting to compare and analyse data from the literature when the assays are not comparable.

5.3. The Effect of Additives on Selected S1, S8 and M4 Keratinases

Various additives were tested on the selected S1, S8 and M4 keratinases-cationic, anionic and neutral detergents, oxidizing agents, reducing agents, mono- and divalent metals. Table 5 contains a summary of additives that had a positive effect on keratinase activity. A positive effect was defined as ≥ 110% activity compared to the control without additive. Some of the papers used keratin as the assay substrate, some used casein and in some cases, both were tested.

Table 5. Influence of additives on the activity of selected S1, S8 and M4 keratinases. Numbers in brackets correspond to the % activity compared to 100% in the absence of the additive.

Protein	Metal ions (%)	Detergents (%)	Reducing Agents (%)	Solvents/Others (%)
S1A Peptidases				
KERDZ	Ca^{2+} (270) Mg^{2+} (180) Fe^{2+} (145)			
KERAK-29	Ca^{2+} (150) Mg^{2+} (110) Mn^{2+} (210)	Zwittergent (114) Tween-20 (130) Triton X-100 (132) Tween-80 (145) TTAB (116) CHAPS (140) Sulfobetaine (135) LAS (118) SDS (115) CTAB (110)	β-ME (102)	H_2O_2 (170)
SFP2	Cu^{2+} (149) Ni^{2+} (116)		DTT (278) β-ME (235)	
NAPase				**Isopropanol (130)**
S8A Peptidases				
SAPDZ	Ca^{2+} (450) Mg^{2+} (195) Mn^{2+} (280) Zn^{2+} (180) Cu^{2+} (110) Co^{2+} (113)			
KerRP	* **Ca^{2+}**			
KerSMD	**Ca^{2+} (112)**		**Na_2SO_3 (116)**	
KerSMF	** **Ca^{2+} (105)**	**Tween-20 (112)**	**Na_2SO_3 (115)** **DTT (115)**	
KerA1	**Ca^{2+} (123)** **Mg^{2+} (199)** **Na^{+} (135)**	**Tween 80 (113)**	β-ME (Casein 100) **(Keratin 192)**	
SAPB	Ca^{2+} (157) Mg^{2+} (112) Na^{+} (118)	LAS (114) Tween 80 (119) Tween 20 (117) SDS (119)	β-ME (192)	Urea (165) H_2O_2 (168)
rK_{27}	**Stability only tested**	**Triton X-100 (677)** **Tween-80 (242)** **Saponin (461)** **Sodium Cholate (276)** **SDS (186)**	**DTT (267)** **β-ME (323)**	**NaClO (276)** **H_2O_2 (275)**
KerQ7	**Ca^{2+} (417)** **Mg^{2+} (175)** **Mn^{2+} (250)** **Ba^{2+} (121)** **Sn^{2+} (115)**			
KBALT	**Ca^{2+} (127)** **Mg^{2+} (134)** **Zn^{2+} (129)** **Ba^{2+} (115)**	**SDS (128)**	**β-ME (102.5)**	
YT06 Protease	**Mg^{2+} (118)** **Mn^{2+} (196)** **Ni^{2+} (120)** **Ba^{2+} (115)**	**Tween-20 (170)**	**β-ME (623)**	
M4 Peptidases				
RecGEOker	**Mg^{2+} (112)** **Mn^{2+} (116)** **Zn^{2+} 1 mM (58); 10 mM (52)** **Ca^{2+} 1 mM (101); 10 mM (66)**	**Triton X-100 (115)** **Tween 40 (180)** **Tween 60 (133)** **Tween 80 (122)** **Triton X-305 (153)**	**DTT (139)**	

Note: Source organism, accession numbers and references can be found in Table 3; Bold = tested on a keratinous substrate; Not bold = tested on casein; * = Only tested for binding; ** included for comparison with KerSMD; underlined = denaturing detergents; DTT = dithiothreitol; β-ME = β-mercaptoethanol; LAS = linear alkylbenzene sulfonate; SDS = sodium dodecyl sulfate; TAED = tetraacetylethylenediamine; TTAB = tetradecyltrimethylammonium bromide; CHAPS = 3-[(3-cholamidopropyl)dimethylammonio]-1-propanesulfonate; CTAB = cetrimonium bromide.

Despite there being large differences in concentrations of metals additives, incubation time and temperature, in general, supplementation with Ca^{2+} showed the largest increase in activity except for KerSMF [101] and kerA1 [97]. In the case of KerSMF, Ca^{2+} had no effect on activity and in the case of kerA1, Mg^{2+} addition increased activity by 199% compared to 123% for Ca^{2+}. In general, Ca^{2+} > Mg^{2+} > Mn^{2+} had a positive effect on all the S1 and S8 keratinases (Table 5). The effect of these divalent metals on M4 metalloproteases is discrete compared to serine proteases. Only the addition of magnesium and manganese ions resulted in keratinase activity slightly above the control without additives. These divalent ions have been described to stabilize the active structure of the peptidases by binding to the metal-binding sites [100]. Other explanations for the higher activity are possible stabilization of enzyme/substrate complex [101] or formation of salt or ion bridges that maintain the enzyme conformation [97,122,128]. Furthermore, these metal ions reduce thermal denaturation [134]. Lin et al. observed that aqualysin, a thermostable peptidase from the S8 family, was only stable at high temperatures in the presence of 1 mM Ca^{2+} [135].

Several studies were carried out on the keratinase activity in the presence of metal ions (Zn^{2+}, Cu^{2+}, Co^{2+}, Ba^{2+}, Sn^{2+} and Ni^{2+}) were carried out (Table 5). The addition of the metal ions improved activity between 10% and 29% except for SAPDZ [100], where Zn^{2+} addition increased activity by 80% and Cu^{2+}addition increased activity of SFP2 [99] by 49%. Li et al. characterized SFP1, a non-keratinolytic peptidase similar to SFP2 and produced by the same organism [99]. It showed increased activity with copper ions, possibly due to the stabilization of the enzyme. Copper ions acting as a stabilizer has rarely been described in previous serine protease studies, and it is not known whether there is a copper-binding site stabilizing the enzyme [136]. In another example, peptidases were more stable in the presence of copper ions, which resulted in a reduction in both autolysis and thermal inactivation rates [137].

Detergents, in general, enable the disruption or formation of hydrophobic and hydrophilic bonds and assist in the extraction of proteins into aqueous media [138]. Detergents can act as denaturing agents on enzymes. Denaturing detergents are anionic (SDS, LAS) or cationic (CTAB, TTAB). They denature proteins by breaking protein–protein interactions. Non-denaturing detergents are non-ionic (Triton X-100, Tweens, cholate, saponin) or zwitterionic (CHAPS, sulfobetaine, zwittergent), and their action is milder and enzyme function is usually maintained. In most cases the addition of denaturing and non-denaturing detergents resulted in an increase in activity (110–150%). However, the addition of the non-ionic detergents to the assay mixture with keratin as substrate of rK_{27} had a dramatic effect on activity compared to the control without detergent [103]. Activities of 677% (Triton X-100), 242% (Tween 80), 461% (saponin) and 276% (cholate) were achieved. The addition of the anionic denaturing detergent, SDS to the assay increased the keratinase activity to 186%. The addition of the non-ionic detergents, Tween 40, Tween 60, Tween 80 and Triton X-305 to the assay mixture with keratin as substrate for the M4 keratinase, RecGEOker [109], showed increased activity to 180%, 133%, 122% and 153%, respectively. This example showed a definite trend of increasing activity with decreasing Tween 80 (monounsaturated C18 derivative) < Tween 60 (saturated C18 derivative) < Tween 40 (saturated C16 derivative). The partial solubilizing action of detergents on the insoluble keratin substrate might explain why both denaturing and non-denaturing detergents have a positive effect on keratin hydrolysis. There are insufficient examples to confirm this Tween effect on keratinases in general.

The reduction of disulphide bonds, destabilizes keratins and acts synergistically with keratin hydrolysis in nature [6,8–10]. Sodium sulphite, dithiothreitol (DTT) and β-mercaptoethanol were tested on some of the keratinases in Table 5. The reducing agents had a positive effect on all S8 keratinases tested with keratin as the substrate. The increase in activity ranged from 115% for Na_2SO_3 (KerSMF [101]) to 623% for β-mercaptoethanol (YT06 protease [98]) except in the case of KBALT [63], where β-mercaptoethanol had no effect on the activity. β-Mercaptoethanol doubled the activity of SAPB [102] when tested with casein as substrate. DTT also increased the activity of the M4 keratinase, RecGEOker (139% [109]), when tested with keratin as substrate. None of the S1 enzymes were tested with keratin and reducing agents. However, the two assays with casein and reducing agent showed on one hand, no effect from β-mercaptoethanol on KERAK-29, [95] and on the other hand, a considerable effect on SFP2 (DTT, 278%; β-mercaptoethanol, 235% [99]). It should be noted that where disulphide bonds present in the enzyme are essential for function the inclusion of reducing agents may negatively affect activity.

Chaotropic agents are comparable to detergents, breaking non-covalent interactions and allowing protein denaturation [139–141]. Urea and isopropanol are chaotropic agents (Table 5). The activity of SAPB [102] was increased to 165% in the presence of urea compared to the control and the activity of NAPase [111] was increased to 130% in the presence of isopropanol [111].

The effect of the oxidizing agents, H_2O_2 and sodium hypochlorite, was also studied on three S8 and S1 keratinases, SAPB [102], rK_{27} [103] and KERAK-29 [95]. Activity was significantly increased in all cases (Table 5).

In most cases the effect of additives like divalent cations, detergents, reducing agents, chaotropic agents and oxidizing agents have a positive effect on keratinase activity. Nearly all compounds capable of disrupting the integrity of the keratin structure without inactivating the keratinase appear to have a positive effect on keratinase activity. The effect of compounds disrupting the keratin structure was, in some cases like rK_{27} [103], significant.

5.4. Substrate Specificity

Table 6 summarizes the substrate specificity data of the selected keratinases from the literature. In general, a variety of keratins and other proteins like gelatin, casein and albumin were tested. To compare the selected keratinases, the values in Table 6 have been normalized using the activity of designated keratin substrates (keratin azure, keratin, feather or wool) as 100% activity.

KerQ7 [104] was the only keratinase in Table 6 tested on multiple types of keratins. KerQ7 showed a preference for the β-keratin-rich feather meal and feathers. The activity on feather meal was only 16% higher than feathers. The activity on rabbit hair, goat hair and bovine hair was 88%, 74% and 50% of the activity on feathers, respectively, whereas activity on wool was only 12%. These substrates are rich in α-keratins [4,9,12,13]. Substrate fibre thickness and fibre surface area may also contribute to the variations in enzyme activity. Nonetheless, the keratinase activity toward various substrates is likely to be multifactorial. KerSMD and KerSMF, from *Stenotrophomonas maltophilia*, showed less activity towards feather powder and wool than keratin azure [101]. KerSMD and KerSMF had similar activity on feather powder (54% and 71%, respectively) and wool (59% and 78%, respectively). However, KerSMD showed an activity of 1589% towards soluble keratin compared to an activity of 126% for KerSMF on the same substrate.

No trends were observable on non-keratin substrates. For example, SAPDZ [100] showed 81% activity on gelatin compared to keratin, whereas the activity of kerA1 [97] and SAPB [102] on gelatin was 22% and 146% compared to keratin, respectively. The same inconsistencies can be seen with casein. The activities of SAPB [102] and KerSMD [101] on casein are 153% and 2800%, respectively, compared to keratin azure, whereas KerSMF [101] has only slightly lower activity on casein (91%) compared to keratin azure.

Keratinases are known for their activity on "hard-to-degrade" proteinogenic substrates. Most of the characterized keratinases in the literature are also capable of degrading collagen, which is an example of another complex and hard-to-degrade substrate [142]. A study in 2008 characterized the first keratinase without collagenase activity [143]. Only three enzymes from the S8 family were tested on collagen or azocoll (azocollagen). Vpr [105] presented collagenase activity (129%) and while SAPDZ [100] did not. Protease C2 [106] showed a surprisingly high activity (24000%) on azocoll compared to keratin azure (100%). KERDZ [94], from the S1 peptidase family, had no activity on collagen, whereas RecGEOker [109], belonging to the M4 metallopeptidase family, was able to hydrolyse collagen. The differences in activity between substrates may be attributed to the specific peptide sequences in the substrates and the sequence specificity of the enzymes.

Table 6. Substrate specificity of the selected S1, S8 and M4 keratinases. Numbers in brackets correspond to the % activity relative to other substrates.

Protein	Keratins (%)	Natural Proteins (%)		Modified Protein (%)	Esters and Others (%)
S1A Peptidases					
KERDZ	Keratin (100) [2]	Gelatin (90) Casein (79) Albumin (75) Elastin (50)	Myoglobin (41) Hemoglobin (20) Collagen type 1/2 (0)	Azocasein (80) Azoalbumin (70)	BAEE (91) TAME (100) BCEE (95) BTEE (0) ATEE (0)
SFP2	Keratin (100)[2]	Casein (111)			
S8A Peptidases					
SAPDZ	Keratin (100)	Gelatin (81) Casein (95) Albumin (72)	Hemoglobin (66) Collagen type 1/2 (0)	Azocasein (91) Keratin Azure (100) [1]	BAEE (0) BTEE (100) BCEE (0) ATEE (95)
KerSMD	Feather powder (54) Soluble keratin (1589) Wool (59)	Casein (2800)		Keratin Azure (100) [1]	
KerSMF	Feather powder (71) Soluble keratin (126) Wool (78)	Casein (92)		Keratin Azure (100) [1]	
KerA1	Keratin (100) [2]	Gelatin (22) Casein (222)	Elastin (54) BSA (97) Egg albumin (4)	Azocasein (177) Azokeratin (92)	
SAPB	Keratin (100)[2]	Gelatin (146) Casein (153)	Bovine serum albumin (80) Egg albumin (18) Gluten (30)	Azocasein (123) Azokeratin (96)	BTEE (109) ATEE (115)
rK_{27}	Powdered chicken feather > haemoglobin > meat protein > hoof keratin > fibrin > elastin > gelatine > casein > BSA > azocasein > keratin azure				
KerQ7	Rabbit hair (88) Wool (12) Goat hair (74) Feather meal (116) Bovine hair (50) Feather (100) [3]				
Vpr	Feather meal (50) Keratin (100) [2]	Gelatin (147) Casein (156)	Fibrin (145) Collagen (129)		
Protease C2	Bovine hair (274) Feather (439)	Albumin (8571) Elastin (11)		Keratin azure (100) [1] Azocasein (102857) Azocoll (24000)	
M4 Peptidases					
RecGEOker	Wool (100) [5]	Gelatin (92) Casein (95)	Albumin (37) Collagen type 1 (98)		

Note: Source organism, accession numbers and references can be found in Table 3; Activity normalized to the following substrates—[1] keratin azure, [2] keratin; [3] feather, [4] wool; BAEE = N-α-benzoyl-L-arginine ethyl ester; TAME = N-α-p-tosyl-L-arginine methyl; BCEE = benzoyl-citrulline ethyl ester; BTEE = N-benzoyl-L-tyrosine ethyl ester; ATEE = N-acetyl-L-tyrosine ethyl ester; DTNB = 5,5′-dithiobis-(2-nitrobenzoic acid); Azocoll = commercially available azocollagen.

Some enzymes also showed esterase activity, which may be of importance for facilitating enzyme access to the substrate. Fatty acids of the cuticle surface are linked via a thioester to the protein matrix below in keratin fibres and feathers [14]. Only three enzymes in Table 6 have been characterized on ester substrates. The two S8 family peptidases—SAPDZ [100] and KerRP [96]—appear to have similar ester substrate affinity with both showing activity against N-benzoyl-L-tyrosine ethyl ester (BTEE) and N-acetyl-L-tyrosine ethyl ester (ATEE). In contrast, KERDZ (S1 family) had no activity towards these substrates but was active towards N-α-benzoyl-L-arginine ethyl ester (BAEE), N-α-p-tosyl-L-arginine methyl (TAME) and benzoyl-citrulline ethyl ester (BCEE) [94].

In general, the peptidases from S1, S8 and M4 families (Table 6) present varied substrate specificities. There are limited examples in the S1 and M4 families to detect trends but even within the S8 peptidases examples there were no obvious substrate preferences.

6. Potential Applications of Keratinases

New keratinases with improved properties for commercialization and the keratin hydrolysates they produce represent an opportunity for adding value to keratin waste.

Commercial keratinases are sold for a variety of applications (Table 7) such as the degradation of infectious prions, as supplements for animal feed to improve its nutritional value, removal of corns and calluses from skin, treatment of acne and nail fungi and, they are also incorporated into cosmetic skin peeling and depilatory creams [6,52,144]. Other applications include the use in cleaning products for unblocking drainpipes and septic tanks.

Table 7. Some examples of commercial keratinases.

Trade Name	Source	EC Number	Substrate or Function	Supplier
Versazyme [1,3]	*Bacillus licheniformis*	3.4.21.62/ S8 family	Improving nutritional value of poultry feed & prions degradation	Bioresource Int'l, Inc.
Ronozyme ProAct [2]	*Nocardiopsis prasina*	3.4.21.-/serine protease	Improving nutritional value animal feed	DSM/Novozymes
Cibenza DP100 [2]	*Bacillus licheniformis* PWD-1	-	Improving nutritional value animal feed	Novus International
Pure Keratinase 100 [3]	*Bacillus licheniformis* PWD-1	-	Prion degradation from medical & dental instruments	Proteus Biotech
BioGuard Plus [3]	Proprietary blend of microorganisms – incl. keratinase producer	-	Cleaning drainpipes, septic tanks & digesters	RuShay Inc.
Keratoclean sensitive PB [3]	*Bacillus licheniformis* (PB333 keratinase)	-	Treatment acne, dead skin removal, promotes cell renewal	Proteus Biotech
Keratoclean Hydra PB [3]	*Bacillus licheniformis*	-	Removal of corns & call uses, acne, Hirsutism, peeling	Proteus Biotech
FixaFungus [3]	-	-	Treatment of toenail fungal infections	Proteus Biotech

Source: [1] [6], [2] [127], [3] [52].

There are also a number of promising applications of keratinases that have not been commercialized to date: dag or manure balls removal from cattle hides and tails [145]; extraction of glucocorticoids from chicken feathers to monitor the stress level in poultry breeding and production programmes [146] extraction of chicken feather cholesterol as a precursor to bile salts that can be used to produce bio-emulsifiers and biosurfactants in the cosmetic industry [18]; selective hydrolysis of wool from wool/polyester or mixed textiles to facilitate textile recycling [147]; and dehairing of hides in the leather industry [64,65,82].

The use of keratinases for the processing of keratinous waste might be advantageous for high value products. The use of enzymes instead of thermochemical methods for keratin hydrolysis reduces chemical modification arising from harsh chemical hydrolysis and might allow a degree of control of the peptide composition that is produced. Keratin hydrolysates are widely used in protein feed supplements [18]. Feather waste, for example, is hydrolysed with saturated steam under high pressure (sometimes with the addition of lime) to produce feather meal, which is used as a feed supplement [148]. These conditions lead to the loss or modification of some of the amino acids, which impacts the nutritional value and digestibility of the feather meal. Hydrolysis with keratinases might offer an alternative, which reduces the energy requirements of the process and enhances the nutritional value of the supplement.

Keratin-derived bioactive peptides have been reported in the literature. These peptides have a range of activities like antimicrobial [149], antihypertensive [150], anti-inflammatory [151–154], antioxidant [149,150,155], inhibition of early stage amyloid aggregation [156], antidiabetic [157] or anti-aging [158–160] depending on the keratin source and the method of preparation. Producing protein feed supplements with antioxidant or anti-inflammatory properties as well as skin and hair products with antioxidant, anti-inflammatory, antimicrobial or anti-aging properties would most likely increase the value of these products.

Keratin peptides and subunits can spontaneously self-assemble [161]. This property can be exploited to form biomaterials like hydrogels, films, sponges, scaffolds and nanofibres for tissue engineering, wound healing, fibroblast cultivation and treatment of burns [161–164]. The production of smart biocomposites is also of interest. An example is the production of transparent plastic film containing citric acid from wool hydrolysate [165]. The plastic has excellent biocidal activity and could be used as a functional packaging for food.

The examples described above demonstrate the commercial potential of keratinases and the large number of opportunities they offer for adding value to keratin waste by producing bioactive protein feed supplements, personal care products and biomaterials from keratin hydrolysates.

7. Discovery and Future Research

The standout problem with the characterization of keratinases, demonstrated by the analysis of the assay conditions in this review, is lack of standardization of the keratinase assay combined with the small number of sequenced peptidases with keratinolytic activity that have been biochemically characterized. Both of these issues hamper the identification and comparison of true keratinases. Current experimental conditions vary in temperature, pH, buffer types and concentration, additives, substrates and their pretreatment biasing possible conclusions. It is unclear whether some proteases are keratinases or whether pretreatment or co-treatment influences their keratinolytic activity to some degree.

The uncertainty in defining keratinases and highly variable characterization of keratinases in the literature increases the challenge of finding new keratinases based on literature data or from sequence databases. However, the discovery of new keratinases is critical for expanding the opportunities for waste keratin valorisation. It would be desirable to identify new keratinases with high activities and specificities enabling control over cleavage sites, peptide molecular weights and amino acid side chain modifications.

Standardized experiments combined with phylogenetic studies and sequence analyses are needed. Standardized experiments, which avoid pre- or co-treatments, would determine the true protease activity on keratin substrates and reduce possible experimental biases. An in-depth phylogenetic analysis would help to clarify the position of keratinases within the phylogenetic trees of the peptidase families in which they are found and may help focus the search for new peptidases with keratinolytic activity. A comprehensive sequence analyses, aimed at the identification of conserved sites between peptidases with keratinolytic activity, as well as the presence of specific domains that possibly contribute to their ability to hydrolyse keratin, may assist in the development of algorithms to search the vast sequence space of the peptidase families.

Author Contributions: Conceptualization, R.S., K.R., J.P.D.O.M., L.N., M.N., G.C. and Z.Z.; investigation, J.P.D.O.M. and K.R.; writing—original draft preparation, J.P.D.O.M. and K.R.; writing—review and editing, R.S., L.N., M.N. and G.C.; supervision, R.S.; project administration, R.S.; funding acquisition, R.S. All authors have read and agreed to the published version of the manuscript.

Acknowledgments: This project is supported by Meat and Livestock Australia through funding from the Australian Government Department of Agriculture as part of its Rural R&D for Profit program and the partners.

References

1. Tesfaye, T.; Sithole, B.; Ramjugernath, D. Valorisation of chicken feathers: A review on recycling and recovery route—current status and future prospects. *Clean Technol. Environ. Policy* **2017**, *19*, 2363–2368. [CrossRef]

2. Thyagarajan, D.; Barathi, M.; Sakthivadivu, R. Scope of poultry waste utilization. *IOSR-JAVS* **2013**, *6*, 29–35.
3. Gooding, C.H.; Meeker, D.L. Review: Comparison of 3 alternatives for large-scale processing of animal carcasses and meat by-products. *PAS* **2016**, *32*, 259–270. [CrossRef]
4. Sinkiewicz, I.; Staroszczyk, H.; Sliwinska, A. Solubilization of keratins and functional properties of their isolates and hydrolysates. *J. Food Biochem.* **2018**, *42*, e12494. [CrossRef]
5. Chojnacka, K.; Gorecka, H.; Michalak, I.M.; Gorecki, H. A review: Valorization of keratinous materials. *Waste Biomass Valoriz.* **2011**, *2*, 3017–3021. [CrossRef]
6. Lange, L.; Huang, Y.; Busk, P.K. Microbial decomposition of keratin in nature—A new hypothesis of industrial relevance. *Appl. Microbiol. Biotechnol.* **2016**, *100*, 2083–2096. [CrossRef]
7. Peng, Z.; Zhang, J.; Du, G.; Chen, J. Keratin waste recycling based on microbial degradation: Mechanisms and prospects. *ACS Sustain. Chem. Eng.* **2019**, *7*, 9727–9736. [CrossRef]
8. Jones, L.N. Hair structure anatomy and comparative anatomy. *Clin. Dermatol.* **2001**, *19*, 95–103. [CrossRef]
9. Bragulla, H.H.; Homberg, D.G. Structure and functions of keratin proteins in simple, stratified, keratinized and cornified epithelia. *J. Anat.* **2009**, *214*, 516–559. [CrossRef]
10. Plowman, J.E. Proteomic database of wool components. *J. Chromatog. B.* **2003**, *787*, 63–76. [CrossRef]
11. Da Silva, R.R. Different processes for keratin degradation: The ways for the biotechnological application of keratinases. *J. Agric. Food Chem.* **2018**, *66*, 9377–9378. [CrossRef] [PubMed]
12. Yang, F.; Zhang, Y.; Rheinstädter, M.C. The structure of people's hair. *PeerJ* **2014**, *2*, e619. [CrossRef] [PubMed]
13. Wang, B.; Yang, W.; McKittrick, J.; Meyers, M.A. Keratin: structure, mechanical properties, occurrence in biological organisms, and efforts of bioinspiration. *Prog. Mater. Sci.* **2016**, *76*, 229–318. [CrossRef]
14. Ganske, F.; Meyer, H.H.; Deutz, H.; Bornscheuer, U. Enzyme-catalysed hydrolysis of 18-methyleicosanoic acid-cysteine thioester. *Eur. J. Lipid Sci. Technol.* **2003**, *105*, 627–632. [CrossRef]
15. Koepke, V.; Nilssen, B. Wool surface properties and their influence on dye uptake—A microscopical study. *J. Text. Inst.* **1960**, *51*, T1398–T1413. [CrossRef]
16. Wolski, T. *Modified Keratin Proteins, Their Physiochemical Properties, Analysis and Application*; Medical Academy: Lublin, Poland, 1985.
17. Xie, H.; Li, S.; Zhang, S. Ionic liquids as novel solvents for the dissolution of and blending of wool keratin fibers. *Green Chem.* **2005**, *7*, 606–608. [CrossRef]
18. Ningthoujam, D.S.; Tamreihao, K.; Mukherjee, S.; Khunjamayum, R.; Devi, L.J.; Asem, R.S. Keratinaceous Wastes and Their Valorization Through Keratinolytic Microorganisms. In *Keratin*; Blumenberg, M., Ed.; IntechOpen: London, UK, 2018; Volume 1, pp. 129–148.
19. Wu, G. *Amino Acids: Biochemistry and Nutrition*, 1st ed.; CRC Press: Boca Raton, FL, USA, 2013.
20. Papadopoulis, M.C. Effect of processing on high-protein feedstuffs: a review. *Biol. Wastes* **1989**, *29*, 123–138. [CrossRef]
21. Latshaw, J.D.; Musharaf, N.; Retrum, R. Processing of feather meal to maximize its nutritional value for poultry. *Animal Feed Sci. Technol.* **1994**, *47*, 179–188. [CrossRef]
22. Schwass, D.E.; Finley, J.W. Heat and alkaline damage to proteins: Racemization and lysinoalanine formation. *J. Ag. Food Chem.* **1984**, *32*, 1377–1382. [CrossRef]
23. Liardon, R.; Hurrell, R.F. Amino acid racemization in heated and alkali-treated proteins. *J. Ag. Food Chem.* **1983**, *31*, 432–437. [CrossRef]
24. Liardon, R.; Ledermann, S. Racemization kinetics of free and protein-bound amino acids under moderate alkaline treatment. *J. Agric. Food Chem.* **1986**, *34*, 557–565. [CrossRef]
25. Vasconcelos, A.; Freddi, G.; Cavaco-Paulo, A. Biodegradable materials based on silk fibroin and keratin. *Biomacromolecules* **2008**, *9*, 1299–1305. [CrossRef] [PubMed]
26. Yamauchi, K.; Yamauchi, A.; Kusonoki, T.; Kohda, A.; Konishi, Y. Preparation of stable aqueous solution of keratins, and physiochemical and degradational properties of films. *J. Biomed. Mater. Res. Part Res.* **1996**, *31*, 439–444. [CrossRef]
27. Wang, S.; Taraballi, F.; Tan, L.P.; Ng, K.W. Human keratin hydrogels support fibroblast attachment and proliferation in vitro. *Cell Tissue Res.* **2012**, *347*, 795–802. [CrossRef]
28. Tonin, C.; Aluigi, A.; Vineis, C.; Varesano, A.; Montarsolo, A.; Ferrero, F. Thermal and structural characterizationof poly (ethylene-oxide)/keratin blend films. *J. Therm. Anal. Calorim.* **2007**, *89*, 601–608. [CrossRef]
29. Wang, K.; Li, R.; Ma, J.H.; Jian, Y.K.; Che, J.N. Extracting keratin from wool using cysteine. *Green Chem.* **2016**, *18*, 476–481. [CrossRef]

30. Torchinski, Y.M. *Sulfur in Proteins*; Pergamon Press: Oxford, UK, 1981.
31. Cardamone, J.M. Investigating the microstructure of keratin extracted from wool: Peptide sequence (MALDI-TOF/TOF) and protein conformation (FTIR). *J. Mol. Struct.* **2010**, *969*, 97–105. [CrossRef]
32. De Guzman, R.C.; Merrill, M.R.; Richter, J.R.; Hamzi, R.I.; Greengauz-Roberts, O.K.; Van Dyke, M.E. Mechanical and biological properties of keratose biomaterials. *Biomaterials* **2011**, *32*, 8205–8217. [CrossRef]
33. Cardamone, J.M.; Nunez, A.; Garcia, R.A.; Aldema-Ramos, M. Characterizing wool keratin. *Lett. Mater. Sci.* **2009**, *2009*, 147175. [CrossRef]
34. Kelly, R.J.; Worth, G.H.; Roddick-Lanzilotta, A.D.; Rankin, D.A.; Ellis, G.D.; Mesman, P.J.R.; Summers, C.G.; Singleton, D.J. The Production of Soluble Keratin Derivatives. U.S. 7148327B2, 2001.
35. Cornfield, M.C.; Robson, A. The amino acid composition of wool. *Biochem. J.* **1955**, *59*, 62–68.
36. Mellet, P. The influence of alkali treatment on native and denatured proteins. *Tex. Res. J.* **1968**, *38*, 977–983. [CrossRef]
37. Masters, P.M.; Friedman, M. Racemization of amino acids in alkali-treated food proteins. *J. Agric. Food Chem.* **1979**, *27*, 507–511. [CrossRef]
38. Provansal, M.M.P.; Cug, J.L.A.; Cheftal, J.C. Chemical and nutritional modifications of sunfower proteins due to alkaline processing. Formation of amino acid cross-links and isomerization of lysine residues. *J. Agric. Food Chem.* **1975**, *23*, 938–943. [CrossRef]
39. Gaidau, C.; Epure, D.-G.; Enascuta, C.E.; Carsote, C.; Sendrea, C.; Proietti, N.C.W.; Gu, H. Wool keratin total solubilisation for recovery and reintegration—An ecological approach. *J. Clean. Prod.* **2019**, *236*, 117586. [CrossRef]
40. Koleva, M.; Danalev, D.; Ivanova, D.; Vezenkov, L.; Vassiliev, N. Synthesis of two peptide mimetics as markers for chemical changes of wool's keratin during skin unhairing process and comparison the wool quality obtaianed by ecological methods for skin unhairing. *Bulg. Chem. Commun.* **2009**, *41*, 160–164.
41. Banga, A.K. *Therapeutic Peptides and Proteins: Formulation, Process and Development Systems*; CRC Press: Boca Raton, FL, USA, 1997.
42. Murray, K.; Rasmussen, P.; Neustaedter, J.; Luck, J.M. The hydrolysis of arginine. *J. Biol. Chem.* **1965**, *240*, 705–709.
43. Daft, F.S.; Coghill, R.D. The alkaline decomposition of serine. *J. Biol. Chem.* **1931**, *90*, 341–350.
44. Hunt, S. The non-protein amino acids. In *Chemistry and Biochemistry of the Amino Acids*; Barrett, G.C., Ed.; Springer: Dordrecht, Germany, 1985.
45. Stapleton, I.; Swan, J. Amino acids and peptides. VI. Studies on cystine and $\alpha\alpha'$-dimethylcystine in relation to the alkaline degradation of protein disulphides. *Aust. J. Chem.* **1960**, *13*, 416–425. [CrossRef]
46. Ohta, T.; Suzuki, S.; Todo, M.; Kurechi, T. The decomposition of tryptophan in acid solutions: specific effect of hydrochloric acid. *Chem. Pharm. Bull.* **1981**, *29*, 1767–1771. [CrossRef]
47. Tung, W.S.; Daoud, W.A. Photocatalytic self-cleaning keratins: A feasibility study. *Acta Biomater.* **2009**, *5*, 50–56. [CrossRef]
48. Kurbanoglu, E.B.; Kurbanoglu, N.I. A new process for the utilization of ram horn waste. *J. Biosci. Bioeng.* **2002**, *94*, 202–206. [CrossRef]
49. Zhang, J.; Li, Y.; Li, J.; Zhao, Z.; Liu, X.; Li, Z.; Han, Y.; Hu, J. Isolation and characterization of biofunctional keratin particles extracted from wool waste. *Powder Technol.* **2013**, *246*, 356–362. [CrossRef]
50. Kurbanoglu, E.B.; Kurbanoglu, N.I. Ram horn hydrolysate as enhancer of xanthan production in batch culture of *Xanthomonas campestris* EBK-4 isolate. *Process Biochem.* **2007**, *42*, 1146–1149. [CrossRef]
51. Brandelli, A. Bacterial keratinases: useful enzymes for bioprocessing agroindustrial wastes and beyond. *Food Bioproc. Technol.* **2008**, *1*, 105–116. [CrossRef]
52. Sharma, R.; Devi, S. Versatility and commercial status of microbial keratinases: A review. *Rev. Environ. Sci. Biotechnol.* **2018**, *17*, 19–45. [CrossRef]
53. Muhsin, T.H.; Aubaid, A.H. Partial purification and some biochemical characteristics of exocellular keratinase from *Trichophyton mentagrophytes* var erinacei. *Mycopathologia* **2001**, *150*, 121–125. [CrossRef]
54. Rai, S.K.; Mukherjee, A.K. Optimization of the production of an oxidant and detergent-stable alkaline beta-keratinase from *Brevibacillus* sp. strain AS-S10-11: application of enzyme in laundry detergent formulations and in the leather industry. *Biochem. Eng. J.* **2011**, *54*, 47–56. [CrossRef]

55. Thankaswamy, S.R.; Sundaramoorthy, S.; Palanivel, S.; Ramudu, K.N. Improved microbial degradation of animal hair waste from leather industry using *Brevibacterium luteolum* (MTCC 5982). *J. Clean. Prod.* **2018**, *189*, 701–708. [CrossRef]
56. Nam, G.-W.; Lee, D.-W.; Lee, H.-S.; Lee, N.-J.; Kim, B.-C.; Choe, E.-A.; Hwang, J.-K.; Suhartono, M.T.; Pyun, Y.-R. Native feather-degradation by *Fervidobacterium islandicum* AW-1, a newly isolated keratinase-producing thermophilic anaerobe. *Arch. Microbiol.* **2002**, *178*, 538–547. [CrossRef]
57. Kshteri, P.; Roy, S.S.; Sharma, S.K.; Singh, T.S.; Ansari, M.A.; Sailo, B.; Singh, S. Feather degrading, phytostimulating, and biocontrol potential of actinobacteria from North Easatern Himalayan Region. *J. Basic Microbiol.* **2018**, *58*, 730–738. [CrossRef]
58. Mamangkey, J.; Suryanto, D.; Munir, E.; Mustopa, A.Z. Isolation, molecular identification and verification of gene encoding bacterial keratinase from crocodile (crocodylus porosus) feces. *IOP Conf. Ser. Earth Environ. Sci.* **2019**, *305*, 012085. [CrossRef]
59. Pernicova, I.; Enev, V.; Marova, I.; Obruca, S. Interconnection oof waste chicken feather biodegradation and keratinase and *mcl*-PHA prooduction employing *Pseudomonas putida* KT2440. *Appl. Food Biotechnol.* **2019**, *6*, 83–90.
60. Fuke, P.; Gujar, V.V.; Khardenavis, A. Genome annotation and vlidation of keratin hydrolyzing proteolytic enzymes from *Serratia marcescens* EGD-HP20. *Appl. Biochem. Biotechnol.* **2018**, *18*, 970–986. [CrossRef] [PubMed]
61. Fuke, P.; Pal, R.R.; Khardenavis, A.A.; Purohit, H. *In silico* characterization of broad range proteases produced by *Serratia marcescens* EGD-HP20. *J. Basic Microbiol.* **2018**, *58*, 492–500. [CrossRef]
62. Bhari, R.; Kaur, M.; Singh, R.S. Thermostable and halotolerant keratinae from *Bacillus aerius* NSMk2 with remarkable dehairing and laundry applications. *J. Basic Microbiol.* **2019**, *59*, 555–558. [CrossRef]
63. Pawar, V.A.; Prajapati, A.S.; Akhani, R.C.; Patel, D.H.; Subramanian, R.B. Molecular and biochemical characteriztion of a thermostable keratinase from *Bacillus altitudinis* RBDV1. *3 Biotech* **2018**, *8*, 107–113. [CrossRef]
64. Cao, S.; Li, D.; Ma, X.X.Q.; Song, J.; Lu, F.; Li, Y. A novel unhairing enzyme produced by heterolgous expression of keratinase gene (*kerT*) in *Bacillus subtilis*. *World J. Microbiol. Biotechnol.* **2019**, *35*, 122–131. [CrossRef]
65. Kumar, M.; Bhatia, D.; Khatak, S.; Kumar, R.; Sharma, A.; Malik, D.K. Optimization and purification of keratinase from *Bacillus anthracis* with dehairing application. *J. Pure Appl. Microbiol.* **2019**, *13*, 585–590. [CrossRef]
66. Kalaikumari, S.S.; Vennila, T.; Monika, V.; Chandraraj, K.; Gunasekaran, P. Bioutilization of poultry feather for keratinase production and its application in leather industry. *J. Clean. Prod.* **2019**, *208*, 44–45. [CrossRef]
67. Hamiche, S.; Mechri, S.; Khelouia, L.; Annane, R.; El Hattab, M.; Badis, A.; Jaouadi, B. Purification and biochemical characterization of two keratinases from *Bacillus amyloliquefaciens* S13 isolated from marine brown algae *Zonaria tournefortii* with potential keratin-biodegradation and hide-unhairing activities. *Int. J. Biol. Macromol.* **2019**, *122*, 758–769. [CrossRef]
68. Devi, C.S.; Shankar, R.S.; Kumar, S.; Mohanasrinivasan, B. Production of keratinase from a newly isolates feather degrading *Bacillus cereus* VITSDVM4 from poultry waste. *Natl. Acad. Sci. Lett.* **2018**, *41*, 307–311. [CrossRef]
69. Arokiyaraj, S.; Varghese, R.; Ahmed, B.A.; Duraipandiyan, V.; Al-Dhabi, N.A. Optimizing the fermentation conditions and enhanced production of keratinase from *Bacillus cereus* isolated from halophilic environment. *Saudi J. Biol. Sci.* **2019**, *26*, 378–381. [CrossRef] [PubMed]
70. Abdel-Fattah, A.M.; El-Gamal, M.S.; Ismail, S.A.; Emran, M.A. Biodegradation of feather waste by keratinase produced from newly isolated *Bacillus licheniformis* ALW1. *JGEB* **2018**, *16*, 311–318. [CrossRef]
71. Cavello, I.; Urbieta, M.S.; Segretin, A.B.; Giaveno, A.; Cavalitto, S.; Donati, E.R. Assessment of keratinase and other hydrolytc enzymes in thermophilic bacteria isolated from geothermal areas in Patagonia Argentina. *Geomicrobiol. J.* **2018**, *35*, 156–165. [CrossRef]
72. Adetunji, C.O.; Adejumo, I.O. Efficacy of crude and immobilized enzymes from *Bacillus licheniformis* for production of biodegraded feather meal and their assessment. *Environ. Technol. Inno.* **2018**, *11*, 116–124. [CrossRef]

73. Hashem, A.M.; Abdel-Fattah, A.M.; Ismail, S.A.; El-Gamal, M.S.; Esawy, M.A.; Emran, M.A. Optimization, characterization and thermodynamic studies on *B. licheniformis* ALW1 keratinase. *Egypt J. Chem.* **2018**, *61*, 591–607. [CrossRef]
74. He, Z.; Sun, R.; Tang, Z.; Bu, T.; Wu, Q.; Li, C.; Chen, H. Biodegradation of feather waste keratin by keratin-degrading strain *Bacillus subtilis* 8. *J. Microbiol. Biotechnol.* **2018**, *28*, 314–322. [CrossRef]
75. Oluwaseun, A.C.; Phazang, P.; Sarin, N.B. Production of ecofriendly biofertilizers produced from crude and immobilized enzymes from *Bacillus subtilis* CH008 and their effect on the growth of *Solanum lycopersicum*. *Plant Arch.* **2018**, *18*, 1455–1462.
76. Nagarajan, S.; Eswaran, P.; Masilamani, R.P.; Natarajan, H. Chicken feather compost to promote the plant growth activity by using keratinolytic bacteria. *Waste Biomass Valor.* **2018**, *9*, 531–538. [CrossRef]
77. Imtiaz, A.; Rehman, A. *Bacillus subtilis* BML5 isolated from soil contaminated with poultry waste has keratinolytic activity. *Pakistan J. Zool.* **2018**, *50*, 143–148. [CrossRef]
78. De Paiva, D.P.; De Oliveira, S.S.A.; Mazotto, A.M.; Vermehlo, A.B.; De Oliveira, S.S. Keratinolytic activity of *Bacillus subtilis* LFB-FIOCRUZ 1266 enhanced by whole-cell mutagenesis. *3 Biotech* **2019**, *9*, 2–13. [CrossRef] [PubMed]
79. Sharma, S.; Prasad, R.K.; Chatterjee, S.; Sharma, A.; Variable, M.G.; Yadav, K.K. Characterization of Bacillus species with keratinase and cellulase properties isolated from feather dumping and cockroach gut. *Proc. Natl. Acad. Sci. India. Sect. B Biol. Sci* **2019**, *89*, 1079–1086. [CrossRef]
80. Suharti, D.R.T.; Nilamsari, N.R. Isolation and characterization of newly keratinase producing *Bacillus* sp. N1 from tofu liquid waste. *IOP Conf. Ser. Earth Environ. Sci.* **2019**, *230*, 012088. [CrossRef]
81. Jin, M.; Chen, C.; He, X.; Zeng, R. Characterization of an extreme alkaline-stable keratinase from the draft genome of feather-degrading *Bacillus* sp. JM7 from deep-sea. *Acta Oceanol. Sin.* **2019**, *38*, 87–95. [CrossRef]
82. Tian, J.; Xu, Z.; Long, X.; Tian, Y.; Shi, B. High-expression keratinase by *Bacillus subtilis* SCK6 for enzymatic dehairing of goatskins. *Int. J. Biol. Macromol.* **2019**, *135*, 119–126. [CrossRef]
83. Koentjoro, M.P.; Prasetyo, E.N.; Rahmatullah, A.M. Optimization of keratinase production by Bacillus SLII-I bacteria in chicken feather waste medium. *ARPN J. Engin. Appl. Sci.* **2018**, *13*, 482–488.
84. Nurkhasanah, U.; Suharti. Preliminary study on keratinase fermentation by *Bacillus* sp. MD24 under solid state fermentation. *IOP Conf. Ser. Earth Environ. Sci.* **2019**, *276*, 012016. [CrossRef]
85. Suharti, S.; Riesmi, M.T.; Hidayati, A.; Zuhriyah, U.F.; Wonorahardjo, S.; Susanti, E. Enzymatic dehairing of goat skin using keratinase from *Bacillus* sp. MD24, a newly isolated soil bacterium. *Pertanika J. Trop. Agric. Sci.* **2018**, *41*, 1449–1461.
86. Ashokkumar, M.; Irudayaraj, G.; Yellapu, N.; Manonmani, A.M. Molecular characterization of *bmyC* gene of the mosquito pupicidal bacterial, *Bacillus amyloliquefaciens* (VCRC B483) and in silico analysis of bacillomycin D snthetase C protein. *World J. Microbiol. Biotechnol.* **2018**, *34*, 116–126. [CrossRef]
87. Yusuf, I.; Ahmad, S.A.; Pang, L.Y.; Yasid, N.A.; Shukor, M.Y. Effective production of keratinase by gellum gum-immobilised *Alcaligenes* sp. AQ05-001 using heavy metal-free and polluted feather wastes as substrates. *3 Biotech* **2019**, *9*, 32–43. [CrossRef]
88. Sultana, N.; Saha, P. Studies on potential application of crude keratinase enzyme from *Stenotrophomonas* sp. for dehairing in leather processing industry. *J. Environ. Biol.* **2018**, *39*, 324–330. [CrossRef]
89. Kang, D.; Herschend, J.; Al-Soud, W.A.; Mortensen, M.S.; Gonzalo, M.; Jacquiod, S.; Sorensen, S.J. Enrichment and characterization of an environmental microbial consortium displaying efficient keratinolytic acitivty. *Bioresour. Technol.* **2018**, *270*, 303–310. [CrossRef] [PubMed]
90. Mercer, D.K.; Stewart, C.S. Keratin hydrolysis by dermophytes. *Med. Mycol. J.* **2019**, *57*, 13–22. [CrossRef]
91. Pillai, P.; Mandge, S.; Archana, G. Statistical optimization of production and tannery application of a keratinolytic serine protease from *Bacillus subtilis* P13. *Process Biochem.* **2011**, *46*, 1110–1117. [CrossRef]
92. Purchase, D. *Microbial Keratinases: Characteristics, Biotechnological Applications and Potential*; CABI: Wallingford, UK, 2016.
93. Gupta, R.; Sharma, R.; Beg, Q.K. Revisiting microbial keratinases: Next generation proteases for sustainable biotechnology. *Crit. Revs. Biotechnol.* **2013**, *33*, 216–228. [CrossRef]
94. Elhoul, M.B.; Jaouadi, N.Z.; Rekik, H.; Benmrad, M.O.; Mechri, S.; Moujehed, E.; Kourdali, S.; Hattab, M.E.; Badis, A.; Bejar, S.; et al. Biochemical and molecular characterization of new keratinoytic protease from *Actinomadura viridilutea* DZ50. *Int. J. Biol. Macromol.* **2016**, *92*, 299–315. [CrossRef] [PubMed]

95. Habbeche, A.; Saoudi, B.; Jaouadi, B.; Haberra, S.; Kerouaz, B.; Boudelaa, M.; Badis, A.; Ladjama, A. Purification and biochemical characterization of a detergent-stable keratinase from a newly thermophilic actinomycete *Actinomadura keratinilytica* strain Cpt29 isolated from poultry compost. *J. Biosci. Bioeng.* **2014**, *117*, 413–421. [CrossRef]
96. Fakhfakh, N.; Kanoun, S.; Manni, L.; Nasri, M. Production and biochemical and molecular characterization of a keratinolytic serine protease from chicken feather-degrading *Bacillus licheniformis* RPk. *Can. J. Microbiol.* **2009**, *55*, 427–436. [CrossRef]
97. Fakhfakh-Zouari, N.; Hmidet, N.; Haddar, A.; Kanoun, S.; Nasri, M. A novel serine metalloprotease from a newly isolated *Bacillus pumilus* A1 grown on chicken feather meal: biochemical and molecular characterization. *Appl. Biochem. Biotechnol.* **2010**, *162*, 329–344. [CrossRef]
98. Wang, L.; Qian, Y.; Cao, Y.; Huang, Y.; Chang, Z.; Huang, H. Production and characterization of keratinolytic proteases by a chicken feather-degrading thermophilic strain, *Thermoactinomyces* sp. YT06. *J. Microbiol. Biotechnol.* **2017**, *27*, 2190–2198. [CrossRef]
99. Li, J.; Shi, P.-J.; Han, X.-Y.; Meng, K.; Yang, P.-L.; Wang, Y.-R.; Luo, H.-Y.; Wu, N.-F.; Yao, B.; Fan, Y.-L. Functional expression of the keratinolytic serine protease gene *sfp2* from *Streptomyces fradiae* var. k11 in *Pichia pastoris*. *Protein Expres. Purif.* **2007**, *54*, 79–86. [CrossRef] [PubMed]
100. Benkiar, A.; Nadia, Z.J.; Badis, A.; Rebzani, F.; Soraya, B.T.; Rekik, H.; Naili, B.; Ferradji, F.Z.; Bejar, S.; Jaouadi, B. Biochemical and molecular characterization of a thermo- and detergent-stable alkaline serin keratinolytic protease from *Bacillus circulans* strain DZ100 for detergent formulations and feather-biodegradation process. *Int. Biodeterior. Biodegradation* **2013**, *83*, 129–138. [CrossRef]
101. Fang, Z.; Zhang, J.; Liu, B.; Jiang, L.; Du, G.; Chen, J. Cloning, heterologous expression and characterization of two keratinases from *Stenotrophomonas maltophilia* BBE11-1. *Proc. Biochem.* **2014**, *49*, 647–654. [CrossRef]
102. Jaouadi, B.; Ellouz-Chaabouni, S.; Rhimi, M.; Bejar, S. Biochemical and molecular characterization of a detergent-stable serine alkaline protease from Bacillus pumilus CBS with high catalytic efficiency. *Biochimie* **2008**, *90*, 1291–1305. [CrossRef]
103. Rajput, R.; Sharma, R.; Gupta, R. Cloning and characterization of a thermostable detergent-compatible recombinant keratinase from *Bacillus pumilus* KS12. *IUBMB* **2011**, *58*, 109–118.
104. Jaouadi, N.Z.; Rekik, H.; Elhoul, M.B.; Rahem, F.Z.; Hila, C.G.; Aicha, H.S.B.; Badis, A.; Toumi, A.; Bejar, S.; Jaouadi, B. A novel keratinase from Bacillus tequilensis strain Q7 with promising potential for the leather bating process. *Int. J. Biol. Macromol.* **2015**, *79*, 952–964. [CrossRef]
105. Ghosh, A.; Chakrabarti, K.; Chattopadhyay, D. Cloning of feather-degrading minor extracellular protease from Bacillus cereus DCUW: dissection of the structural domains. *Microbiology* **2009**, *155*, 2049–2057. [CrossRef]
106. Wang, L.; Cheng, G.; Ren, Y.; Dai, Z.; Zhao, Z.-S.; Liu, F.; Li, S.; Wei, Y.; Xiong, J.; Tang, X.-F.; et al. Degradation of intact chicken feathers by *Thermoactinomyces* sp. CDF and characterization of its keratinolytic protease. *Appl Microbiol. Biotechnol.* **2015**, *99*, 3949–3959. [CrossRef]
107. Wu, W.-L.; Chen, M.-Y.; Tu, I.-F.; Lin, Y.-C.; Kumar, N.E.; Chen, M.-Y.; Ho, M.-C.; Wu, S.-H. The discovery of novel heat-stable keratinases from *Meiothermus taiwanensis* WR-220 and other extremophiles. *Sci. Rep.* **2017**, *7*, 4658–4669. [CrossRef]
108. Liang, X.; Bian, Y.; Tang, X.-F.; Xiao, G.; Tang, B. Enhancement of keratinolytic activity of a thermophilic subtilase by improving its autolysis resistance and thermostability under reducing conditions. *Appl. Microbiol. Biotechnol.* **2010**, *87*, 999–1006. [CrossRef]
109. Gegeckas, A.; Gudiukaite, R.; Debski, J.; Citavicius, D. Keratinous waste decomposition and peptide production by keratinase from *Geobacillus stearothermophilus* AD-11. *Int. J. Biol. Macromol.* **2015**, *75*, 158–165. [CrossRef] [PubMed]
110. Sharma, R.; Murty, N.A.R.; Gupta, R. Molecular characterization of N-terminal pro-sequence of keratinase ker P from *Pseudomonas aeruginosa*: identification of region with chaperone activity. *Appl. Biochem. Biotechnol.* **2011**, *165*, 892–901. [CrossRef] [PubMed]
111. Mitsuiki, S.; Ichikawa, M.; Oka, T.; Sakai, M.; Moriyama, Y.; Sameshima, Y.; Goto, M.; Furukawa, K. Molecular characterization of a keratinolytic enzyme from an alkaliphilic *Nocardiopsis* sp. TOA-1. *Enzyme Microb. Technol.* **2004**, *34*, 482–489. [CrossRef]
112. Ghosh, A.; Chakrabarti, K. Degradtion of raw feather by a novel high molecular weight extracellular protease from newly isolated *Bacillus cereus* DCUW. *J. Ind. Microbiol. Biotechnol.* **2008**, *35*, 825–834. [CrossRef]

113. Tripathi, L.P.; Sowdhamini, R. Genome-wide survey of prokaryotic serine proteases: analysis of distribution and domain architectures of five serine protease families in prokaryotes. *BMC Genomics* **2008**, *9*, 549–576. [CrossRef]
114. Rawlings, N.D.; Waller, M.; Barrett, A.J.; Bateman, A. MEROPS: the database of proteolytic enzymes, their substrates and inhibitors. *Nucleic Acids Res.* **2014**, *42*, D503–D509. [CrossRef]
115. Siezen, R.J. Homology modelling and protein engineering strategy of subtilases, the family of subtilisin-like serine proteinases. *Protein Eng.* **1991**, *4*, 719–737. [CrossRef]
116. Page, M.J.; Di Cera, E. Serine peptidases: classification, structure and function. *Cell Mol. Life Sci.* **2008**, *65*, 1220–1236. [CrossRef]
117. Donlon, J.; Polaina, J.; MacCabe, A.P. (Eds.) *Industrial Enzymes*; Springer: Dordrecht, Germany, 2007.
118. Jongeneel, C.V.; Bouvier, J.; Bairoch, A. A unique signature identifies a family of zinc-dependent metallopeptidases. *FEBS Lett.* **1989**, *242*, 211–214. [CrossRef]
119. De Kreij, A.; Venema, G.; Van den Burg, B. Substrate specificity in the highly heterogeneous M4 peptidase family is determined by a small subset of amino acids. *JBC* **2000**, *275*, 31115–31120. [CrossRef]
120. Gregg, K.A. A comparison of genomic coding sequences for feather and scale keratins: Structural and evolutionary implications. *EMBO J.* **1984**, *3*, 175–178. [CrossRef] [PubMed]
121. Gradisar, H.; Friedrich, J.; Krizaj, I.; Jerala, R. Similarities and specificities of fungal keratinolytic proteases: comparison of keratinases of *Paecilomyces marquandii* and *Doratomyces microsporus* to some known proteases. *Appl. Environ. Microbiol.* **2005**, *71*, 3420–3426. [CrossRef]
122. Brandelli, A.; Daroit, D.J.; Riffel, A. Biochemical features of microbial keratinases and their production and applications. *Appl Microbiol. Biotechnol.* **2010**, *85*, 1735–1750. [CrossRef] [PubMed]
123. Di Cera, E. Serine proteases. *IUBMB Life* **2009**, *61*, 510–515. [CrossRef] [PubMed]
124. Wu, J.W.; Chen, X.L. Extracellular metalloproteases from bacteria. *Appl. Microbiol. Biotechnol.* **2011**, *92*, 253–262. [CrossRef]
125. Wainwright, M. A new method for determining the microbial degradation of keratin in soils. *Experientia* **1982**, *38*, 243–244. [CrossRef]
126. Riffel, A.; Lucas, F.; Heeb, P.; Brandelli, A. Characterization of a new keratinolytic bacterium that completely degrades native feather keratin. *Arch. Microbiol.* **2003**, *179*, 258–265. [CrossRef]
127. Navone, L.; Speight, R. Understanding the dynamics of keratin weakening and hydrolysis by proteases. *PLoS ONE* **2018**, *13*, e0202608. [CrossRef]
128. Farag, A.M.; Hassan, M.A. Purification, characterization and immobilization of a keratinase from *Aspergillus orizae*. *Enzyme Microb. Technol.* **2004**, *34*, 85–93. [CrossRef]
129. Grazziotin, A.; Pimentel, F.A.; De Jong, E.V.; Brandelli, A. Nutritional improvement of feather protein by treatment with microbial keratinase. *Animal Feed Sci. Technol.* **2006**, *126*, 135–144. [CrossRef]
130. Gill, S.C.; Von Hippel, P.H. Calculation of protein extinction coefficients from amino acid sequence data. *Anal. Biochem.* **1989**, *182*, 319–326. [CrossRef]
131. Daroit, D.J.; Corrêa, A.P.F.; Brandelli, A. Keratinolytic potential of a novel *Bacillus* sp. P45 isolated from the Amazon basin fish *Piaractus mesopotamicus*. *Int. Biodeter. Biodegr.* **2009**, *63*, 358–363. [CrossRef]
132. Pereira, J.Q.; Lopes, F.C.; Petry, M.V.; Da Costa Medina, L.F.; Brandelli, A. Isolation of three novel Antarctic psychrotolerant feather-degrading bacteria and partial purification of keratinolytic enzyme from Lysobacter sp. A03. *Int. Biodeter. Biodegr.* **2014**, *88*, 1–7. [CrossRef]
133. Wawrzkiewicz, K.; Wolski, T.; Lobarzewski, J. Screening the keratinoltic activity of dermatophytes in vitro. *Mycopathologia* **1991**, *114*, 1–8. [CrossRef] [PubMed]
134. Bressollier, P.; Letourneau, F.; Urdaci, M.; Verneuil, B. Purification and characterization of a keratinolytic serine proteinase from *Streptomyces albidoflavus*. *Appl. Environ. Microbiol.* **1999**, *65*, 2570–2576. [CrossRef] [PubMed]
135. Lin, S.J.; Yoshimura, E.; Sakai, H.; Wakagi, T.; Matsuzawa, H. Weakly bound calcium ions involved in the thermostability of aqualysin I, a heat-stable subtilisin-type protease of *Thermus aquaticus* YT-1. *BBA-Protein Struct. M.* **1999**, *133*, 132–138. [CrossRef]
136. Meng, K.; Li, J.; Cao, Y.; Shi, P.; Wu, B.; Han, X.; Bai, Y.; Wu, N.-F.; Yao, B. Gene cloning and heterologous expression of a serine protease from *Streptomyces fradiae* var.k11. *Can. J. Microbiol.* **2007**, *53*, 186–195. [CrossRef]

137. Öztürk, N.Ç.; Kazan, D.; Denizci, A.A.; Erarslan, A. The influence of copper on alkaline protease stability toward autolysis and thermal inactivation. *Eng. Life Sci.* **2012**, *12*, 662–671. [CrossRef]
138. Walker, J.M. *The Proteins Protocols Handbook*, 2nd ed.; Humana Press: Totowa, NJ, USA, 1996.
139. Fershi, A.R. *Structure and Mechanism in Protein Science: A Guide to Enzyme Catalysis and Protein Folding*; Freeman W. H. & Co.: New York, NY, USA, 1999.
140. Kunugi, S.; Tanaka, N. Cold denaturation of proteins under high pressure. *Biochim. Biophys. Acta* **2002**, *1595*, 309–314. [CrossRef]
141. Salvi, G.; De Los Rios, P.; Vendruscolo, M. Effective interactions between chaotropic agents and proteins. *Proteins Struct. Funct. Bioinf.* **2005**, *61*, 429–499. [CrossRef]
142. Suzuki, Y.; Tsujimoto, Y.; Matsui, H.; Watanabe, K. Decomposition of extremely hard-to-degrade animal proteins by thermophilic bacteria. *J. Biosci. Bioeng.* **2006**, *102*, 73–81. [CrossRef] [PubMed]
143. Macedo, A.J.; Beys da Silva, W.O.; Termignoni, C. Properties of a non collagen-degrading *Bacillus subtilis* keratinase. *Can. J. Microbiol.* **2008**, *54*, 180–188. [CrossRef] [PubMed]
144. Gupta, R.; Rajput, R.; Sharma, R.; Gupta, N. Biotechnological applications and prospective market of microbial keratinases. *Appl. Microbiol. Biotechnol.* **2013**, *97*, 9931–9940. [CrossRef] [PubMed]
145. Navone, L.; Speight, R.E. Enzyme systems for effective dag removal from cattle hides. *Anim. Prod. Sci.* **2019**, *59*, 1387–1398. [CrossRef]
146. Alba, A.C.; Strauch, T.A.; Keisler, D.H.; Wells, K.D.; Kesler, D.C. Using a keratinase to degrade chicken feathers for improved extraction of glucocorticoids. *Gen. Comp. Endocr.* **2019**, *270*, 35–40. [CrossRef]
147. Navone, L.; Moffitt, K.; Hansen, K.-A.; Blinco, J.; Payne, A.; Speight, R. Closing the textile loop: enzymatic fibre spearation and recycling of wool/polyester fabric blends. *J. Waste Manag.* **2020**, *102*, 149–160. [CrossRef]
148. El Boushy, A.R.Y.; Van der Poel, F.B. Poultry by-Products. In *Handbook of Poultry Feed from Waste*; Springer: Dordrecht, Germany, 2000; pp. 90–152. [CrossRef]
149. Sundaram, M.; Legadavi, R.; Banu, N.A.; Gayathri, V.; Palanisammy, A. A study of antibacterial activity of keratin nanoparticles from chicken feather waste against *Staphylococcus aureus* (Bovine mastitis bacteria) and its antioxidant activity. *Eur. J. Biotechnol. Biosci.* **2015**, *6*, 1–5.
150. Ohba, R.; Deguchi, T.; Kishikawa, M.; Arsyad, F.; Morimura, S.; Kida, K. Physiological function of enzymatic hydrolysates of collagen or keratin contained in livestock or fish waste. *Food Sci. Technol. Res.* **2003**, *9*, 91–93. [CrossRef]
151. Li, L.; Wang, W.; Shi, B.-H.; Li, J.-C.; Zhang, R.-Z.; Liu, S.-T.; Chen, R.-M.; Gao, W.-H.; Chen, G.-R.; Zheng, Y.-Q.; et al. *Anti-Inflammatory Activity of Antelope Horn Keratin and its Tryptic Hydrolysate*; Nutrition Press Inc.: Trumbull, CT, USA, 1999.
152. Kelly, R.; Ellis, G.; Macdonald, R.; McPherson, R.; Middlewood, P.; Nuthall, M.; Rao, G.-F.; Roddick-Lanzilotta, A.; Sigurjonsson, G.; Singleton, D. Keratin and Soluble Derivatives Thereof for a Nutraceutical and to Reduce Oxidative Stress and to Reduce Inflammation and to Promote Skin Health. U.S. Patent 0065506, 22 March 2007.
153. Kelly, R.J.; Ellis, G.D.; Macdonald, R.J.; McPherson, R.A.; Middlewood, P.G.; Nuthall, M.G.; Rao, G.-F.; Roddick-Lanzilotta, A.D.; Sigurjonsson, G.F.; Singleton, D.J. Nutraceutical Composition Comprising Soluble Keratin or Derivatives Thereof. U.S. Patent 7579317, 25 August 2009.
154. Cutler, P. *Protein Purification Protocols*; Humana Press: Totowa, NJ, USA, 2004.
155. Zeng, W.-C.; Zhang, W.-C.; Zhang, W.-H.; Shi, B. Antioxidant activity and characterization of bioactive polypeptides from bovine hair. *Funct. Polym.* **2013**, *73*, 573–578. [CrossRef]
156. Jones, L.N.; Sinclair, R.D.; Ecroyd, H.; Lui, Y.; Bennett, L.E. Bioprospecting keratinous material. *Int. J. Trichol.* **2010**, *2*, 47–49. [CrossRef]
157. Fontoura, R.; Daroit, D.J.; Correa, A.P.F.; Meira, S.M.M.; Mosquera, M.; Brandelli, A. Production of feather hydrolysates with antioxidant, angiotensin-1 converting enzyme- and dipeptidyl peptidase-IV-inhibitory activities. *New Biotechnol.* **2014**, *31*, 506–513. [CrossRef] [PubMed]
158. Yeo, I.; Lee, Y.-J.; Song, K.; Jin, H.-S.; Lee, J.-E.; Kim, D.; Lee, D.-W.; Kang, N.J. Low molecular weight keratins with anti-skin aging activity produced by anaerobic digestion of poultry feathers with Fervidobacterium islandicum AW-1. *J. Biotechnol.* **2018**, *271*, 17–25. [CrossRef]
159. Jin, H.-S.; Song, K.; Baek, J.-H.; Lee, J.-E.; Kim, D.J.; Nam, G.-W.; Kang, N.J.; Lee, D.-L. Identification of matrix metalloproteinasae-1-suppressive peptides in feather hydrolysates. *J. Agric. Food Chem.* **2018**, *66*, 12719–12729. [CrossRef] [PubMed]

160. Jin, H.-S.; Park, S.Y.; Kim, J.-Y.; Lee, J.-E.; Lee, H.-S.; Kang, N.J.; Lee, D.-W. Fluorescence-based quantification of bioactive keratin peptides from feathers for optimising large-scale anaerobic fermentation and purification. *Biotechnol. Bioproc. Eng.* **2019**, *2*, 240–249. [CrossRef]
161. Rouse, J.G.; Van Dyke, M.E. A review of keratin-based biomaterials for biomedical applications. *Materials* **2010**, *3*, 999–1014. [CrossRef]
162. Tachibana, A.; Furuta, Y.; Takesima, H.; Tanabe, T.; Yamauchi, K. Fabrication of wool keratin sponge scaffolds for long-term cell cultivation. *J. Biotechnol.* **2002**, *93*, 165–170. [CrossRef]
163. Tang, L.; Ollague, S.J.; Kelly, R.; Kirsner, R.S.; Li, J. Wool-derived keratin stimulates uman keratinocyte migration and types IV and VIIcollagen expression. *Exp. Dermatol.* **2012**, *21*, 458–460. [CrossRef]
164. Loan, F.; Marsh, C.; Cassidy, S.; Simcock, J. Keratin-based products for effective wound care management in superficial and partial thickeness burns injuries. *Burns* **2016**, *4*, 541–547. [CrossRef]
165. Sanchez Ramirez, D.O.; Carletto, R.A.; Tonetti, C.; Giachet, F.T.; Varesano, A.; Vineis, C. Wool keratin film plasticized by citric acid for food packaging. *Food Pack. Shelf Life* **2017**, *12*, 100–110. [CrossRef]

12

Hydrolysis of Glycosyl Thioimidates by Glycoside Hydrolase Requires Remote Activation for Efficient Activity

Laure Guillotin [1], Zeinab Assaf [1], Salvatore G. Pistorio [2], Pierre Lafite [1], Alexei V. Demchenko [2] and Richard Daniellou [1,*]

1 Institut de Chimie Organique et Analytique, ICOA, Université d'Orléans, CNRS, UMR 7311, BP6759 Rue de Chartres, F-45067 Orléans CEDEX 2, France; guillotin.laure@gmail.com (L.G.); assaf_zeinab@hotmail.fr (Z.A.); pierre.lafite@univ-orleans.fr (P.L.)

2 Department of Chemistry and Biochemistry, University of Missouri-St. Louis, One University Boulevard, St. Louis, MO 63121, USA; salvopistorio@hotmail.com (S.G.P.); demchenkoa@umsl.edu (A.V.D.)

* Correspondence: richard.daniellou@univ-orleans.fr

Abstract: Chemoenzymatic synthesis of glycosides relies on efficient glycosyl donor substrates able to react rapidly and efficiently, yet with increased stability towards chemical or enzymatic hydrolysis. In this context, glycosyl thioimidates have previously been used as efficient donors, in the case of hydrolysis or thioglycoligation. In both cases, the release of the thioimidoyl aglycone was remotely activated through a protonation driven by a carboxylic residue in the active site of the corresponding enzymes. A recombinant glucosidase (*Dt*Gly) from *Dictyoglomus themophilum*, previously used in biocatalysis, was also able to use such glycosyl thioimidates as substrates. Yet, enzymatic kinetic values analysis, coupled to mutagenesis and in silico modelling of *Dt*Gly/substrate complexes demonstrated that the release of the thioimidoyl moiety during catalysis is only driven by its leaving group ability, without the activation of a remote protonation. In the search of efficient glycosyl donors, glycosyl thioimidates are attractive and efficient. Their utility, however, is limited to enzymes able to promote leaving group release by remote activation.

Keywords: glycoside hydrolase; thioglycosides; biocatalysis

---

1. Introduction

Enzymes proved to be efficient synthetic tools for the eco-compatible synthesis of many classes of compounds. Non-organic solvents, mild experimental conditions and high regio- or stereo- specificity inherent to biocatalyzed reactions have increased the added value of enzymes in transformation processes, from the laboratory bench to the industrial scale [1]. Moreover, genetic modifications of recombinant enzymes are now powerful tools to easily alter versatility and properties of the engineered proteins. Rational mutagenesis, directed evolution, or even de novo design have dramatically broadened the applicability of enzymes in biocatalysis [2].

In the glycochemistry field, a vast array of carbohydrate-metabolizing enzymes (CAZYmes), including glycoside hydrolases (GH) or glycosyltransferases (GT), has been engineered and used for the chemo-enzymatic synthesis of glycosides [3]. The corresponding methodologies have proved useful in numerous applications ranging from glycosylated natural products to pharmaceuticals [4,5]. However, only few examples in the literature have been describing the use of CAZYmes for the preparation of synthetic thioglycosides that exhibit a sulphur atom linking the glycone and aglycone counterparts instead of more conventional oxygen or nitrogen atoms [6]. Interestingly, when compared to the corresponding *O*-glycosides, *S*-glycosides are highly stable towards enzymatic and acidic hydrolyses.

As a result, thioglycosides have been used as substrate analogues or inhibitors of *O*-GH involved in many diseases including cancer, lysosomal storage disorder, viral and bacterial infections [7,8].

Activated glycosyl donors have been used for a long time, especially in chemoenzymatic synthesis of oligosaccharides [9–11]. In retaining GH, where the stereochemistry of the anomeric carbon is conserved, these activated donors are of high interest because they enable the formation of the glycosyl-enzyme intermediate through the release of the leaving group (Figure 1). This first step is common to all enzymatic activities (hydrolase [12], transglycosidase [13], halogenase [12] and thioligase [14]) because the final outcome of the reaction only depends on the nature of the nucleophile that will attack the glycosyl-enzyme intermediate in the second step. Depending on the reaction and the substrate employed, this step can be rate-determining.

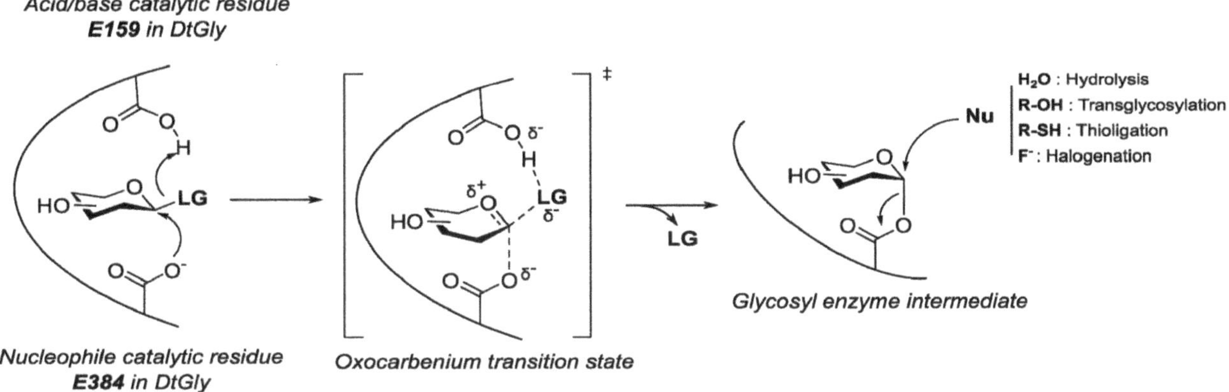

Figure 1. Schematic mechanism of the first step involving the glycosyl-enzyme intermediate formation in retaining GH. The leaving group (LG) release can also be catalysed through another catalytic residue according to its nature. Depending on the nucleophile (Nu) attacking the intermediate, three reactions can take place—hydrolysis, transglycosylation or thioligation.

In addition to the well characterized *O*-glycosides bearing a potent leaving group, some activated S-glycosides have been reported as efficient substrates for thioligases [15] or glycoside hydrolases [16]. This latter hydrolytic activity is peculiar as very few examples of *S*-glycosides hydrolysis by glycoside hydrolases have been reported in literature [16–26]. Among those examples, putting aside Glc*N*Acase, GH4 and myrosinase that do not operate through the canonical GH mechanism, only almond β-glucosidase GH1, *Aspergillus niger* GH3 [16,22,27], *Micromonospora viridifaciens* sialidase [21], *Caldocellum saccharolyticum* glucosidase [24] and *Oryza sativa* Os4BGlu12 [23] have been isolated and identified as thioglycoside hydrolases (Table 1).

Table 1. Comparison of *S*- and *O*-glycoside hydrolysis by GH.

Enzyme	Organism	Substrates Tested	Relative Activity *S*-vs. *O*-(%)	Ref
β-D-Glucosidase	Sweet almond	pNPSGlc	0.13 [a]	[22]
		pNPSGal	0.07 [a]	
		pNPSFuc	0.06 [a]	
		Glc*S*Biz	80 [a]	[16]
		Glc*S*(*N*-Me)Biz	10 [a]	
		Glc*S*Box	5 [a]	
β-D-Glucosidase	*A. niger*	Glc*S*Biz	5 [a]	[16]
		Glc*S*(*N*-Me)Biz	1 [a]	
Sialidase	*M. viridifaciens*	Substituted pNPSNeuAc	0.01–60	[21]
Os4BGlu12	*O. sativa*	pNPSGlc	0.5 [a]	[23]
		OctylSGlc	0.1 [b]	

[a] ratio of k_{cat}/K_M for thioglycoside substrate vs. corresponding *para*-nitrophenyl glycoside. [b] ratio of k_{cat}/K_M for octyl *S*-glucoside vs. octyl *O*-glucoside.

In most cases, *S*-substrate hydrolysis is much less efficient than the rate observed for the corresponding *O*-substrate. Indeed, thioglycosides are less efficient substrates because no general acid/base catalysis is available [28]. Yet, a new class of reactive thioglucosides (Figure 2, Table 1) bearing a thioimidoyl moiety was reported, which were efficiently hydrolysed by almond GH1, as well as *A. niger* GH3 [16]. In both cases, the authors demonstrated that benzoxazolyl 1-thio-β-D-glucopyranoside (Glc*S*Box) and benzimidazolyl 1-thio-β-D-glucopyranoside (Glc*S*Biz) hydrolyses were catalysed by remote activation of the C-S bond through protonation of the ring nitrogen in the aglycone. Such remote activation was also described in the case of Araf51 [15], which was able to use similar arabinofuranosyl thioimidates as glycosyl donors in thioglycoligation reaction [29–35]. In the context of chemoenzymatic synthesis of glycosides, these substrates are attractive because of their high stability towards chemical hydrolysis in aqueous solutions, as well as efficient leaving group ability [15].

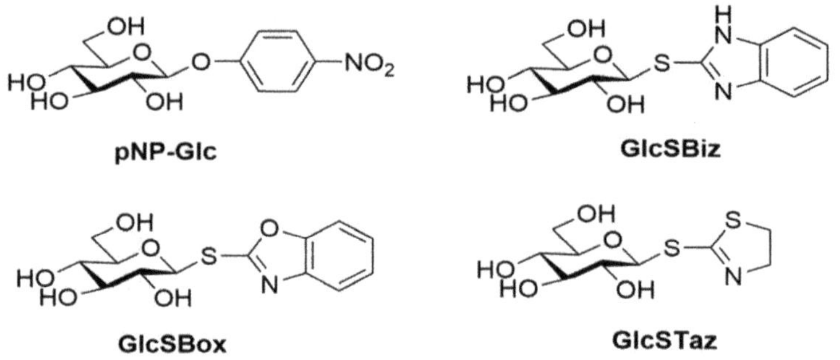

pNP-Glc

GlcSBiz

GlcSBox

GlcSTaz

Figure 2. Substrates used in this study.

In this work, we demonstrated that *Dt*Gly, a GH previously used in chemoenzymatic synthesis of *O*-glycosides, was able to hydrolyse these glycosyl thioimidates. Combined in vitro enzymatic analysis with in silico modelling of the enzyme-substrate interaction have helped us to decipher the molecular mechanism of this rare hydrolysis.

2. Results

2.1. DtGly Can Hydrolyze Thioglycosides

*Dt*Gly (uniprot B5YCI2_DICT6) is encoded by *dicth_0359* gene in the thermophile *Dictyoglomus thermophilum* genome. We have recently reported the cloning, expression and purification of this protein [36]. As many other *D. thermophilum* proteins [37–41], *Dt*Gly was found to be thermostable and also exhibited a wide substrate specificity, as it is able to hydrolyse *p*NP β-D-glucoside, *p*NP β-D-galactoside and *p*NP β-D-fucoside. Moreover, our previous study demonstrated that *Dt*Gly could be used in chemoenzymatic synthesis of glycosides, thereby serving as an attractive biocatalyst that needed to be assayed for other substrates [36].

In this context, we have focused on thioglycoside hydrolysis, as few examples of *S*-GH are available in literature. Three *S*-containing substrates were tested, namely Glc*S*Biz, Glc*S*Box and Glc*S*Taz that bear benzimidazolyl, benzoxazolyl and thiazolinyl aglycones, respectively (Figure 2).

Unlike *p*NP-Glc, wherein the hydrolysis can be easily monitored by quantification of the released *p*NP group, hydrolysis rates of the *S*-containing substrates were determined by quantification of the released glucose. This was achieved by monitoring *o*-dianisidine oxidation enzymatically coupled to glucose production [42]. This methodology applied to *p*NP-Glc hydrolysis gave similar kinetic values to those previously reported using *p*NP quantification (*data not shown*).

All three *S*-containing substrates were hydrolysed by *Dt*Gly (Table 2), with K_M values higher but in the same order of magnitude, as those observed for *p*NP-Glc (2- to 5-fold increase). However, the catalytic rate k_{cat} was decreased by one order of magnitude, indicating that the reaction is dramatically slowed in the case of *S*-containing substrates. Therefore, the catalytic efficiencies of

Glc*S*Biz, Glc*S*Box and Glc*S*Taz were found to be 20 to 40 times lower than the value determined for *p*NP-Glc.

Table 2. Kinetic parameters of WT and acid/base E159Q mutant of *Dt*Gly. pNP-Glc hydrolysis activity was measured by *p*NP release quantification. Other substrate hydrolysis activities were determined by quantification of the released glucose. All experiments were done in three independent replicates and are expressed as Mean ± SD.

Enzyme	Substrate	K_M (μM)	k_{cat} (s^{-1})	k_{cat}/K_M ($s^{-1}.mM^{-1}$)
WT	*p*NP-Glc [a]	460 ± 40	31 ± 0.7	67
WT	Glc*S*Biz	1533 ± 114	0.23 ± 0.01	0.15
WT	Glc*S*Box	2246 ± 289	0.38 ± 0.03	0.17
WT	Glc*S*Taz	880 ± 52	0.31 ± 0.01	0.35
E159Q	*p*NP-Glc	200 ± 20	0.20 ± 0.01	1.0
E159Q	Glc*S*Box	445 ± 40	0.06 ± 0.01	0.13

[a] Previously reported data [36].

Glc*S*Biz, Glc*S*Box and Glc*S*Taz have previously been used as substrates for sweet almond and *A. niger* β-glucosidases [16], yet with a much different behaviour. Glc*S*Biz was hydrolysed by this enzyme as efficiently as *p*NP-Glc. Kinetics analysis proved that Glc*S*Biz was efficiently hydrolysed by those glucosidases thanks to the remote protonation of the imidazole ring nitrogen. A much lower activity was observed for Glc*S*Box and no activity could be observed for Glc*S*Taz.

To better understand the chemistry underlying the thioglucoside hydrolysis by *Dt*Gly, we first investigated whether these substrates were efficiently binding in the active site, because low GH activities can arise from a second binding mode of substrates, as already reported [43]. Inhibition of *p*NP-Glc hydrolysis by Glc*S*Biz demonstrated that the latter is a competitor in the active site to *p*NP-Glc (Figure 3). Moreover, it efficiently binds into the active site, as an inhibitory constant K_i of 177 ± 11 μM was calculated from the inhibition curves.

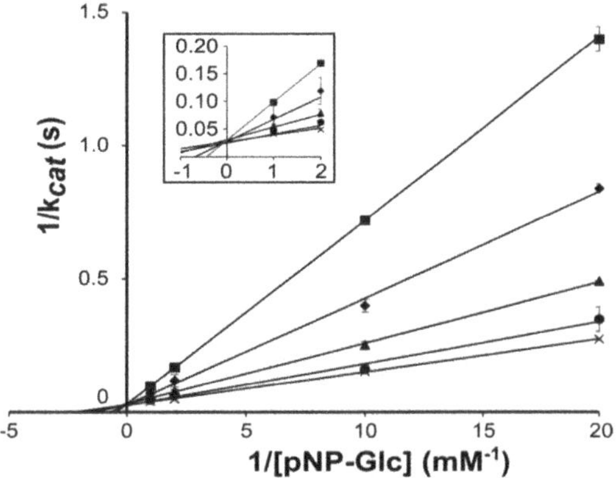

Figure 3. Lineweaver plot of wild-type *Dt*Gly inhibition with increasing concentrations of Glc*S*Biz. Data are expressed as mean ± SD from three independent experiments. Inhibitor concentrations are respectively depicted as crosses (0 μM), circles (100 μM), triangles (250 μM), diamonds (500 μM) and squares (1000 μM). Inset: 2X zoom on axes origin highlighting the intersection of fitted lines on *y*-axis.

2.2. Identification of Residues Surrounding the Thioglycoside Substrates in DtGly Active Site

Structural analysis of *Dt*Gly was carried out to identify potential residues that might be involved in *S*-containing substrate hydrolysis mechanism. Despite our efforts to crystallize *Dt*Gly, no diffracting crystal could be obtained, thus we decided to build a homology model of the enzyme. To do so, a 3D structure of β-glycosidase from *Pyrococcus horikoshii* was chosen because of its high sequence identity (resp. homology) with *Dt*Gly of 45% (resp. 63%) [44].

An initial model of residues 2-416 was built using a ModWeb server (ModPipe Protein Quality Score of 1.6, considered as reliable); missing residues were then added and the overall model was equilibrated by several cycles of energy minimization and molecular dynamics (Figure 4A).

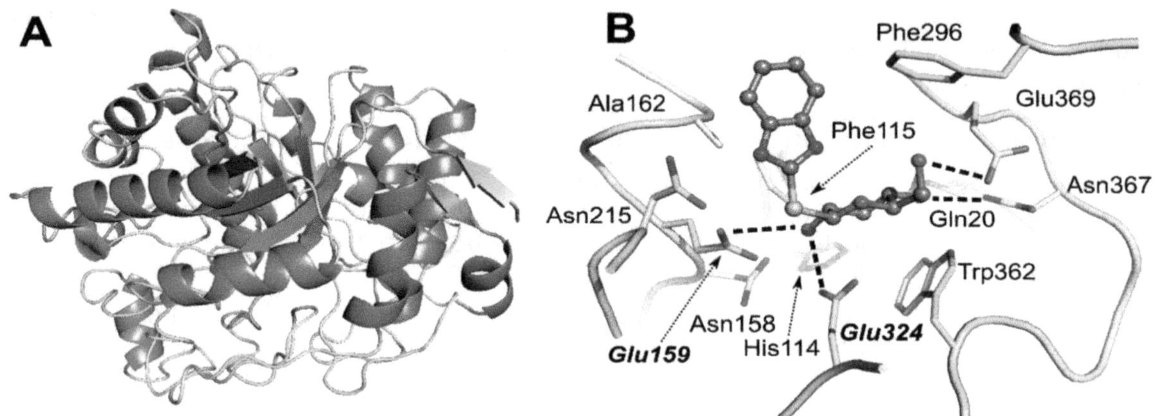

Figure 4. (**A**) Overall representation of *Dt*Gly model. Helices and sheets are respectively coloured in blue and orange. (**B**) Model of docked Glc*S*Biz in *Dt*Gly active site. Residues surrounding the ligand binding pocket are depicted as sticks. For clarity purposes, hydrogens are not represented. Catalytic residues Glu159 (acid/base) and Glu324 (nucleophile) are highlighted in bold. H-bonds are indicated as dashed lines.

In order to evaluate potential roles of active site residues in sulphur-containing substrate hydrolysis, modelling of substrate-bound *Dt*Gly were done by molecular docking. Using the conformation of glucosides in other closely related GH1 x-ray structures (β-glucosidase from *Thermotoga maritima* PDB 1OIM and 1OIF [45]), Glc*S*Biz, Glc*S*Box and Glc*S*Taz were independently docked into the *Dt*Gly active site. Figure 4B depicts the residues surrounding Glc*S*Biz, as well as the network of H-bonds between the sugar moiety and several polar residues (Gln20, Glu159, Glu324, Asn367, Glu369). An additional H-π interaction between glucose and Trp362 is also visible, as already seen for other GH [40]. The same interactions were found for other substrates or conformations (see the Supplementary Materials).

In the context of identification of potential residues involved in the *S*-glycoside activation during hydrolysis this model confirms that no acidic residue except Glu159 was close enough to remotely protonate aglycone moieties of substrates, as expected considering in vitro assays.

2.3. DtGly Hydrolysis of S-Glycosides Does Not Involve General Acid/Base Catalysis

In our model, the catalytic glutamate Glu159, which acts as the acid/base residue in retaining GH mechanism [46], is the only one close enough to activate thioglycoside hydrolysis. Although direct protonation on sulphur cannot occur in the case of thiogycosides [28], examples of distant protonation of the aglycone by a catalytic residue have been reported [15]. We have thus generated two mutants, namely Glu159Ala and Glu159Gln to assess the potential role of this residue in the thioglycoside hydrolysis. Unlike Glu159Gln that could be purified to homogeneity, Glu159Ala mutant was not soluble after cell lysis and thus could not be purified. This mutant was left aside for further experiments.

Glu159Gln mutation led to a dramatic decrease of catalytic efficiency for *p*NPGlc, as shown in Table 2. K_M values for this substrate are lower but in the same order of magnitude (200 μM vs. 460 μM), which can be explained by conservation of the active site structure in the mutant and decreased k_{cat} value by a factor of 150 because K_M is related to k_{cat}. This loss of hydrolytic activity upon acid/base mutation is usual, as reported in many other studies, especially those concerning thioligase generation [14,40].

When using Glc*S*Box as a substrate, *Dt*Gly Glu159Gln exhibits a reduced k_{cat} value (0.38 to 0.06 s^{-1}), as expected because nucleophile water attack is not activated by deprotonation. However, second order rate constant remains unchanged, indicating that the first step of the reaction is not compromised by the removal of the glutamate residue. Thus, the release of the thiol leaving group is not activated by Glu159 and is only dictated by its leaving capability (i.e., pKa).

3. Discussion

We have previously used *D. thermophilum* *Dt*Gly as a versatile tool for synthesis of glycosides and looked for alternate substrates for this enzyme. Thioimidate glycosyl donors have been used for a long time in organic synthesis to generate a wide range of glycosides and glycans [34,47–49]. In this context, we tested previously reported glycosyl thioimidates as substrates for almond GH1 and *A. niger* GH3 [16]. Examples of cloning and characterization of thioglycoside hydrolases are scarcely available in literature and even fewer studies on mechanism underlying the thioglycoside hydrolysis by GH have been published.

*Dt*Gly is able to hydrolyse *S*-glycosides, with lower activities than those observed for *O*-glycosides. This hydrolytic activity is rate-limited by release of the thiol-containing leaving group and not water nucleophilic attack, unlike generally accepted mechanism for the *O*-glycoside hydrolysis [19,28]. The modelling of substrate-*Dt*Gly complexes as well as mutagenesis of the acid/base residue also demonstrated that no residue was able to remotely protonate the benzimidazole group nor the sulphur atom. If *Dt*Gly is able to hydrolyse *S*-containing substrates without general acid catalysis, the hydrolysis rate is limited by the leaving group capability, as no remote activation is possible. The pKa of leaving groups 2-mercaptobenzimidazole (for Glc*S*Biz) and 2-mercaptobenzoxazole (for Glc*S*Box) have been experimentally determined at 5.8 [50] and 6.58 [51]. To our knowledge, no value is available for 2-mercaptothiazoline (for Glc*S*Taz).

In the case of almond and *A. niger* glycosidases, a remote protonation occurring on a nitrogen atom of the benzimidazole moiety of Glc*S*Biz was shown to accelerate the leaving group release, thus increasing the catalytic rate to a value close to those observed for *O*-glycosides. Another GH exhibiting thioglycosidase activity on 2′-thio-benzimidazolyl arabinosides activated by remote deprotonation, namely Ara*f*51 [15], was also reported. The modelling of Ara*f*51/substrate complex demonstrated that the nucleophile catalytic residue was responsible for the remote protonation on imidazole nitrogen, mostly because a furanosyl ring is much more flexible than a pyranosyl ring and allows the nucleophile residue to interact with the aglycone.

4. Materials and Methods

4.1. Materials

para-Nitrophenyl β-D-glucopyranoside (*p*NP-Glc) was purchased from Carbosynth (Oxford, UK). 2-benzoxazolyl 1-thio-β-D-glucopyranoside (Glc*S*Box) [52], 2-benzimidazolyl 1-thio-β-D-glucopyranoside (Glc*S*Biz) [16] and 2-thiazolinyl 1-thio-β-D-glucopyranoside (Glc*S*Taz) [49] were prepared as previously described. Otherwise specified, all other chemicals were purchased from Thermo Fisher Scientific (Waltham, MA, USA) and were of purest quality available. Mutagenic primers were purchased from Eurofins Genomics (Ebersberg, Germany) and WT *Dt*Gly coding expression plasmid (pET28a-*dtgly*) was prepared as previously described [36].

4.2. Production of WT and E159Q DtGly

pET28a-*dtgly* was used as a template for mutagenic PCR in the Quikchange Site-directed mutagenesis kit (Agilent, Les Ulis, France). Primers containing desired mutation on acid/base residue position (E159) were constituted of a pair of complement oligonucleotides designed using Agilent tools website (www.genomics.agilent.com, mutated codons are highlighted in bold): *Dt*Gly E159A: 5′-gaattactggatgactataaat**gcg**cccaatgcttatgccttt-3′ and *Dt*Gly E159Q: 5′-atcttgtgaattactggatgactataaat**cag**cccaatgcttatg-3′. Mutagenesis procedure was performed according to the kit procedure. Sequences of pET28a-*dtgly*E159A and pET28a-*dtgly*E159Q were verified by Sanger sequencing at Eurofins Genomics (Ebersberg, Germany).

Production and purification of *Dt*Gly variants was done as previously reported [36]. Briefly, *Escherishia coli* Rosetta(DE3) transformed with expression plasmids were grown in LB medium supplemented with chloramphenicol (34 μg/mL) and kanamycin (30 μg/mL) at 37 °C until OD_{600} reached 0.6. Induction was then done by addition of 1 mM IPTG and incubated overnight a 25 °C. Cells were harvested, lyzed by freeze-thaw cycles and sonication and supernatant was clarified by heat treatment for 15 min at 70 °C before centrifugation. Finally, supernatant was loaded on a Nickel column (HisPure, Thermo Scientific) and purified by elution with lysis buffer containing 500 mM imidazole.

4.3. *p*NP Release Quantification Assay

*Dt*Gly variants (WT or mutants) activity towards *p*NP-Glc hydrolysis was determined at 37 °C in a 200 μL incubation containing 5 ng of the enzyme, 0.01–10 mM *p*NP-Glc, Citrate-Phosphate buffer (20 mM, pH 6) and 0.1–1 mM GlcBox for inhibition studies. After 20 min of incubation, 100 μL of Na_2CO_3 1 M was added and released *p*NP was quantified at 405 nm (ε_{405} = 19,500 $cm^{-1}.M^{-1}$). Prism 4 (GraphPad) was used to fit data according to Michaelis-Menten model, or competitive inhibition model and retrieve kinetic parameters.

4.4. Glucose Release Assay

To determine Glc*S*Box, Glc*S*Taz and Glc*S*Biz hydrolysis rate by DtGly variants, produced glucose was quantified using a continuous coupled enzyme assay [42]. Incubations were similar as those for *p*NP-Glc hydrolysis with the addition of glucose oxidase from *Aspergillus niger* (Sigma-Aldrich, Saint Louis, MO, USA, 0.4 u), horseradish peroxidase (Sigma-Aldrich, 0.4 u) and *o*-dianisidine (Sigma-Aldrich, 100 μM). Dianisidine oxidation coupled to glucose production was monitored at 442 nm during 30 min. Prism 4 (GraphPad) was used to fit data according to Michaelis-Menten model and retrieve kinetic parameters.

4.5. Computational Studies

The structure of β-glycosidase from Pyrococcus horikoshii [44] (PDB 1VFF, 45%/63% sequence identity/homology) was used as a template for homology model building using a ModWeb server from the A. Sali Laboratory (https://modbase.compbio.ucsf.edu/modweb/). The resulting model was prepared with AmberTools [53] and equilibrated using NAMD software [54] and Amber fb15 force field [55] (3 cycles of 10,000 minimization steps and 0.5 ns dynamics at 100 K).

Docking of Glc*S*Box, Glc*S*Taz and Glc*S*Biz substrates into *Dt*Gly active site model was done by firstly applying AM1-BCC charges on ligands [56]. Then each substrate was placed 10 Å away, facing the active site (according to PDB 1VFF). *Dt*Gly-substrate complexes were formed using steered molecular dynamics [57] at 100 K using the structural alignment of glucose moiety in its binding pocket as the final orientation according to closest structures bearing a ligand in their active site (β-glucosidase from *Thermotoga maritima* PDB 1OIM, 1OIF and 1OIF) [45]. *Dt*Gly backbone was kept constrained during the whole procedure. Finally, protein-ligand complexes models were equilibrated by releasing substrate constraints and applying several cycles of energy minimization (10,000 steps, steepest descent) followed by molecular dynamics (100 K, 1 ns). Final complex models were obtained by a final energy minimization. For each substrate, several initial conformations were tested, mostly by rotation of the glycosidic bond. All structural figures were drawn using the PyMOL Molecular Graphics system 1.8 (www.pymol.org).

5. Conclusions

This study demonstrates that glycosyl thioimidates are not universal glycosyl donors for chemoenzymatic syntheses. While the above examples of efficient enzymatic activities using such substrates were reported in literature, they rely on an activation by protonation of the aglycone moiety, either with a distant carboxylic acid residue (almond GH1) or the catalytic nucleophile (Ara*f*51, Figure 5). *Dt*Gly seems to be the paradigm of the general case of an enzyme that can use those substrates without acid catalysis, yet with a much lower activity. This study paves the way for broadening *Dt*Gly applications in biocatalysis. Identification of efficient substrates and mutation into a thioligase are currently under further investigation.

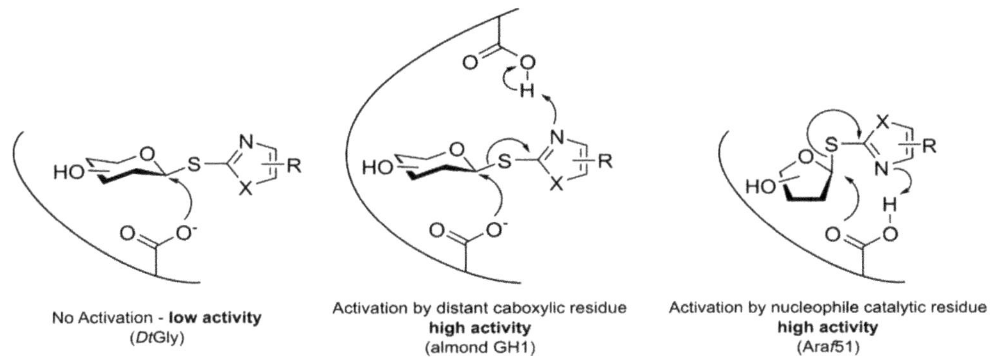

Figure 5. Glycosyl thioimidates require remote activation to promote the leaving group release.

Author Contributions: Conceptualization, R.D. and A.V.D.; Enzymatic studies, L.G., P.L. and Z.A.; data curation and modelling, P.L.; chemical synthesis of substrates, S.G.P. writing—original draft preparation, P.L.; writing—review and editing, P.L., R.D. and A.V.D.; project administration, R.D.; funding acquisition, R.D. and A.V.D.

References

1. Sheldon, R.A.; Woodley, J.M. Role of Biocatalysis in Sustainable Chemistry. *Chem. Rev.* **2018**, *118*, 801–838. [CrossRef] [PubMed]
2. Quin, M.B.; Schmidt-Dannert, C. Engineering of Biocatalysts: From Evolution to Creation. *ACS Catal.* **2011**, *1*, 1017–1021. [CrossRef] [PubMed]
3. Hancock, S.M.; Vaughan, M.D.; Withers, S.G. Engineering of glycosidases and glycosyltransferases. *Curr. Opin. Chem. Biol.* **2006**, *10*, 509–519. [CrossRef] [PubMed]
4. Li, T.-L.; Liu, Y.-C.; Lyu, S.-Y. Combining biocatalysis and chemoselective chemistries for glycopeptide antibiotics modification. *Curr. Opin. Chem. Biol.* **2012**, *16*, 170–178. [CrossRef] [PubMed]
5. Ati, J.; Lafite, P.; Daniellou, R. Enzymatic synthesis of glycosides: From natural O- and N-glycosides to rare C- and S-glycosides. *Beilstein J. Org. Chem.* **2017**, *13*, 1857–1865. [CrossRef]
6. Guillotin, L.; Lafite, P.; Daniellou, R. Chapter 10 Enzymatic thioglycosylation: Current knowledge and challenges. In *Carbohydrate Chemistry*; The Royal Society of Chemistry: London, UK, 2014; Volume 40, pp. 178–194. ISBN 978-1-84973-965-8.
7. Driguez, H. Thiooligosaccharides as Tools for Structural Biology. *Chembiochem* **2001**, *2*, 311–318. [CrossRef]
8. Wardrop, D.J.; Waidyarachchi, S.L. Synthesis and biological activity of naturally occurring α-glucosidase inhibitors. *Nat. Prod. Rep.* **2010**, *27*, 1431–1468. [CrossRef]
9. O'Neill, E.C.; Field, R.A. Enzymatic synthesis using glycoside phosphorylases. *Carbohydr. Res.* **2015**, *403*, 23–37. [CrossRef]
10. Lairson, L.L.; Henrissat, B.; Davies, G.J.; Withers, S.G. Glycosyltransferases: Structures, Functions and Mechanisms. *Annu. Rev. Biochem.* **2008**, *77*, 521–555. [CrossRef]
11. De Bruyn, F.; Maertens, J.; Beauprez, J.; Soetaert, W.; De Mey, M. Biotechnological advances in UDP-sugar based glycosylation of small molecules. *Biotechnol. Adv.* **2015**, *33*, 288–302. [CrossRef]
12. Rye, C.S.; Withers, S.G. Glycosidase mechanisms. *Curr. Opin. Chem. Biol.* **2000**, *4*, 573–580. [CrossRef]
13. Wang, L.-X.; Huang, W. Enzymatic transglycosylation for glycoconjugate synthesis. *Curr. Opin. Chem. Biol.* **2009**, *13*, 592–600. [CrossRef] [PubMed]
14. Jahn, M.; Marles, J.; Warren, R.A.J.; Withers, S.G. Thioglycoligases: Mutant Glycosidases for Thioglycoside Synthesis. *Angew. Chem. Int. Ed.* **2003**, *42*, 352–354. [CrossRef] [PubMed]
15. Almendros, M.; Danalev, D.; François-Heude, M.; Loyer, P.; Legentil, L.; Nugier-Chauvin, C.; Daniellou, R.; Ferrières, V. Exploring the synthetic potency of the first furanothioglycoligase through original remote activation. *Org. Biomol. Chem.* **2011**, *9*, 8371–8378. [CrossRef]

16. Avegno, E.A.-B.; Hasty, S.J.; Parameswar, A.R.; Howarth, G.S.; Demchenko, A.V.; Byers, L.D. Reactive thioglucoside substrates for β-glucosidase. *Arch. Biochem. Biophys.* **2013**, *537*, 1–4. [CrossRef] [PubMed]
17. Goodman, I.; Fouts, J.R.; Bresnick, E.; Menegas, R.; Hitchings, G.H. A Mammalian Thioglycosidase. *Science* **1959**, *130*, 450–451. [CrossRef]
18. Meulenbeld, G.H.; Hartmans, S. Thioglucosidase activity from *Sphingobacterium sp.* strain OTG1. *Appl. Microbiol. Biotechnol.* **2001**, *56*, 700–706. [CrossRef] [PubMed]
19. Macauley, M.S.; Stubbs, K.A.; Vocadlo, D.J. O-GlcNAcase catalyzes cleavage of thioglycosides without general acid catalysis. *J. Am. Chem. Soc.* **2005**, *127*, 17202–17203. [CrossRef] [PubMed]
20. Cetinbaş, N.; Macauley, M.S.; Stubbs, K.A.; Drapala, R.; Vocadlo, D.J. Identification of Asp174 and Asp175 as the key catalytic residues of human O-GlcNAcase by functional analysis of site-directed mutants. *Biochemistry* **2006**, *45*, 3835–3844. [CrossRef]
21. Narine, A.A.; Watson, J.N.; Bennet, A.J. Mechanistic requirements for the efficient enzyme-catalyzed hydrolysis of thiosialosides. *Biochemistry* **2006**, *45*, 9319–9326. [CrossRef]
22. Shen, H.; Byers, L.D. Thioglycoside hydrolysis catalyzed by beta-glucosidase. *Biochem. Biophys. Res. Commun.* **2007**, *362*, 717–720. [CrossRef] [PubMed]
23. Sansenya, S.; Opassiri, R.; Kuaprasert, B.; Chen, C.J.; Cairns, J.R.K. The crystal structure of rice (*Oryza sativa L.*) Os4BGlu12, an oligosaccharide and tuberonic acid glucoside-hydrolyzing beta-glucosidase with significant thioglucohydrolase activity. *Arch. Biochem. Biophys.* **2011**, *510*, 62–72. [CrossRef] [PubMed]
24. Wathelet, J.-P.; Iori, R.; Leoni, O.; Rollin, P.; Mabon, N.; Marlier, M.; Palmieri, S. A recombinant β-O-glucosidase from *Caldocellum saccharolyticum* to hydrolyse desulfo-glucosinolates. *Biotechnol. Lett.* **2001**, *23*, 443–446. [CrossRef]
25. Reese, E.T.; Clapp, R.C.; Mandels, M. A thioglucosidase in fungi. *Arch. Biochem. Biophys.* **1958**, *75*, 228–242. [CrossRef]
26. Yip, V.L.Y.; Withers, S.G. Family 4 glycosidases carry out efficient hydrolysis of thioglycosides by an alpha,beta-elimination mechanism. *Angew. Chem. Int. Ed.* **2006**, *45*, 6179–6182. [CrossRef] [PubMed]
27. Niemec-Cyganek, A.; Szeja, W. Heteroaryl thioglycosides, a new class of substrated for glycosidases. *Pol. J. Chem.* **2003**, *77*, 969–973.
28. Jensen, J.L.; Jencks, W.P. Hydrolysis of benzaldehyde O,S-acetals. *J. Am. Chem. Soc.* **1979**, *101*, 1476–1488. [CrossRef]
29. Hanessian, S.; Bacquet, C.; Lehong, N. Chemistry of the glycosidic linkage. Exceptionally fast and efficient formation of glycosides by remote activation. *Carbohydr. Res.* **1980**, *80*, C17–C22. [CrossRef]
30. Ferrières, V.; Blanchard, S.; Fischer, D.; Plusquellec, D. A novel synthesis of D-galactofuranosyl, D-glucofuranosyl and D-mannofuranosyl 1-phosphates based on remote activation of new and free hexofuranosyl donors. *Bioorg. Med. Chem. Lett.* **2002**, *12*, 3515–3518. [CrossRef]
31. Demchenko, A.V.; Malysheva, N.N.; De Meo, C. S-Benzoxazolyl (SBox) Glycosides as Novel, Versatile Glycosyl Donors for Stereoselective 1,2-Cis Glycosylation. *Org. Lett.* **2003**, *5*, 455–458. [CrossRef]
32. Demchenko, A.V.; Pornsuriyasak, P.; De Meo, C.; Malysheva, N.N. Potent, Versatile and Stable: Thiazolyl Thioglycosides as Glycosyl Donors. *Angew. Chem. Int. Ed.* **2004**, *43*, 3069–3072. [CrossRef] [PubMed]
33. Euzen, R.; Guégan, J.-P.; Ferrières, V.; Plusquellec, D. First *O*-Glycosylation from Unprotected 1-Thioimidoyl Hexofuranosides Assisted by Divalent Cations. *J. Org. Chem.* **2007**, *72*, 5743–5747. [CrossRef] [PubMed]
34. Kamat, M.N.; De Meo, C.; Demchenko, A.V. S-Benzoxazolyl as a Stable Protecting Moiety and a Potent Anomeric Leaving Group in Oligosaccharide Synthesis. *J. Org. Chem.* **2007**, *72*, 6947–6955. [CrossRef] [PubMed]
35. Hasty, S.J.; Demchenko, A.V. Glycosyl Thioimidates as Versatile Building Blocks for Organic Synthesis. *Chem. Heterocycl. Compd.* **2012**, *48*, 220–240. [CrossRef] [PubMed]
36. Guillotin, L.; Cancellieri, P.; Lafite, P.; Landemarre, L.; Daniellou, R. Chemo-enzymatic synthesis of 3-O-(β-d-glycopyranosyl)-sn-glycerols and their evaluation as preservative in cosmetics. *Pure Appl. Chem.* **2017**, *89*, 11302. [CrossRef]
37. Fukusumi, S.; Kamizono, A.; Horinouchi, S.; Beppu, T. Cloning and nucleotide sequence of a heat-stable amylase gene from an anaerobic thermophile, *Dictyoglomus thermophilum*. *Eur. J. Biochem.* **1988**, *174*, 15–21. [CrossRef] [PubMed]
38. Shi, R.; Li, Z.; Ye, Q.; Xu, J.; Liu, Y. Heterologous expression and characterization of a novel thermo-halotolerant endoglucanase Cel5H from *Dictyoglomus thermophilum*. *Bioresour. Technol.* **2013**, *142*, 338–344. [CrossRef] [PubMed]

39. Nakajima, M.; Imamura, H.; Shoun, H.; Horinouchi, S.; Wakagi, T. Transglycosylation Activity of *Dictyoglomus thermophilum* Amylase A. *Biosci. Biotechnol. Biochem.* **2004**, *68*, 2369–2373. [CrossRef]
40. Guillotin, L.; Richet, N.; Lafite, P.; Daniellou, R. Is the acid/base catalytic residue mutation in β-d-mannosidase Dt Man from *Dictyoglomus thermophilum* sufficient enough to provide thioglycoligase activity? *Biochimie* **2017**, *137*, 190–196. [CrossRef] [PubMed]
41. Guillotin, L.; Kim, H.; Traore, Y.; Moreau, P.; Lafite, P.; Coquoin, V.; Nuccio, S.; De Vaumas, R.; Daniellou, R. Biochemical Characterization of the α-L-Rhamnosidase Dt Rha from *Dictyoglomus thermophilum*: Application to the Selective Derhamnosylation of Natural Flavonoids. *ACS Omega* **2019**, *4*, 1916–1922. [CrossRef]
42. Huggett, A.S.G.; Nixon, D.A. Use of glucose oxidase, peroxidase and o-dianisidine in determination of blood and urinary glucose. *Lancet* **1957**, *270*, 368–370. [CrossRef]
43. Guillotin, L.; Lafite, P.; Daniellou, R. Unraveling the Substrate Recognition Mechanism and Specificity of the Unusual Glycosyl Hydrolase Family 29 BT2192 from *Bacteroides thetaiotaomicron*. *Biochemistry* **2014**, *53*, 1447–1455. [CrossRef] [PubMed]
44. Akiba, T.; Nishio, M.; Matsui, I.; Harata, K. X-ray structure of a membrane-bound β-glycosidase from the hyperthermophilic archaeon *Pyrococcus horikoshii*. *Proteins Struct. Funct. Bioinform.* **2004**, *57*, 422–431. [CrossRef] [PubMed]
45. Zechel, D.L.; Boraston, A.B.; Gloster, T.; Boraston, C.M.; Macdonald, J.M.; Tilbrook, D.M.G.; Stick, R.V.; Davies, G.J. Iminosugar Glycosidase Inhibitors: Structural and Thermodynamic Dissection of the Binding of Isofagomine and 1-Deoxynojirimycin to β-Glucosidases. *J. Am. Chem. Soc.* **2003**, *125*, 14313–14323. [CrossRef] [PubMed]
46. Zechel, D.L.; Withers, S.G. Dissection of nucleophilic and acid-base catalysis in glycosidases. *Curr. Opin. Chem. Biol.* **2001**, *5*, 643–649. [CrossRef]
47. Zhu, X.; Schmidt, R.R. New Principles for Glycoside-Bond Formation. *Angew. Chem. Int. Ed.* **2009**, *48*, 1900–1934. [CrossRef] [PubMed]
48. Codée, J.D.C.; Litjens, R.E.J.N.; Van den Bos, L.J.; Overkleeft, H.S.; Van der Marel, G.A. Thioglycosides in sequential glycosylation strategies. *Chem. Soc. Rev.* **2005**, *34*, 769–782. [CrossRef] [PubMed]
49. Pornsuriyasak, P.; Demchenko, A.V. S-Thiazolinyl (STaz) Glycosides as Versatile Building Blocks for Convergent Selective, Chemoselective and Orthogonal Oligosaccharide Synthesis. *Chem.-Eur J.* **2006**, *12*, 6630–6646. [CrossRef] [PubMed]
50. Jerez, G.; Kaufman, G.; Prystai, M.; Schenkeveld, S.; Donkor, K.K. Determination of thermodynamic p K_a values of benzimidazole and benzimidazole derivatives by capillary electrophoresis. *J. Sep. Sci.* **2009**, *32*, 1087–1095. [CrossRef]
51. Akhond, M.; Ghaedi, M.; Tashkhourian, J. Development of a New Copper(II) Ion-selective Poly(vinyl chloride) Membrane Electrode Based on 2-Mercaptobenzoxazole. *Bull. Korean Chem. Soc.* **2005**, *26*, 882–886.
52. Kamat, M.N.; Rath, N.P.; Demchenko, A. V Versatile Synthesis and Mechanism of Activation of S-Benzoxazolyl Glycosides. *J. Org. Chem.* **2007**, *72*, 6938–6946. [CrossRef] [PubMed]
53. Case, D.A.; Darden, T.A.; III, T.E.C.; Simmerling, C.L.; Wang, J.; Duke, R.E.; Luo, R.; Walker, R.C.; Zhang, W.; Merz, K.M.; et al. *AMBER 12 2012*; University of California: San Francisco, CA, USA, 2012.
54. Phillips, J.C.; Braun, R.; Wang, W.; Gumbart, J.; Tajkhorshid, E.; Villa, E.; Chipot, C.; Skeel, R.D.; Kale, L.; Schulten, K. Scalable molecular dynamics with NAMD. *J. Comput. Chem.* **2005**, *26*, 1781–1802. [CrossRef] [PubMed]
55. Wang, L.-P.; McKiernan, K.A.; Gomes, J.; Beauchamp, K.A.; Head-Gordon, T.; Rice, J.E.; Swope, W.C.; Martínez, T.J.; Pande, V.S. Building a More Predictive Protein Force Field: A Systematic and Reproducible Route to AMBER-FB15. *J. Phys. Chem. B* **2017**, *121*, 4023–4039. [CrossRef] [PubMed]
56. Jakalian, A.; Jack, D.B.; Bayly, C.I. Fast, efficient generation of high-quality atomic charges. AM1-BCC model: II. Parameterization and validation. *J. Comput. Chem.* **2002**, *23*, 1623–1641. [CrossRef] [PubMed]
57. Lafite, P.; André, F.; Zeldin, D.C.; Dansette, P.M.; Mansuy, D. Unusual Regioselectivity and Active Site Topology of Human Cytochrome P450 2J2. *Biochemistry* **2007**, *46*, 10237–10247. [CrossRef] [PubMed]

13

Characterization of the Novel Ene Reductase Ppo-Er1 from *Paenibacillus Polymyxa*

David Aregger, Christin Peters and Rebecca M. Buller *

Competence Center for Biocatalysis, Institute of Chemistry and Biotechnology, Department of Life Sciences and Facility Management, Zurich University of Applied Sciences, Einsiedlerstrasse 31, 8820 Waedenswil, Switzerland; David.Aregger@zhaw.ch (D.A.); Christin.peters@zhaw.ch (C.P.)

* Correspondence: rebecca.buller@zhaw.ch

Abstract: Ene reductases enable the asymmetric hydrogenation of activated alkenes allowing the manufacture of valuable chiral products. The enzymes complement existing metal- and organocatalytic approaches for the stereoselective reduction of activated C=C double bonds, and efforts to expand the biocatalytic toolbox with additional ene reductases are of high academic and industrial interest. Here, we present the characterization of a novel ene reductase from *Paenibacillus polymyxa*, named Ppo-Er1, belonging to the recently identified subgroup III of the old yellow enzyme family. The determination of substrate scope, solvent stability, temperature, and pH range of Ppo-Er1 is one of the first examples of a detailed biophysical characterization of a subgroup III enzyme. Notably, Ppo-Er1 possesses a wide temperature optimum (T_{opt}: 20–45 °C) and retains high conversion rates of at least 70% even at 10 °C reaction temperature making it an interesting biocatalyst for the conversion of temperature-labile substrates. When assaying a set of different organic solvents to determine Ppo-Er1's solvent tolerance, the ene reductase exhibited good performance in up to 40% cyclohexane as well as 20 vol% DMSO and ethanol. In summary, Ppo-Er1 exhibited activity for thirteen out of the nineteen investigated compounds, for ten of which Michaelis–Menten kinetics could be determined. The enzyme exhibited the highest specificity constant for maleimide with a k_{cat}/K_M value of 287 $mM^{-1}\ s^{-1}$. In addition, Ppo-Er1 proved to be highly enantioselective for selected substrates with measured enantiomeric excess values of 92% or higher for 2-methyl-2-cyclohexenone, citral, and carvone.

Keywords: biocatalysis; ene reductase; enzyme sourcing; old yellow enzyme; solvent stability

---

1. Introduction

Many bioactive molecules contain at least one chiral center rendering the development of effective asymmetric synthesis methods essential for the chemical industry. Besides the well-established metal- and organocatalytic approaches [1], biocatalytic strategies offer an interesting alternative to install chirality into small molecules. To date, industrial biocatalysis has mastered a range of enzyme families including ketoreductases [2], transaminases [3], and imine reductases [4]. Looking forward, the increasing power of genomic mining and enzyme engineering will allow industrial access to even more enzyme families leading to an expansion of the available biocatalytic toolbox [5].

The families of enzymes collectively known as ene reductases (ERs) catalyze the stereoselective trans- and, more rarely, cis-hydrogenation of activated alkenes [6–9]. Thus, ene reductases offer a valuable access route to asymmetric compounds, which is complementary to the chemical cis-hydrogenation catalyzed by chiral rhodium or ruthenium phosphine catalysts [10,11]. Today, ene reductases are classified into five enzyme groups, which differ in structure, reaction mechanism, substrate spectrum, and stereoselectivity (Figure 1) [12]. While enoate reductases, medium- and short-chain dehydrogenases/reductases (MDR and SDR), as well as the recently discovered quinone reductase-like ene reductases [13], are currently being investigated in terms of their industrial potential [14], enzymes stemming from the old yellow enzyme (OYE) family are established members of the biocatalytic toolbox and are the best characterized and most extensively employed ene reductases today [6].

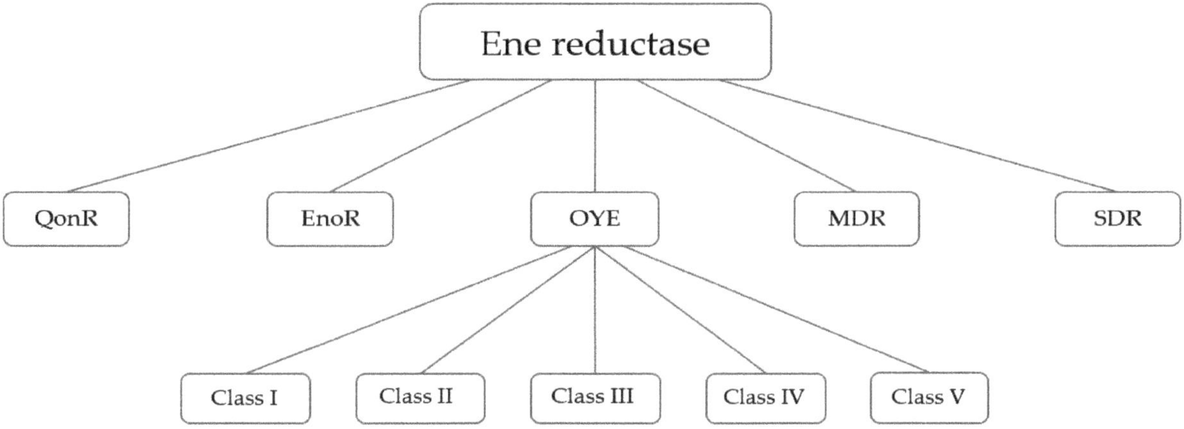

Figure 1. Overview of the classification within the ene reductase family [15]. QnoR (NADPH-dependent quinone reductase like ene-reductases), EnoR (enoate reductase), OYE (old yellow enzyme), MDR (medium-chain dehydrogenase/reductase), and SDR (short-chain dehydrogenase/reductase); Class I (classical OYE); Class II (thermophilic-like OYE) and Class V (fungal OYE).

Isolated in 1932 by Warburg and Christian from bottom-fermented brewer's yeast (*Saccharomyces pastorianus*), the first such ene reductase was named "yellow enzyme" [16]. After the discovery of several additional members belonging to the same enzyme family the "yellow enzyme" was renamed to "old yellow enzyme" (OYE1) [17]. OYEs preferentially accept α,β-unsaturated ketones, aldehydes, nitroalkenes, and some carboxylic acids as substrates [7]. In the last decade, the catalytic mechanism of OYEs has been exhaustively investigated and its general principle is well understood: The enzymes follow a bi–bi ping–pong mechanism, which can be divided into a reductive and an oxidative half reaction [18]. In the reductive half-reaction, flavin mononucleotide (FMN) is reduced through hydride transfer from NAD(P)H, whereas in the oxidative half reaction a hydride is transferred from the reduced flavin to the C_β of the activated alkene. The missing proton for the C_α is transferred via a tyrosine residue from the opposite site [18,19], ultimately leading to an anti-addition hydrogenation.

The catalytic machinery of OYE enzymes is supported by a typical (α,β) 8-barrel (TIM-barrel) fold with additional secondary structural elements present (e.g., four β-strands and five α-helices in OYE1 [20]; six β-strands and two α-helices in 12-oxophytodienoate reductase OPR [18]). The folded domain is known to occur in different oligomeric states, such as monomers (PETN reductase) [21], dimers (OYE1) [20], tetrameters (dimers of dimers such as YqjM [22] or TOYE [23]), octamers, and dodecamers [23]. The oligomerization state is described to be often governed by the position and amino acid composition of surface loops [7]. In addition, the constitution of the loops can have an influence on thermostability [23].

Notably, amino acid sequence alignments of OYE homologs show high conservation in specific regions of the proteins, such as residues involved in catalysis, FMN, and substrate binding [7,15,23]. To account for these differences in sequence and the resulting structural features, the old yellow enzyme family can be further divided into five subclasses [15]. While enzyme members of the subclass I, also termed "classical" old yellow enzymes, and class II, introduced by Scrutton's group in 2010 and dubbed "thermophilic-like" [23], have been well explored [7,14], the recently described class III–V are less well investigated [15,24].

Synthetic applications of ene reductases are manifold and range from the preparation of profens [25–27] and chiralγ-amino acids [28–30] to the synthesis of chiral phosphonates [31] and nitroalkanes [32], precursors in the synthesis of pharmaceutically active ingredients. To further promote an off-the-shelve synthetic use of ene reductases, which can reduce the time and cost of the implementation of a biocatalytic step into a process significantly, we set out to expand the available biocatalytic toolbox [15]. In this context, not only the discovery and engineering of novel ene reductases is of great utility [33], but also a careful characterization of the new biocatalysts is needed as it may lead to the construction of a more targeted enzyme library associated with reduced screening time and costs.

Herein, we showcase the detailed characterization of Ppo-Er1 from *Paenibacillus polymyxa*, an OYE subclass III enzyme, and highlight the enzyme's substrate scope, kinetic parameters, solvent tolerance, as well as pH and temperature profile. The data presented may facilitate future screening and engineering studies and, in selected cases, thus, lead to the faster adoption of an ene reductase in chemical process development.

2. Results and Discussion

The enzyme Ppo-Er1 from *P. polymyxa* was discovered during the screening of 19 bacterial wild-type strains from the Culture Collection of Switzerland, as previously described [15]. Ppo-Er1 (41.3 kDa) is characterized by a substantial sequence similarity with the old yellow enzyme YqiG from *Bacillus subtilis* (50%) [34], Bac-OYE2 from *Bacillus* sp. (50%) [35], Lla-Er from *Lactococcus lactis* (39%) [15], and LacER from *Lactobacillus paracasei* (47%) [36], all of which belong to the subclass III of the OYE family. In detail, Ppo-Er1 contains a specific combination of motifs known from the classical and thermophilic-like groups that has been found to be characteristic for class III enzymes [15]: Gln104 and Arg228 predicted to interact with the pyrimidine ring of FMN [22], His 171, and Asn 175 proposed to interact with N1 and N3 of FMN [22,37]; Thr30 suggested to interact with isoalloxazine ring O4 of FMN [38]; and Met29, Leu324, and Arg321, which presumably interact with the dimethyl benzene moiety of FMN. As expected, subclass III old yellow enzyme Ppo-Er1 is thus phylogenetically positioned between classical and thermophilic-like OYEs.

2.1. Expression and Characterization of Ppo-Er1

The ready-to-use plasmid consisting of pET-28b(+) vector and the Ppo-Er1 sequence was assembled by Twist Bioscience and a C-terminal His_6 tag for protein purification by affinity chromatography was included. The soluble recombinant expression of Ppo-Er1 in *Escherichia coli* BL21 (DE3) was achieved in terrific broth (TB) medium at 25 °C. Ppo-Er1 was purified by affinity chromatography using Ni-NTA resin (Figure S1) and the cofactor FMN was reconstituted before further analysis. FMN reconstitution (100 μM) proved necessary to obtain a fully active enzyme as without this step the enzyme preparation only exhibited 8% (0.05 U/mg for cyclohexanone) of the expected activity (0.61 U/mg for cyclohexanone). This effect was also described for the OYEs LacER [36] and Lla-Er [15]. In the case of LacER, for example, the addition of FMN after purification by DEAE ion exchange chromatography increased the activity by a factor of 92 from 0.0018 to 0.168 U/mg for the substrate *trans*-2-hexen-1-al. This observation suggests that—similar to other known OYEs—the binding affinity of Ppo-Er1 to FMN under purification conditions is low, a fact that has to be kept in mind for any following activity analysis. The storage stability of the purified Ppo-Er1 proved to be very good, boding well for the enzyme's incorporation in potential enzyme screens: At −20 °C and in the presence of 20% glycerol, the enzyme did not lose any activity even when stored for an extended period of time (one week), whereas an activity drop of approximately 20% was observed after incubation for 10 days at 4 °C (no additives). In contrast to a number of reported OYEs [15,39], we found that NADPH and NADH are equally preferred physiological cofactors of Ppo-ER1 (Figure S14) allowing for maximum flexibility in the choice of recycling system during process development. Both, the coupled-enzyme approach [40] or the use of alternative hydride sources [41,42] will thus be conceivable options to avoid having to add stoichiometric amounts of the coenzymes.

The oligomeric state of Ppo-Er1 was determined via gel filtration by correlation with a commercial gel filtration standard containing proteins of specific size. Based on this comparison, Ppo-Er1 mostly occurs as a monomer (Figure S2) as do for example PETN from *Enterobacter cloacae* [21] and RmER from *Ralstonia metallidurans* [43], both thermophilic-like ene reductases.

Further relevant parameters for application such as optimum pH, optimum temperature, and long-term temperature stability were determined using the substrate cyclohexenone. The pH profile of Ppo-Er1 was measured in Davies buffer covering pH 5 to pH 10 [44], in which the enzyme reached about 50% of the activity observed in 50 mM phosphate buffer (Figure S3). The pH profile was found to be bell-shaped, exhibiting a narrow optimum at pH 6.5–7.5 (Figure 2). Beyond this range, enzyme activity decreases rapidly, especially when the enzyme was pre-incubated for a longer time period (24 h) in the measurement buffers (Figure 2). In the case of other characterized class III OYEs such as LacER [36] and YqiG [15,34], a similar pH profile was determined albeit with a wider pH working range as indicated by the reported optimum activities in the range of pH_{opt} 8–9 and pH_{opt} 6–9, respectively. Notably, OYE enzymes belonging to other subclasses exhibit similar pH profiles as reported for Ppo-Er1, e.g., the "classical" XenB [45] and NemA [45] with a pH_{opt} of 6–7.5, the "thermophilic-like" YqjM [46] and Chr-OYE3 [47] with a pH_{opt} of 6–8, and the class IV enzyme Ppo-Er3 [15] with a pH_{opt} of 7–8.5.

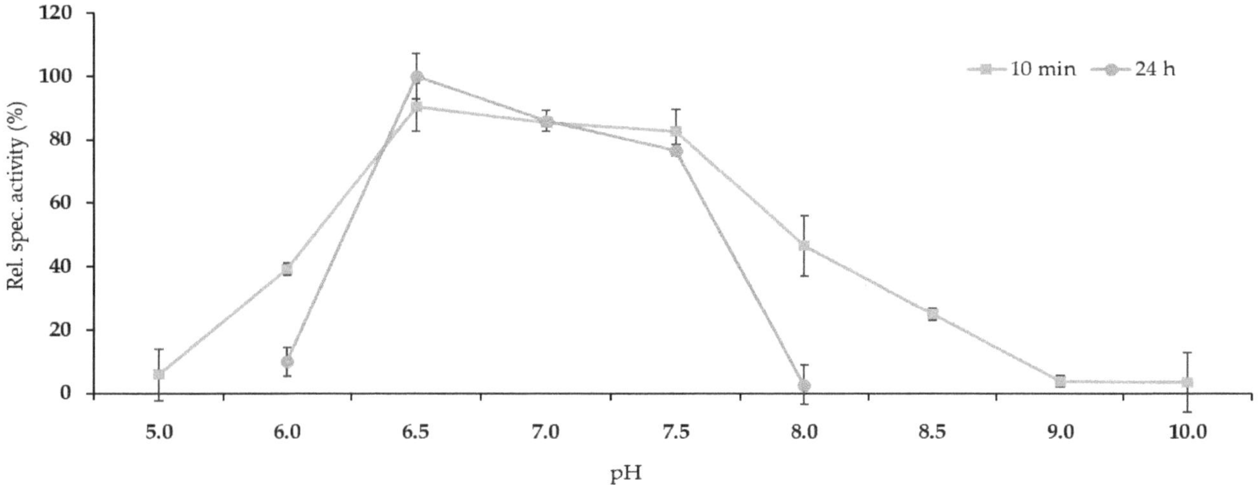

Figure 2. pH profile of Ppo-Er1 measured between pH 5 and pH 10 in Davies buffer [44]. The enzyme was preincubated at 25 °C in the respective measurement buffer solution for 10 min and 24 h, respectively, to determine the stability and activity of Ppo-Er1 in dependence of pH. Relative specific activity corresponds 100% to an activity of 0.41 U/mg for cyclohexenone. The error bars show the standard deviation of triplicates.

In terms of thermal robustness, Ppo-Er1 possesses interesting long-term stability. After 24 h incubation at 20 °C, enzyme activity toward cyclohexenone remained virtually unchanged, whereas residual activity of approximately 70% was detected after an equally long incubation time at 30 °C. Furthermore, short-term exposure of Ppo-Er1 to 45 °C led to only a marginal loss in activity (<10%) allowing the enzyme to be used for applications that require higher temperatures (Figure 3). These results are in line with data obtained for other class III and IV enzymes such as YqiG and Ppo-Er3, which have reported T_{opt} values of 25–40 °C [15,34]. Strikingly, Ppo-Er1 retained a relative specific activity of >70% at temperatures as low as 10 °C making the enzyme an interesting candidate to be used for the transformation of thermolabile substrates such as aldehydes (Figure 3). Overall, our Ppo-Er1 data confirm that the temperature profile of class III enzymes resembles those of their mesophilic counterparts of class I, for example NemA [45] with a reported T_{opt} of 30–50 °C and OYE2p [48] with a T_{opt} of 25–40 °C. Finally, we employed the *Thermo*FAD technique to determine the melting temperature of Ppo-Er1 and found that the ene reductase unfolds at T_m = 46.5 ± 1 °C (Figure S15).

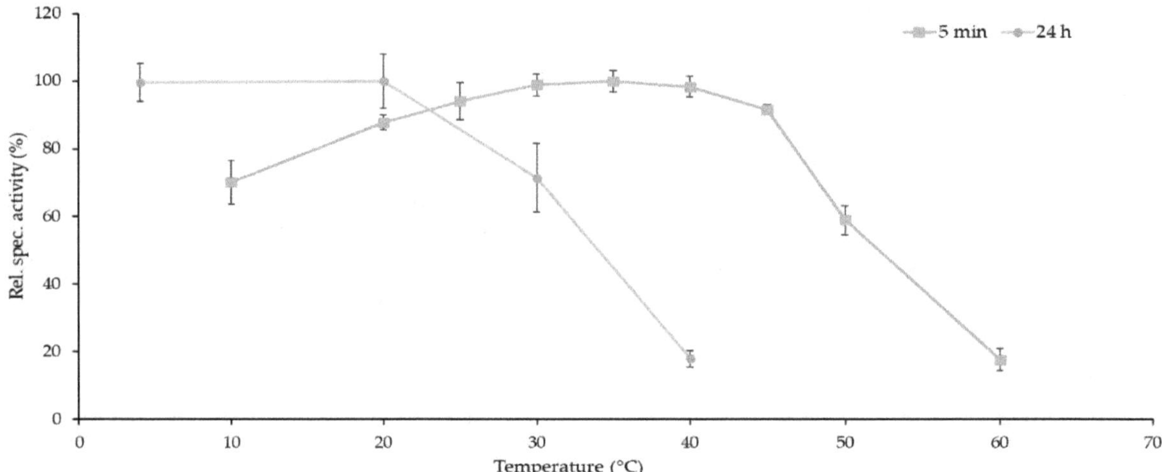

Figure 3. The temperature profile and the temperature stability of Ppo-Er1. For the temperature profile Ppo-Er1 was incubated for 5 min at different temperatures (10–60 °C) and directly measured for the conversion of substrate cyclohexenone (1 mM). For the temperature stability measurement, Ppo-Er1 was incubated at four different temperatures (4–40 °C) and measured after 24 h at 25 °C. The error bars show the standard deviation of triplicates. Relative specific activity corresponds 100% to an activity of 0.52 U/mg for cyclohexenone.

The use of cosolvents is often a "must" in biocatalytic processes due to the presence of high concentrations of various organic substrates. Consequently, in many instances the solvent stability of enzymes needs to be optimized by enzyme engineering to generate catalysts that are compatible with the process conditions [49]. To verify the stability of Ppo-Er1 in the presence of a set of typical solvents, we thus determined the enzymatic activity over a concentration range of 10–40% of DMSO, DMF, cyclohexane, ethanol, and ethyl acetate. The enzyme performed best in cyclohexane (assayed substrate: 1 mM hexenal), which did not cause a significant loss in activity even when supplemented to a final volume of up to 40% in the assay. Alternatively, DMSO could be considered as a viable cosolvent for Ppo-Er1 as the enzyme was virtually unaffected up to a concentration of 20% *v*/*v*. Even at a concentration of 30% *v*/*v* DMSO, Ppo-Er1 retained a relative activity of approximately 80% (assay substrate: 1 mM cyclohexenone). The solvent ethanol was shown to also be a suitable choice for this enzyme, as it was tolerated well up to a concentration of 10% *v*/*v*. DMF or ethyl acetate, however, should not be used in combination with Ppo-Er1 as their presence was found to be detrimental for enzymatic activity. Already at a concentration of 10% *v*/*v* activity drops of 30% and 85% were observed, respectively (Figure 4).

In comparison to most known old yellow enzymes, Ppo-Er1 exhibits similar solvent resistance: The thermophilic-like OYE YqjM [46] has been reported to remain active in an analogous concentration range of DMSO, DMF, and ethyl acetate as Ppo-Er1. However, an ethanol concentration of 10% *v*/*v* led to a strong reduction of the half-life of YqjM, which we did not observe in the case of Ppo-Er1. TOYE [23], another thermophilic-like OYE, was reported to exhibit a 50% loss of activity at an ethanol concentration of 45% corresponding to a higher stability toward this solvent compared to Ppo-Er1, whereas the classical PETNR [50] already lost 50% activity in the presence of an ethanol concentration of 20% *v*/*v*. In this context, it should be noted that organic-solvent-tolerant ene reductases have also been reported: FOYE1, originating from an acidophilic iron oxidizer, was shown to perform well in many solvent systems with up to 20% *v*/*v* solvents (ethanol, methanol, acetone, isopropanol, DMSO, THF) clearly outperforming all abovementioned ene reductases in terms of solvent stability [51].

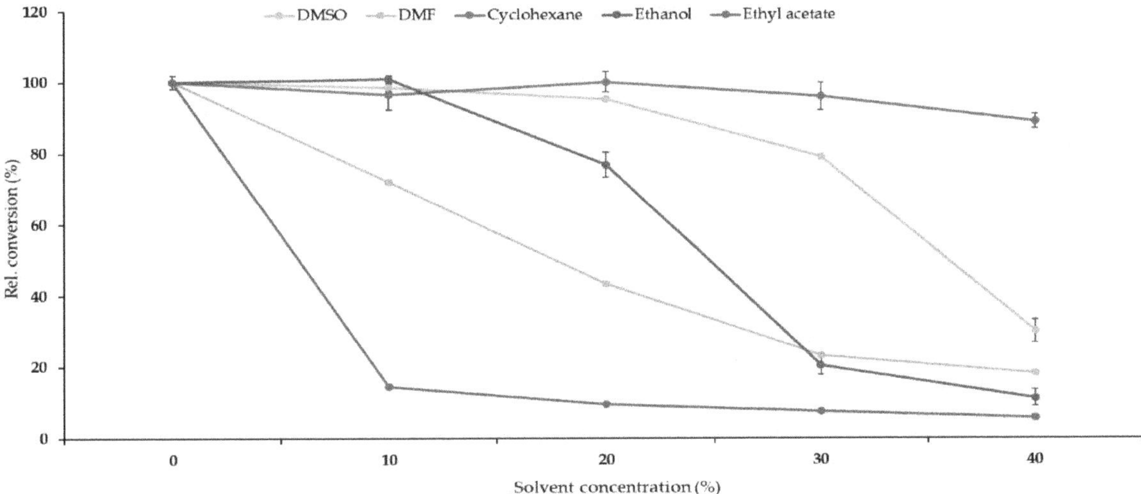

Figure 4. Overview of the solvent stability of Ppo-Er1 in DMSO (dimethyl sulfoxide), DMF (dimethyl formamide), cyclohexane, ethanol, and ethyl acetate in a concentration range of 10%–40% *v*/*v*. The standard enzyme assay was performed while the concentration of solvents was varied (substrate for cyclohexane: 1 mM hexenal, all other solvents: 1 mM cyclohexenone). Data are shown as values relative to an enzyme assay without cosolvent in which 100% relative conversion corresponds to the production of 0.84 mM cyclohexanone or 0.49 mM hexanal, respectively. The error bars show the standard deviation of triplicates, except for the 30% *v*/*v* cyclohexane point for which only two measurements were available.

2.2. Substrate Scope, Determination of Michaelis–Menten Parameters, and Stereoselectivity

To determine the substrate profile of Ppo-Er1, the enzyme was tested for the conversion of nineteen structurally diverse aliphatic and cyclic alkenes bearing ketone, aldehyde, nitro, carboxylic acid, or ester moieties as electron-withdrawing groups. For thirteen substrates, product formation by Ppo-Er1 could be detected. Cyclohexenone, hexenal, 2-methyl-2-pentenal, 4-phenyl-3-buten-2-one, cinnamic aldehyde, maleimide, and carvone (at 5 mM concentration) were converted especially well, and >99% conversion was obtained within 4 h (Table 1). Substrates not accepted by Ppo-Er1 included α,β-unsaturated carboxylic acids such as butenic acid, cinnamic acid, and citraconic acid as well as the ketones 3-methyl-2-cyclohexenone and 3-methyl-2-cyclopentenone, which are characterized by an additional methyl group in the β-position. The α,β-unsaturated ester ethyl crotonate was also not converted.

Based on the obtained data, it can be concluded that the overall substrate profile of Ppo-Er1 resembles that of other subclass III enzymes such as YqiG [15,34] and Lla-Er [15]. For example, 5 mM of cinnamic aldehyde and cyclohexenone are also well converted by Lla-Er [15] (65% ± 4.2% and 23% ± 3.1%) and YqiG [15] (58% ± 2.4% and 55% ± 6.1%) after 1 h at 30 °C. Notably, however, marked differences in substrate acceptance by class III enzymes occur for some of the investigated substrates highlighting the importance of an in-depth substrate profiling: Whereas carvone and maleimide are very well converted by Ppo-Er1 (both: >99%), Lla-Er, for example, accepts this compound only poorly (carvone: 2.6% ± 0.1%, maleimide: not converted) [15]. Diethylbenzylidenemalonate conversion by YqiG [15,34] (11% ± 1.3%), on the other hand, significantly exceeded the detected product formations achieved by Lla-Er (<1%) [15] and Ppo-Er1 (1.2%). Moreover, 3-methyl-2-cyclopentenone, which is not converted by Ppo-Er1, Lla-Er [15], and YqiG [15,34], has been shown to be accepted by LacER [36]. Generally, we noted that Ppo-Er1 has a restricted substrate acceptance for cyclic β-methylated substrates such as 3-methyl-2-cyclohexenone and 3-methyl-2-cyclopentenone, which possibly results from a difficulty in accepting substituents at the C_β position of cyclic compounds in the active site in analogy to other class II, III, and IV enzymes [15,39]. In addition, carboxylic acids and esters seem to be non-optimal alkene activating groups for this enzyme as conversion of the corresponding substrates was low or not detectable.

Table 1. Conversion, steady state kinetics,[a] and enantiomeric excess (ee) of various substrates converted with purified enzymes as determined after 4 h at 20 °C (n.d.: not detected; n.s.: not soluble). The given uncertainties show the standard deviation of triplicates.

Substrate		**Conversion**	**ee**	k_{cat}/K_m	K_m	k_{cat}
Name	**Structure**	**(%)**	**(%)**	**(mM^{-1} s^{-1})**	**(mM)**	**(s^{-1})**
Maleimide	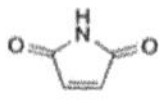	≥99 ± 3.7		287.8 ± 0.12	0.10 ± 0.01	28.78 ± 0.62
trans-β-Methyl-β-nitrostyrene	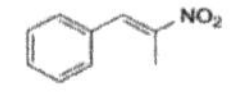	81 ± 1.0		41.4 ± 0.23	0.12 ± 0.03	4.97 ± 0.36
2-Methyl-2-pentenal	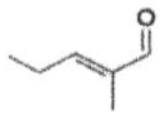	≥99 ± 7.4	(*S*) 63	15.3 ± 0.09	0.41 ± 0.04	6.27 ± 0.11
Cinnamaldehyde	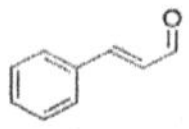	≥99 ± 1.5		14.6 ± 0.14	0.36 ± 0.05	5.27 ± 0.18
Hexenal		≥99 ± 3.4		3.3 ± 0.10	2.22 ± 0.21	7.42 ± 0.01
Carvone	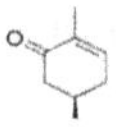	≥99 ± 2.1	(*R*) 98	0.5 ± 0.16	4.35 ± 0.69	2.20 ± 0.08
Cyclohexenone	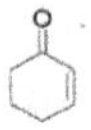	≥99 ± 0.5		0.4 ± 0.08	13.42 ± 1.0	5.25 ± 0.1
Citral	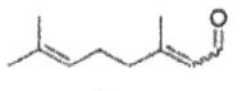	29 ± 1.4	(*S*) 94	0.2 ± 0.93	1.12 ± 1.0	0.17 ± 0.04
2-Methyl-2-cyclohexenone	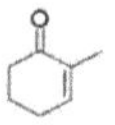	76.2 ± 0.4	(*R*) 92	0.1 ± 0.23	14.93 ± 3.3	1.30 ± 0.08
Cyclopentenone	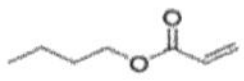	59 ± 1.7		0.03 ± 0.17	57.24 ± 9.4	1.75 ± 0.16
4-Phenyl-3-buten-2-one		≥99 ± 1.0			n.s.	
Butylacrylate	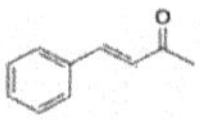	22 ± 6.5				
Diethyl benzyldienemalonate	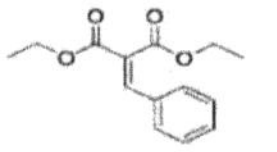	1.2 ± 0.0				
3-Methyl-2-cyclohexenone	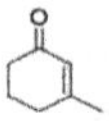	n.d.				
3-Methyl-2-cyclopentenone	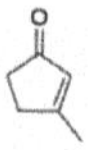	n.d.				
Etylcrotonate	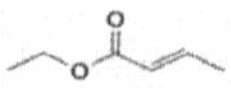	n.d.				
Butenoic acid	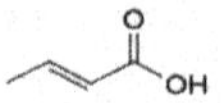	n.d.				
Cinnamic acid	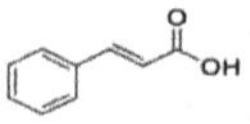	n.d.				
Citraconic acid	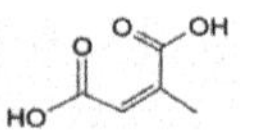	n.d.				

[a] Reactions (1 mL) were performed in potassium phosphate buffer (50 mM, pH 7.0) containing NADPH (175 μM) and substrate (20 μM–80 mM), depending on substrate, Ppo-Er1 (0.61 μM), and DMSO to solubilize the substrates. The reactions were followed continuously by monitoring NADPH oxidation at 340 nm for 90 sec at 25 °C.

To complement the substrate acceptance profile, Michaelis–Menten parameters of Ppo-Er1 for ten diverse substrates were determined (Table 1, Figures S4–S13). Within the tested substrate range, Ppo-Er1 showed the highest catalytic efficiency for maleimide (k_{cat}/K_m = 287 mM^{-1} s^{-1}) followed by trans-β-methyl-β-nitrostyrene (k_{cat}/K_m = 41 mM^{-1} s^{-1}). In combination with the conversion data, the measured kinetic parameters (Table 1) indicate a general preference for alkenes carrying a phenyl substituent at the C_β position of the substrates. Overall, Ppo-Er1's specific activity for other typical ene reductase substrates such as carvone (k_{cat}/K_m = 0.5 mM^{-1} s^{-1}) and cyclohexanone (k_{cat}/K_m = 0.4 mM^{-1} s^{-1}) was found to be in a similar range as those described for other well-known OYEs such as the classical PETNR (carvone: k_{cat}/K_m = 2 mM^{-1} s^{-1}; cyclohexanone: k_{cat}/K_m = 5 mM^{-1} s^{-1}) [50] and the thermophilic-like YqjM (cyclohexanone: k_{cat}/K_m = 6.4 mM^{-1} s^{-1}) [46] (Table 2). Maleimide, however, is better converted by ene reductases from photosynthetic extremophiles such as CtOYE (k_{cat}/K_m = 1940 mM^{-1} s^{-1}) or GsOYE (k_{cat}/K_m = 399 mM^{-1} s^{-1}) [52] the thermophilic-like OYERo2 (k_{cat}/K_m = 10,800 mM^{-1} s^{-1}) [53] or the class III OYE YqiG (k_{cat}/K_m = 800 mM^{-1} s^{-1}) (Table 2).

Table 2. Comparison of the catalytic efficiencies (mM^{-1} s^{-1}) of a range of known old yellow enzymes (OYEs) (YqiG [34], PETNR [50], YqjM [46], TOYE [23], DrER [43], RmER [43], and OYERo2 [53]) from class I–III.

Substrate	Class I	Class II					Class III	
	PETNR	**YqjM**	**TOYE**	**DrER**	**RmER**	**OYERo2**	**Ppo-Er1**	**YqiG**
Cyclohexenone	5	6.4	0.5	2.1	0.7		0.4	22
2-Methyl-cyclohexenone	4	1.0					0.1	
Cyclopentenone	<0.5	1.9	0.6				0.03	
Hexenal		0.60					3.3	
Citral	9	0.02	0.05				0.2	6.7
2-Methyl-2-pentenal	61		0.14				15.3	18
Cinnamaldehyde	8						14.6	
Carvone	2		1.5				0.5	7.5
Maleimide						10,800	287.8	800
trans-β-Methyl-β-nitrostyrene							41.4	

In addition to determining the steady-state kinetic parameters, we also investigated the stereopreference of Ppo-Er1. Based on our results with four selected substrates, Ppo-Er1 displays a similar stereopreference to other reported OYE class III enzymes (Table 3), preferentially forming the *S*-product when converting 2-methy-2-pentenal and citral and forming the *R*-product when transforming carvone and 2-methyl-2-cyclohexenone. Notably, the detected ee values of Ppo-Er1 are generally superior to values determined for YqiG and Lla-Er [15] with the only exception being the enantiomeric excess reported for the conversion of carvone by Lla-Er (>99.9% ee). It should be noted, however, that Lla-Er displayed a low conversion of 2.6% of 5 mM substrate after 1 h at 30 °C compared to the >99% conversion of 5 mM substrate by Ppo-Er1 after 4 h at 20 °C.

Table 3. The enantiomeric excess of some selected OYEs (YqiG [15], Lla-Er [15], Ppo-Er3 [15], OPR1 [54], OPR3 [54], PETNR [50], YqjM [54], TOYE [23]) from classes I–IV. The values presented for YqjM were measured as a reference for Ppo-Er1 and compared with the literature [54].

Substrate	Class I			Class II		Class III			Class IV
	OPR1	**OPR3**	**PETNR**	**YqjM**	**TOYE**	**Ppo-Er1**	**YqiG**	**Lla-ER**	**Ppo-Er3**
2-Methyl-2-pentenal	(*R*) 47	(*S*) 78		(*R*) 20	(*S*) 55	(*S*) 63	(*S*) 33	(*S*) 5	(*S*) 67
Carvone			(*R*) 95	(*R*) 82	(*R*) 95	(*R*) 98	(*R*) 89	(*R*) >99.9	(*R*) 91
2-Methyl-2-cyclohexenone	(*R*) 77	(*R*) 62		(*R*) 81		(*R*) 92	(*R*) 83	(*R*) 11	(*R*) 86
Citral	(*S*) >95	(*S*) >95		(*S*) 95	(*S*) 91	(*S*) 94			

3. Materials and Methods

3.1. Materials

All chemicals were purchased from Merck (Darmstadt, Germany), VWR (Hannover, Germany), or Carl Roth (Karlsruhe, Germany). The purchased chemicals were of the highest available purity or of analytical grade and were used without further purification unless otherwise specified. NADPH tetrasodium salt was ordered from Oriental Yeast Co. Ltd. (Tokyo, Japan). The plasmid (pET 28b(+)

incl. Ppo-Er1) was ordered from Twist Bioscience (San Francisco, CA, USA). The HisTrap FF and the HiTrap Desalting columns were ordered from GE Healthcare (Uppsala, Sweden).

3.2. Plasmid

Twist Bioscience (San Franscisco, CA, USA) cloned the synthetic gene of the codon optimized Ppo-Er1 (Accession Nr: WP_013369181) with *Nde*I and *Xho*I in the commercial pET28b(+) vector.

3.3. Bacterial Strains and Culture Conditions

E. coli BL21 (DE3) [*fhuA2 [lon] ompT gal (λ DE3) [dcm] ΔhsdS*] was purchased from New England Biolabs (Beverly, MA, USA). *E. coli* strains were cultured routinely in Lysogeny broth (LB) or TB media and were supplemented with kanamycin (50 μg mL^{-1}). Bacterial cultures were incubated in baffled Erlenmeyer flasks in a New Brunswick Innova 42 orbital shaker at 200 rpm and 37 °C. Bacteria on agar plates were incubated in a HERATherm Thermo Scientific incubator under air. All materials and biotransformation media were sterilized by autoclaving at 121 °C for 20 min. Aqueous stock solutions were sterilized by filtration through 0.22 μm syringe filters. Agar plates were prepared with LB medium supplemented by 1.5% (w/v) agar.

3.4. Expression

The expression of Ppo-Er1 in *E. coli* BL21 (DE3) was performed by inoculation of TB media (400 mL) supplemented with kanamycin (50 μg mL^{-1}) with an overnight culture (4 mL; 1:100). The culture was incubated at 37 °C and 180 rpm until optical density OD_{600} = 0.5–0.8 was reached. Afterward expression was induced by the addition of 100 μM IPTG, and incubation was continued at 25 °C for 18 h. Cells were harvested by centrifugation at 4500× *g* for 10 min at 4 °C and either used directly or the pellet was stored by freezing at −20 °C.

3.5. Enzyme Purification

The cell disruption was performed by resuspending the pellet from a 400 mL culture in 20 mL buffer (100 mM sodium phosphate buffer pH 7.5, 300 mM NaCl, supplemented by 30 mM imidazole) and a single passage through a French press (2000 psi). The crude extract was separated from the cell debris by centrifugation at 8000× *g* for 45 min. Purification was achieved by affinity chromatography exploiting the C-terminal His-Tag using an automated Äkta purifier system. The crude extract was filtered (0.45 μm) and applied to a pre-equilibrated 5 mL HisTrap FF column. The unbound protein was washed with five column volumes of buffer supplemented with 45 mM imidazole. The elution of Ppo-Er1 was accomplished by a three-column volume of buffer supplemented with 300 mM of imidazole. The resulting fractions were collected and analyzed by SDS-PAGE. The fractions with a high content of Ppo-Er1 were pooled and desalted using 50 mM sodium phosphate buffer (pH 7.5) to remove the imidazole. This step was performed employing the Äkta purifier system using three coupled 5 mL HiTrap desalting columns. After the system was equilibrated, the Ppo-Er1-containing sample was applied and fractioned. The protein fractions were analyzed via the integrated online absorption measurement at 280 nm. The protein content of the pooled purified sample was determined by measuring the adsorption with a NanoDrop One (Thermo Fisher Scientific) system and using the molecular weight (41.3 kDa) and extinction coefficient ($\epsilon_{\lambda = 280\,nm}$ = 38′390 M^{-1} cm^{-1}) of Ppo-Er1 for the calculation. The extinction coefficient was obtained by using the online calculation tool Prot pi [55].

3.6. Activity Assay

The activity measurements were recorded spectrophotometrically by observing NADPH consumption at 340 nm for 60–90 s in a 1 mL (1 cm) plastic cuvette in the Lambda 465 (PDA UV/VIS) system from Perkin Elmer. The biocatalytic experiments to obtain the pH and the temperature profile were conducted in sodium phosphate buffer (50 mM, pH 7.5) using 175 μM NADPH, 1 mM cyclohexenone, and 0.61 μM purified Ppo-Er1. For the determination of the Michaelis–Menten parameters, the substrate concentration was varied in the range of 20 μM–80 mM depending on the substrate while the enzyme concentration was kept constant at 0.61 μM. For the pH profile, Davies buffer [44] was used. All measurements were done in triplicates. Background NADPH consumption was determined in assays in which either the enzyme or the substrate had been eliminated. The substrates were solubilized as 1 M stock in DMSO.

3.7. Biocatalysis Reaction

The in vitro biocatalysis reaction were performed by using desalted Ppo-Er1 (with a concentration of 12.1 μM), 5 mM substrate (1 M stock in DMSO) supplemented with 100 μM NADPH, 10 mM glucose, and 5 μL GDH (20% w/v cell suspension). The reaction volume was adjusted to 1 mL in a glass vial by using sodium phosphate buffer (200 mM, pH 7.0) and incubated for 4 h at 20 °C and 1000 rpm. To determine the solvent stability of Ppo-Er1, the biocatalysis reaction conditions were adapted to include 2.4 μM Ppo-Er1 and 0%–40% v/v solvent (ethanol, ethyl acetate, DMSO, DMF, cyclohexane) in a total reaction volume of 1 mL for 50 min at 20 °C and 1000 rpm. All biocatalysis reactions were done in triplicate, biocatalysis results were verified by control reactions omitting the enzyme.

3.8. GC-Analysis

One milliliter biocatalysis reactions were extracted once with 500 μL methyl *tert*-butyl ether (incl. 1 g/L 1-octanol as internal standard). The phase separation was achieved by centrifugation of the biphasic sample, and the organic phase was separated and subjected to GC analysis (Table S1).

3.9. Gel Filtration

For the determination of the oligomeric state of Ppo-Er1, the Äkta purifier system employing a HiLoad 16/600 Superdex 75 pg column (GE Healthcare (Uppsala, Sweden)) and sodium phosphate buffer (50 mM, pH 7.5) was used. In a first step, the system was calibrated by using the gel filtration standard from Bio Rad (1.35–670 kDa Prod. no.: #1511901). Then flavin-saturated Ppo-Er1 was applied to system under identical conditions.

3.10. Melting Temperature

The unfolding temperature was determined by a *Thermo*FAD assay [56] using Rotor-Gene Q RT-PCR machine. Protein samples (0.5–0.3 mg/mL) in 20 μL sodium phosphate buffer pH 7 were measured using a temperature gradient from 25 to 90 °C, performing fluorescence measurements every 0.5 °C increase after a 10 s delay for signal stabilization. The measurements were performed in triplicates using 470 nm excitation wavelength and 510 nm emission wavelength.

4. Conclusions

Ppo-Er1 is a well-expressed, easy to purify, old yellow enzyme belonging to the recently introduced subclass III designation. In terms of cofactor preference, the enzyme accepts NADPH and NADH equally well, whereas pH and optimum temperature resemble those of previously described OYEs. Notably, the enzyme exhibits only slightly reduced performance (>70% conversion of 1 mM cyclohexenone) at lowered temperatures (10 °C) making it a possible candidate for the transformation of labile substrates such as some aldehydes. In addition, the enzyme was shown to have noteworthy stability in the presence of the solvents cyclohexane (up to at least 40% v/v), DMSO, and ethanol (up to 20% v/v).

The substrate profile analysis with a set of 19 representative alkenes allowed the establishment of Ppo-Er1's substrate scope highlighting its acceptance of a variety of linear and cyclic compounds with often excellent transformation efficiencies and exquisite stereoselectivity (e.g., 98% ee for carvone). Complementing this analysis with the determination of steady-state kinetics for ten of the substrates allowed us to conclude that Ppo-Er1 classifies well with other subgroup III old yellow enzymes.

In summary, our in-depth characterization of Ppo-Er1 allows the enlargement of the available panel of ene reductases with a versatile biocatalyst having interesting synthetic properties. Its introduction in the biocatalytic toolbox may further facilitate academic and industrial efforts when screening for biocatalysts capable of asymmetric double bond reduction. Looking forward, Ppo-Er1's performance could be further optimized via enzyme and process engineering.

Supplementary Materials
Figure S1: SDS-PAGE of the different purification steps for the ene reductase Ppo-ER1; Figure S2: Gel filtration of Ppo-ER1; Figure S3: Activity of Ppo-ER1 in the two used buffers; Table S1: Overview of the used GC-methods; Figure S4: Michaelis–Menten kinetic for maleimide; Figure S5: Michaelis–Menten kinetic for trans-β-methyl-β-nitrostyrene; Figure S6: Michaelis–Menten kinetic for cyclohexanone; Figure S7: Michaelis–Menten kinetic for cinnamaldehyde; Figure S8: Michaelis–Menten kinetic for 2-methyl-2-pentenal; Figure S9: Michaelis–Menten kinetic for carvone; Figure S10: Michaelis–Menten kinetic for citral; Figure S11: Michaelis–Menten kinetic for 2-methyl-2-cyclohexenone; Figure S12: Michaelis–Menten kinetic for cyclopentenone; Figure S13: Michaelis–Menten kinetic for hexenal; Figure S14: Comparison conversion with NADH and NADPH; Figure S15: Melting curve.

Author Contributions: Conceptualization: C.P. and R.M.B.; experimental work: D.A.; writing: D.A., C.P., and R.M.B. All authors have read and agreed to the published version of the manuscript.

References

1. List, B.; Yang, J.W. Chemistry. The organic approach to asymmetric catalysis. *Science* **2006**, *313*, 1584–1586. [CrossRef] [PubMed]
2. Huffman, M.A.; Fryszkowska, A.; Alvizo, O.; Borra-Garske, M.; Campos, K.R.; Canada, K.A.; Devine, P.N.; Duan, D.; Forstater, J.H.; Grosser, S.T.; et al. Design of an in vitro biocatalytic cascade for the manufacture of islatravir. *Science* **2019**, *366*, 1255–1259. [CrossRef] [PubMed]
3. Savile, C.K.; Janey, J.M.; Mundorff, E.C.; Moore, J.C.; Tam, S.; Jarvis, W.R.; Colbeck, J.C.; Krebber, A.; Fleitz, F.J.; Brands, J.; et al. Biocatalytic asymmetric synthesis of chiral amines from ketones applied to sitagliptin manufacture. *Science* **2010**, *329*, 305–309. [CrossRef] [PubMed]
4. Schober, M.; MacDermaid, C.; Ollis, A.A.; Chang, S.; Khan, D.; Hosford, J.; Latham, J.; Ihnken, L.A.F.; Brown, M.J.B.; Fuerst, D.; et al. Chiral synthesis of lsd1 inhibitor gsk2879552 enabled by directed evolution of an imine reductase. *Nat. Catal.* **2019**, *2*, 909–915. [CrossRef]
5. Adams, J.P.; Brown, M.J.B.; Diaz-Rodriguez, A.; Lloyd, R.C.; Roiban, G.D. Biocatalysis: A pharma perspective. *Adv. Synth. Catal.* **2019**, *361*, 2421–2432. [CrossRef]
6. Toogood, H.S.; Scrutton, N.S. Discovery, characterisation, engineering and applications of ene reductases for industrial biocatalysis. *ACS Catal.* **2019**, *8*, 3532–3549. [CrossRef]
7. Toogood, H.S.; Gardiner, J.M.; Scrutton, N.S. Biocatalytic reductions and chemical versatility of the old yellow enzyme family of flavoprotein oxidoreductases. *Chemcatchem* **2010**, *2*, 892–914. [CrossRef]
8. Winkler, C.K.; Faber, K.; Hall, M. Biocatalytic reduction of activated cc-bonds and beyond: Emerging trends. *Curr. Opin. Chem. Biol.* **2018**, *43*, 97–105. [CrossRef]
9. Shimoda, K.; Ito, D.I.; Izumi, S.; Hirata, T. Novel reductase participation in the syn-addition of hydrogen to the c=c bond of enones in the cultured cells of *Nicotiana tabacum*. *J. Chem. Soc. Perkin Trans.* **1996**, *1*, 355–358. [CrossRef]
10. Knowles, W.S. Asymmetric hydrogenations (nobel lecture). *Angew. Chem. Int. Ed. Engl.* **2002**, *41*, 1999–2007.
11. Noyori, R. Asymmetric catalysis: Science and opportunities (nobel lecture). *Angew. Chem. Int. Ed.* **2002**, *41*, 2008–2022. [CrossRef]
12. Knaus, T.; Toogood, H.S.; Scrutton, N.S. Ene-reductases and their applications. In *Green Biocatalysis*; John Wiley & Sons, Inc.: Hoboken, NJ, USA, 2016; pp. 473–488.
13. Steinkellner, G.; Gruber, C.C.; Pavkov-Keller, T.; Binter, A.; Steiner, K.; Winkler, C.; Lyskowski, A.; Schwamberger, O.; Oberer, M.; Schwab, H.; et al. Identification of promiscuous ene-reductase activity by mining structural databases using active site constellations. *Nat. Commun.* **2014**, *5*, 4150. [CrossRef] [PubMed]
14. Hecht, K.; Buller, R. Ene-reductases in pharmaceutical chemistry. In *Pharmaceutical Biocatalysis: Chemoenzymatic Synthesis of Active Pharmaceutical Ingredients*; Grunwald, P., Ed.; Jenny Stanford Publishing: Singapore, 2019.

15. Peters, C.; Frasson, D.; Sievers, M.; Buller, R. Novel old yellow enzyme subclasses. *Chembiochem* **2019**, *20*, 1569–1577. [CrossRef] [PubMed]
16. Warburg, O.; Christian, W. Ein zweites sauerstoffübertragendes ferment und sein absorptionsspektrum. *Naturwissenschaften* **1932**, *20*, 688. [CrossRef]
17. Haas, E. Isolierung eines neuen gelben ferments. *Biochem. Z* **1938**, *298*, 369–390.
18. Breithaupt, C.; Strassner, J.; Breitinger, U.; Huber, R.; Macheroux, P.; Schaller, A.; Clausen, T. X-ray structure of 12-oxophytodienoate reductase 1 provides structural insight into substrate binding and specificity within the family of oye. *Structure* **2001**, *9*, 419–429. [CrossRef]
19. Kohli, R.M.; Massey, V. The oxidative half-reaction of old yellow enzyme: The role of tyrosine 196. *J. Biol. Chem.* **1998**, *273*, 32763–32770. [CrossRef]
20. Fox, K.M.; Karplus, P.A. Old yellow enzyme at 2 å resolution: Overall structure, ligand binding, and comparison with related flavoproteins. *Structure* **1994**, *2*, 1089–1105. [CrossRef]
21. Barna, T.M.; Khan, H.; Bruce, N.C.; Barsukov, I.; Scrutton, N.S.; Moody, P.C.E. Crystal structure of pentaerythritol tetranitrate reductase: "Flipped" binding geometries for steroid substrates in different redox states of the enzyme. *J. Mol. Biol.* **2001**, *310*, 433–447. [CrossRef]
22. Kitzing, K.; Fitzpatrick, T.B.; Wilken, C.; Sawa, J.; Bourenkov, G.P.; Macheroux, P.; Clausen, T. The 1.3 å crystal structure of the flavoprotein yqjm reveals a novel class of old yellow enzymes. *J. Biol. Chem.* **2005**, *280*, 27904–27913. [CrossRef]
23. Adalbjörnsson, B.V.; Toogood, H.S.; Fryszkowska, A.; Pudney, C.R.; Jowitt, T.A.; Leys, D.; Scrutton, N.S. Biocatalysis with thermostable enzymes: Structure and properties of a thermophilic 'ene'-reductase related to old yellow enzyme. *ChemBioChem* **2010**, *11*, 197–207. [CrossRef] [PubMed]
24. Nizam, S.; Verma, S.; Borah, N.N.; Gazara, R.K.; Verma, P.K. Comprehensive genome-wide analysis reveals different classes of enigmatic old yellow enzyme in fungi. *Sci. Rep.* **2014**, *4*, 4013. [CrossRef] [PubMed]
25. Pietruszka, J.; Schölzel, M. Ene reductase-catalysed synthesis of (r)-profen derivatives. *Adv. Synth. Catal.* **2012**, *354*, 751–756. [CrossRef]
26. Li, Z.N.; Wang, Z.X.; Meng, G.; Lu, H.; Huang, Z.D.; Chen, F.E. Identification of an ene reductase from yeast *Kluyveromyces marxianus* and application in the asymmetric synthesis of (R)-profen esters. *Asian. J. Org. Chem.* **2018**, *7*, 763–769. [CrossRef]
27. Waller, J.; Toogood, H.S.; Karuppiah, V.; Rattray, N.J.W.; Mansell, D.J.; Leys, D.; Gardiner, J.M.; Fryszkowska, A.; Ahmed, S.T.; Bandichhor, R.; et al. Structural insights into the ene-reductase synthesis of profens. *Org. Biomol. Chem.* **2017**, *15*, 4440–4448. [CrossRef] [PubMed]
28. Fryszkowska, A.; Fisher, K.; Gardiner, J.M.; Stephens, G.M. A short, chemoenzymatic route to chiral β-aryl-γ-amino acids using reductases from anaerobic bacteria. *Org. Biomol. Chem.* **2010**, *8*, 533–535. [CrossRef] [PubMed]
29. Winkler, C.K.; Clay, D.; Davies, S.; O'Neill, P.; McDaid, P.; Debarge, S.; Steflik, J.; Karmilowicz, M.; Wong, J.W.; Faber, K. Chemoenzymatic asymmetric synthesis of pregabalin precursors via asymmetric bioreduction of β-cyanoacrylate esters using ene-reductases. *J. Org. Chem.* **2013**, *78*, 1525–1533. [CrossRef]
30. Winkler, C.K.; Clay, D.; Turrini, N.G.; Lechner, H.; Kroutil, W.; Davies, S.; Debarge, S.; O'Neill, P.; Steflik, J.; Karmilowicz, M.; et al. Nitrile as activating group in the asymmetric bioreduction of beta-cyanoacrylic acids catalyzed by ene-reductases. *Adv. Synth. Catal.* **2014**, *356*, 1878–1882. [CrossRef]
31. Janicki, I.; Kiełbasiński, P.; Turrini, N.G.; Faber, K.; Hall, M. Asymmetric bioreduction of β-activated vinylphosphonate derivatives using ene-reductases. *Adv. Synth. Catal.* **2017**, *359*, 4190–4196. [CrossRef]
32. Bertolotti, M.; Brenna, E.; Crotti, M.; Gatti, F.G.; Monti, D.; Parmeggiani, F.; Santangelo, S. Substrate scope evaluation of the enantioselective reduction of β-alkyl-β-arylnitroalkenes by old yellow enzymes 1–3 for organic synthesis applications. *ChemCatChem* **2016**, *8*, 577–583. [CrossRef]
33. Dobrijevic, D.; Benhamou, L.; Aliev, A.E.; Méndez-Sánchez, D.; Dawson, N.; Baud, D.; Tappertzhofen, N.; Moody, T.S.; Orengo, C.A.; Hailes, H.C.; et al. Metagenomic ene-reductases for the bioreduction of sterically challenging enones. *RSC Adv.* **2019**, *9*, 36608–36614. [CrossRef]
34. Sheng, X.; Yan, M.; Xu, L.; Wei, M. Identification and characterization of a novel old yellow enzyme from *Bacillus subtilis* str.168. *J. Mol. Catal. B Enzym.* **2016**, *130*, 18–24. [CrossRef]
35. Zhang, H.; Gao, X.; Ren, J.; Feng, J.; Zhang, T.; Wu, Q.; Zhu, D. Enzymatic hydrogenation of diverse activated alkenes. Identification of two bacillus old yellow enzymes with broad substrate profiles. *J. Mol. Catal. B Enzym.* **2014**, *105*, 118–125. [CrossRef]

36. Gao, X.; Ren, J.; Wu, Q.; Zhu, D. Biochemical characterization and substrate profiling of a new nadh-dependent enoate reductase from *Lactobacillus casei*. *Enzyme Microb. Technol.* **2012**, *51*, 26–34. [CrossRef] [PubMed]
37. Brown, B.J.; Deng, Z.; Karplus, P.A.; Massey, V. On the active site of old yellow enzyme: Role of histidine 191 and asparagine 194. *J. Biol. Chem.* **1998**, *273*, 32753–32762. [CrossRef]
38. Spiegelhauer, O.; Dickert, F.; Mende, S.; Niks, D.; Hille, R.; Ullmann, M.; Dobbek, H. Kinetic characterization of xenobiotic reductase a from *Pseudomonas putida* 86. *Biochemistry* **2009**, *48*, 11412–11420. [CrossRef]
39. Scholtissek, A.; Tischler, D.; Westphal, A.; van Berkel, W.; Paul, C. Old yellow enzyme-catalysed asymmetric hydrogenation: Linking family roots with improved catalysis. *Catalysts* **2017**, *7*, 130. [CrossRef]
40. Hummel, W.; Gröger, H. Strategies for regeneration of nicotinamide coenzymes emphasizing self-sufficient closed-loop recycling systems. *J. Biotechnol.* **2014**, *191*, 22–31. [CrossRef]
41. Knaus, T.; Paul, C.E.; Levy, C.W.; de Vries, S.; Mutti, F.G.; Hollmann, F.; Scrutton, N.S. Better than nature: Nicotinamide biomimetics that outperform natural coenzymes. *J. Am. Chem. Soc.* **2016**, *138*, 1033–1039. [CrossRef]
42. Lee, S.H.; Choi, D.S.; Pesic, M.; Lee, Y.W.; Paul, C.E.; Hollmann, F.; Park, C.B. Cofactor-free, direct photoactivation of enoate reductases for the asymmetric reduction of c=c bonds. *Angew. Chem. Int. Ed.* **2017**, *56*, 8681–8685. [CrossRef]
43. Litthauer, S.; Gargiulo, S.; van Heerden, E.; Hollmann, F.; Opperman, D.J. Heterologous expression and characterization of the ene-reductases from *Deinococcus radiodurans* and *Ralstonia metallidurans*. *J. Mol. Catal. B Enzym.* **2014**, *99*, 89–95. [CrossRef]
44. Davies, M.T. A universal buffer solution for use in ultra-violet spectrophotometry. *Analyst* **1959**, *84*, 248–251. [CrossRef]
45. Peters, C.; Kölzsch, R.; Kadow, M.; Skalden, L.; Rudroff, F.; Mihovilovic, M.D.; Bornscheuer, U.T. Identification, characterization, and application of three enoate reductases from *Pseudomonas putida* in in vitro enzyme cascade reactions. *ChemCatChem* **2014**, *6*, 1021–1027. [CrossRef]
46. Pesic, M.; Fernández-Fueyo, E.; Hollmann, F. Characterization of the old yellow enzyme homolog from *Bacillus subtilis* (yqjm). *ChemistrySelect* **2017**, *2*, 3866–3871. [CrossRef]
47. Xu, M.-Y.; Pei, X.-Q.; Wu, Z.-L. Identification and characterization of a novel "thermophilic-like" old yellow enzyme from the genome of *Chryseobacterium* sp. Ca49. *J. Mol. Catal. B Enzym.* **2014**, *108*, 64–71. [CrossRef]
48. Zheng, L.; Lin, J.; Zhang, B.; Kuang, Y.; Wei, D. Identification of a yeast old yellow enzyme for highly enantioselective reduction of citral isomers to (R)-citronellal. *Bioresour. Bioprocess.* **2018**, *5*, 1–12. [CrossRef]
49. Rudroff, F.; Mihovilovic, M.D.; Gröger, H.; Snajdrova, R.; Iding, H.; Bornscheuer, U.T. Opportunities and challenges for combining chemo- and biocatalysis. *Nat. Catal.* **2018**, *1*, 12–22. [CrossRef]
50. Fryszkowska, A.; Toogood, H.; Sakuma, M.; Gardiner, J.M.; Stephens, G.M.; Scrutton, N.S. Asymmetric reduction of activated alkenes by pentaerythritol tetranitrate reductase: Specificity and control of stereochemical outcome by reaction optimisation. *Adv. Synth. Catal.* **2009**, *351*, 2976–2990. [CrossRef]
51. Tischler, D.; Gadke, E.; Eggerichs, D.; Gomez Baraibar, A.; Mugge, C.; Scholtissek, A.; Paul, C.E. Asymmetric reduction of (r)-carvone through a thermostable and organic-solvent-tolerant ene-reductase. *Chembiochem* **2019**. [CrossRef]
52. Robescu, M.S.; Niero, M.; Hall, M.; Cendron, L.; Bergantino, E. Two new ene-reductases from photosynthetic extremophiles enlarge the panel of old yellow enzymes: Ctoye and gsoye. *Appl. Microbiol. Biotechnol.* **2020**, *104*, 2051–2066. [CrossRef]
53. Riedel, A.; Mehnert, M.; Paul, C.E.; Westphal, A.H.; van Berkel, W.J.; Tischler, D. Functional characterization and stability improvement of a 'thermophilic-like' ene-reductase from *Rhodococcus opacus* 1cp. *Front. Microbiol.* **2015**, *6*, 1073. [CrossRef] [PubMed]
54. Hall, M.; Stueckler, C.; Ehammer, H.; Pointner, E.; Oberdorfer, G.; Gruber, K.; Hauer, B.; Stuermer, R.; Kroutil, W.; Macheroux, P.; et al. Asymmetric bioreduction of c=c bonds using enoate reductases opr1, opr3 and yqjm: Enzyme-based stereocontrol. *Adv. Synth. Catal.* **2008**, *350*, 411–418. [CrossRef]
55. Josuran, R. Prot pi. Available online: https://www.protpi.ch/ (accessed on 22 January 2020).
56. Forneris, F.; Orru, R.; Bonivento, D.; Chiarelli, L.R.; Mattevi, A. Thermofad, a thermofluor-adapted flavin ad hoc detection system for protein folding and ligand binding. *FEBS J.* **2009**, *276*, 2833–2840. [CrossRef] [PubMed]

Highly Selective Oxidation of 5-Hydroxymethylfurfural to 5-Hydroxymethyl- 2-Furancarboxylic Acid by a Robust Whole-Cell Biocatalyst

Ran Cang [1,†], Li-Qun Shen [1,†], Guang Yang [1], Zhi-Dong Zhang [3], He Huang [1,2,*] and Zhi-Gang Zhang [1,*]

1 School of Pharmaceutical Sciences, Nanjing Tech University, 30 Puzhu Road(S), Nanjing 211816, China; crnj123@126.com (R.C.); shenliqun163@126.com (L.-Q.S.); qwe88227518@njtech.edu.cn (G.Y.)
2 State Key Laboratory of Materials-Oriented Chemical Engineering, Nanjing Tech University, 30 Puzhu Road(S), Nanjing 211816, China
3 Institute of Microbiology, Xinjiang Academy of Agricultural Sciences, 403 Nanchang Rd, Wulumuqi 830091, China; zhangzheedong@sohu.com
* Correspondence: huangh@njtech.edu.cn (H.H.); zhangzg@njtech.edu.cn (Z.-G.Z.)
† These authors contributed equally to this work.

Abstract: Value-added utilization of biomass-derived 5-hydroxymethylfurfural (HMF) to produce useful derivatives is of great interest. In this work, extremely radiation resistant *Deinococcus wulumuqiensis* R12 was explored for the first time as a new robust biocatalyst for selective oxidation of HMF to 5-hydroxymethylfuroic acid (HMFCA). Its resting cells exhibited excellent catalytic performance in a broad range ofpH and temperature values, and extremely high tolerance to HMF and the HMFCA product. An excellent yield of HMFCA (up to 90%) was achieved when the substrate concentration was set to 300 mM under the optimized reaction conditions. In addition, 511 mM of product was obtained within 20 h by employing a fed-batch strategy, affording a productivity of 44 g/L per day. Of significant synthetic interest was the finding that the *D. wulumuqiensis* R12 cells were able to catalyze the selective oxidation of other structurally diverse aldehydes to their corresponding acids with good yield and high selectivity, indicating broad substrate scope and potential widespread applications in biotechnology and organic chemistry.

Keywords: biocatalysis; extremophile; 5-hydroxymethylfurfural; 5-hydroxymethylfuroic acid; platform chemicals; whole cells

---

1. Introduction

The production of bio-fuels and chemicals from carbon-neutral and renewable biomass is attracting increasing interest [1–5]. Biomass is regarded as a sustainable resource from which some platform chemicals can be manufactured [6,7]. 5-hydroxymethylfurfural (HMF), derived from lignocellulosic materials via dehydration of carbohydrates, is one of the most important platform chemicals [8–10]. It has been listed as one of "Top 10+4" bio-based chemicals by the U.S. Department of Energy (DOE) [11], being applied in the synthesis of a variety of value-added pharmaceutical and biomaterial intermediates [12]. Due to its high reactivity, HMF is a versatile molecule that can be converted into various useful furan derivatives [12–14]. Its structure comprises a furan ring, an aldehyde group and a hydroxymethyl group which can be subjected to upgrading processes by selective redox reactions, leading to 5-hydroxymethylfuroic acid (HMFCA), 2,5-diformylfuran (DFF), 5-formylfuroic acid (FFCA), 2,5-furandicarboxylic acid (FDCA), maleic anhydride (MA) and 2,5-bis (hydroxymethyl) furan (BHMF) (Scheme 1). Among these HMF derivatives, the completely oxidized product FDCA displays very promising application potential and may serve as a "greener" substitute for terephthalate in the manufacture of polyester and polyamide materials [15,16]. HMFCA is the oxidation product of the aldehyde group in HMF and a promising starting material for the synthesis of various polyesters [17]. It was reported that HMFCA can also be used as an antitumor agent and interleukin inhibitor [18,19].

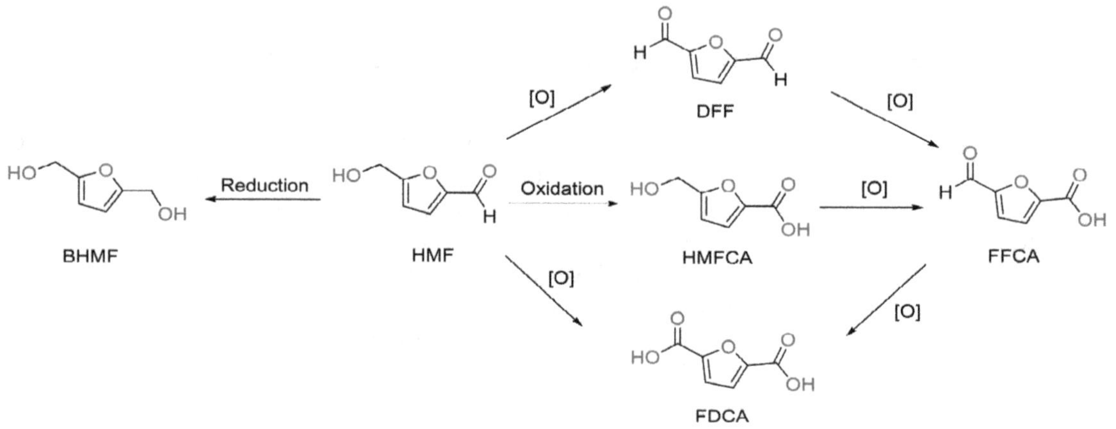

Scheme 1. Catalytic biotransformation of 5-hydroxymethylfurfural (HMF) into high value derivatives.

In order to form HMFCA, selective oxidation of the aldehyde group in HMF is required, while the alcohol group is left intact. Chemoselective oxidation methods are mainly used in the synthesis of HMFCA from HMF, in which noble metal catalysts are generally used [12,20–22]. Recently, HMF was selectively oxidized to HMFCA by an immobilized molybdenum complex in toluene within 3 h, with a yield of approximately 87% [23]. Han et al. reported a selective and mild photocatalytic method for HMFCA synthesis from HMF under ultraviolet and visible light conditions with a yield of 90–95% [24]. In addition, the conversion of HMF to HMFCA via the Cannizzaro reaction is of great value [25,26]. However, the maximal selectivity of HMFCA was 50% due to the formation of an equimolar by-product.

Biocatalytic oxidation of HMF to HMFCA represents a promising alternative to chemical methods [14,27]. Biocatalysis offers many advantages, such as mild, environmentally friendly reaction conditions and often excellent selectivity, as well as high efficiency. However, compared to chemical methods, there are only a few reports on biotransformation of HMF to selectively form HMFCA in the literature [28–32]. In seminal work, Sheldon et al. reported the chloroperoxidase-catalyzed oxidation of HMF affording HMFCA with a selectivity of 25–40% [31]. Krystof et al. reported lipase-mediated and peracid-assisted oxidation of the HMF process to produce HMFCA [32]. Recently, Li and co-workers made use of a molybdenum-dependent enzyme—xanthine oxidase from *Escherichia coli*—for the biocatalytic oxidation of HMF to form HMFCA, with 94% yield and 99% selectivity [29].

Relative to the use of isolated enzymes, we believe that, in HMF oxidation, whole-cell biocatalysts have advantages. They are not only inexpensive and relatively stable, but they also do not require cofactor regeneration [27,33]. Biocatalysis is more efficient when recombinant whole cells that overexpress the enzyme(s) important for catalysis are used [34]. However, employing whole-cell biocatalysts for HMF oxidation is still challenging due to the well-known toxicity of HMF to microbial cells [30]. In addition, due to the variety of enzymes in microbial cells many side reactions are likely to occur during the process of HMF oxidation with formation of HMFCA [28]. Hence, exploring highly tolerant and selective microbial strains is crucial for the biotransformation of HMF into value-added derivative. To our knowledge, there are only a few studies on whole-cell-catalyzed selective oxidation of HMF to form HMFCA in the literature [28,30]; processes that are accompanied by a certain amount of HMF derivatives as byproducts. It was reported that some *Pseudomonas* strains have an HMF degradation pathway, in which HMF is converted to HMFCA as an intermediate [35–37]. A careful literature search did not reveal any studies describing the use of this system for the production of HMFCA. In 2010, Koopman et al. reported the production of 2,5-furandicarboxylic acid (FDCA) from HMF by using recombinant *P. putida* S12_hmfH. As part of this biotransformation, HMFCA hardly accumulated, leading to a mixture of other metabolites [38,39]. Therefore, in the challenging quest to obtain large amounts of pure HMFCA, the use of the *Pseudomonas* strain metabolic pathway is not feasible. Moreover, long standing issues still exist, such as low substrate loading, substrate toxicity and insufficient selectivity, etc. Therefore, searching for new and robust biocatalytic systems with high selectivity is a demanding task.

Extremophiles are organisms that have evolved to thrive under one or more extreme adverse environmental conditions where other organisms cannot survive [40,41]. They are regarded as an ideal and valuable source of biocatalysts, allowing biotransformation under relatively harsh industrial conditions [42–44]. Nevertheless, employing whole-cells or isolated enzymes derived from extremophiles for biocatalysis in a general manner is just beginning to be implemented experimentally. Recently, a *Deinococcus sp*, designated as *Deinococcus wulumuqiensis* R12, was isolated from radiation-polluted soil [45,46]. Previous studies showed that it is phylogenetically more closely related to a prototype strain of the *Deinococcus* genus, namely *Deinococcus radiodurans* R1 [47]. It was found that this strain was capable of producing carotenoids with good yield, and related biosynthesis genes were subsequently cloned and heterogeneously expressed in *E. coli.* by Xu et al. [48]. Furthermore, its whole genome was sequenced by Huang et al. [49]. Recently, genes encoding heat shock proteins from *D. wulumuqiensis* R12 were introduced into *Clostridium acetobutylicum* ATCC824 in order to improve the robustness and butanol titers of host cells [50]. Considering the robustness of *D. wulumuqiensis* R12, it would be of great interest to explore the catalytic properties of its whole cells in biotransformation or bioconversion.

In this study, we report that the radiation resistant strain *D. wulumuqiensis* R12 that can indeed be used as a whole-cell biocatalyst in HMFCA synthesis by selective oxidation of HMF (Scheme 2). The catalytic properties of this strain were evaluated in the transformation of HMF, and the reaction conditions were optimized. In addition, the substrate scope of this new whole-cell biocatalyst was also investigated.

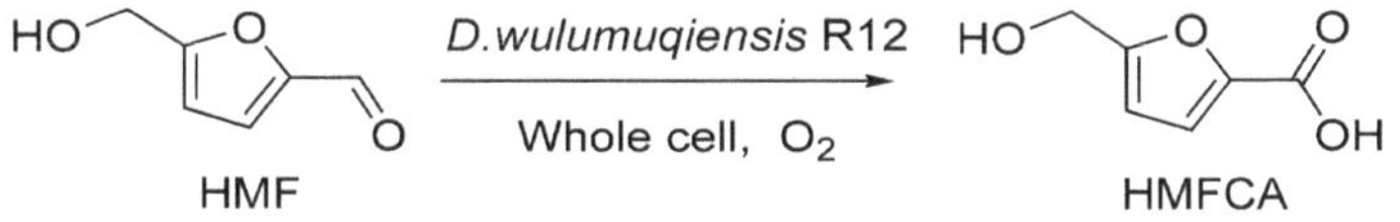

HMF —(*D.wulumuqiensis* R12; Whole cell, O_2)→ HMFCA

Scheme 2. Chemoselective oxidation of MF to 5-hydroxymethylfuroic acid (HMFCA) with whole *D. wulumuqiensis* R12 cells.

2. Results and Discussion

2.1. Growing and Resting Deinococcus Cells as Catalysts in HMF Oxidation with Selective Formation of HMFCA

Similar to the prototype strain of the *Deinococcus* genus, *D. radiodurans* R1, *D. wulumuqiensis* R12 is also well known for its excellent ability to resist extremely high doses of gamma and UV radiation [45]. In order to explore its potential applications in biocatalysis, growing and resting cells of this strain were applied as biocatalysts in the conversion of HMF to form HMFCA. As shown in Figure 1a, 100 mM of the HMF substrate were converted almost completely within 12 h using resting cells, whereas growing cells gave only a 32% yield at a prolonged reaction time of 36 h. Resting cells enabled a much higher yield with more than 98% of HMFCA and a trace amount of 2,5-bis (hydroxymethyl) furan (BHMF) as sole byproduct, indicating excellent chemoselectivity in this biocatalytic process. Increasing substrate concentration further did not affect the selectivity of the resting cells (Figure S2). Considering the reported degradation mechanism of HMF in microbial cells [51], it is reasonable to speculate that the intermediate HMF alcohol (from HMF reduction) was almost completely oxidized in a very short time to form the final HMFCA, or the HMF substrate was oxidized directly—which constitutes a different mechanistic hypothesis. However, to validate this inference, more efforts need to be invested.

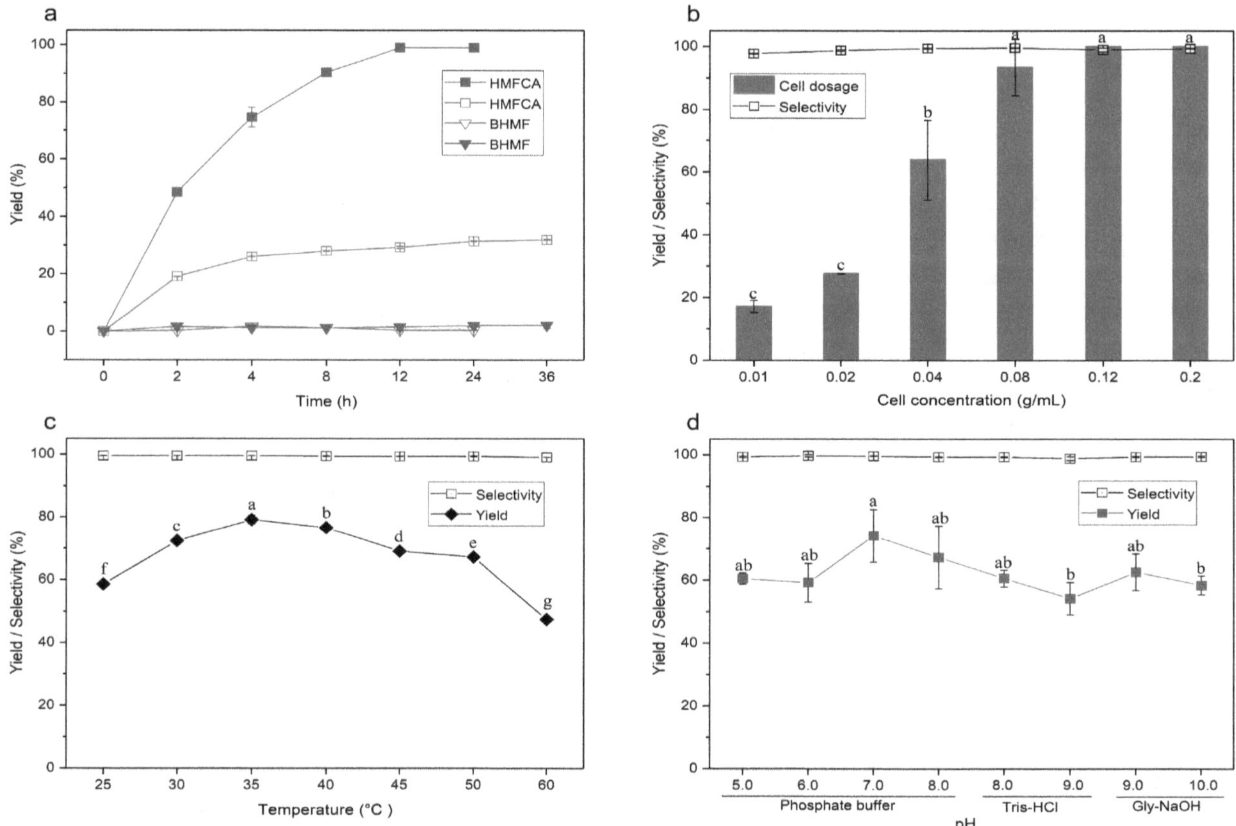

Figure 1. The influence of various factors on HMFCA synthesis by HMF. General conditions unless otherwise stated: 100 mM HMF, 0.12 g/mL microbial cells, 100 mM,pH 7.4, phosphate buffer, 850 rpm, 30 °C, 4h reaction time. (**a**) Resting (solid symbols) and growing cells (open symbols), wherein tryptone glucose yeast extract (TGY) culture was used when growing cells were employed under 200 rpm, 30 °C; (**b**) cell concentration; (**c**) temperature; and (**d**)pH values. Time courses of HMF biotransformation using resting (solid symbols) and growing cells (open symbols).

In addition, we were pleased to discover the performance of two other *Deinococcus* strains stored in our lab, *D. radiodurans* R1 and *Deinococcus xibeiensis* R13, which were also used in the biocatalytic oxidation of HMF to HMFCA. It was found that both radiation resistant strains selectively oxidized HMF with formation of HMFCA. The conversions achieved by *D. radiodurans* R1 and *D. xibeiensis* R13 were slightly lower than that of *D. wulumuqiensis* R12 under the same reaction conditions (Figure S3). These results suggest that the *D. wulumuqiensis* R12 cells act as a catalytic system with high activity and excellent chemoselectivity in the oxidation of HMF to HMFCA. Its catalytic properties were subsequently investigated in greater detail (Figure 1).

2.2. *Effect of Cell Dosage in the Reaction System for HMFCA Synthesis*

Figure 1b shows the influence of microbial cell dosage on HMFCA synthesis based on selective oxidation of HMF. The yield of HMFCA increased steadily from 18% to 99% with increasing cell dosage in the presence of 100 mM of HMF substrate. The maximal yield of 99% was achieved when the cell dosage reached 0.12 g/mL, and further increasing did not improve the HMFCA yield, indicating that the biocatalyst was potentially saturated by substrate under the given reaction conditions. Our results imply that the conversion of HMF to HMFCA correlates with the cell dosage employed in the biocatalytic system. A small amount of cell dosage was required to reach maximal conversion when the substrate concentration decreased to 40 mM under the same reaction conditions (Figure S4). In addition, cell dosage had no significant effect on the selectivity of the reactions (>98%). A higher cell dosage may result in higher viscosity, however, which could impact mass transfer of the reaction mixture. Thus, the optimal cell dosage of 0.12 g/mL wet cells was used in subsequent experiments.

2.3. *Effect of Temperature and pH on HMFCA Synthesis*

The influence of temperature and pH on HMFCA synthesis in the whole-cell catalyzed oxidation of HMF was also studied. As shown in Figure 1c, the effect of reaction temperature on HMF selective oxidation was determined by performing the transformation at different temperatures. Remarkably, the microbial cell biocatalyst showed considerable activity at a broad temperature range, from 25 to 60 °C. The maximal substrate conversion of 79% was obtained at 35 °C after 4hinthepresence of 100 mM HMF substrate.

In addition, even at 50 °C, 67% of the HMF substrate was converted to HMFCA, which is in accord with an early report that *D. wulumuqiensis* R12 has a broad growth temperature range [45]. Slightly decreased conversion is possibly due to the inactivation of the enzymes in the microbial cells at 60 °C. It should be mentioned that HMFCA was obtained as essentially the only oxidative product —with a yield of 99%—in the reaction within the temperature range of 25 °C t o 6 0 °C, indicating excellent catalytic selectivity of the whole-cell biocatalyst. Considering the thermostability of cells and energy efficiency, a temperature of 35 °C was set for subsequent experiments.

In addition, we further studied the pH profile of the whole-cell catalyst in HMFCA synthesis via selective oxidation of HMF (Figure 1d). It was found that the microbial cells had a broadpH activity profile and exhibited a particularly good catalytic performance in the pH range of 5.0 to 10.0. The best yield of 81% was achieved in 100 mM phosphate buffer at a pH 7.0 after 4 h. Interestingly, a conversion percentage of 60% and 58% was obtained in phosphate buffer ofpH 5.0 and Gly-NaOH buffer ofpH 10.0, respectively, after a reaction time of 4 h.

In addition, it appeared that the buffer types had a moderate influence on the conversion of HMF, as a yield of 54% was obtained in Tris-HCl buffer (pH 9.0), compared to 63% in Gly-NaOH buffer at the samepH . One should not be surprised that *D. wulumuqiensis* R12 cells are able to resist such harsh reaction conditions with extremepH values. In their studies, Wang et al. reported that the *D. wulumuqiensis* R12 strain is able to grow in a widepH range from 5.0 to 12.0 [45]. Compared to *Comamonas testosterone* SC1588, which has been applied in HMFCA synthesis from HMF [28], *D. wulumuqiensis* R12 cells showed higher tolerance to extremepH values. Therefore, the optimalpH value of 7.0 was selected for all subsequent experiments.

2.4. *Inhibitory and Toxic Effect of Substrate*

HMF is a well-known toxic inhibitor of microbial cells, inhibiting their growth and hindering their upgrading of HMF by whole-cell biocatalysis [52]. The catalytic performance of *D. wulumuqiensis* R12 cells towards HMF under varying concentrations was therefore tested. As shown in Figure 2a, HMFCA was synthesized in 99% yield within 12 h when the substrate concentration was 150 mM. However, the yield decreased slightly to 81% at the substrate concentration of 200 mM. The yield decreased gradually in the substrate concentration range of 250–1000 mM. Remarkably, 41% yield of HMFCA was obtained when the substrate concentration reached 500 mM, and a yield of 23% was observed at the substrate concentration of 1000 mM. The essentially complete selectivity remained almost constant at these varying substrate concentrations.

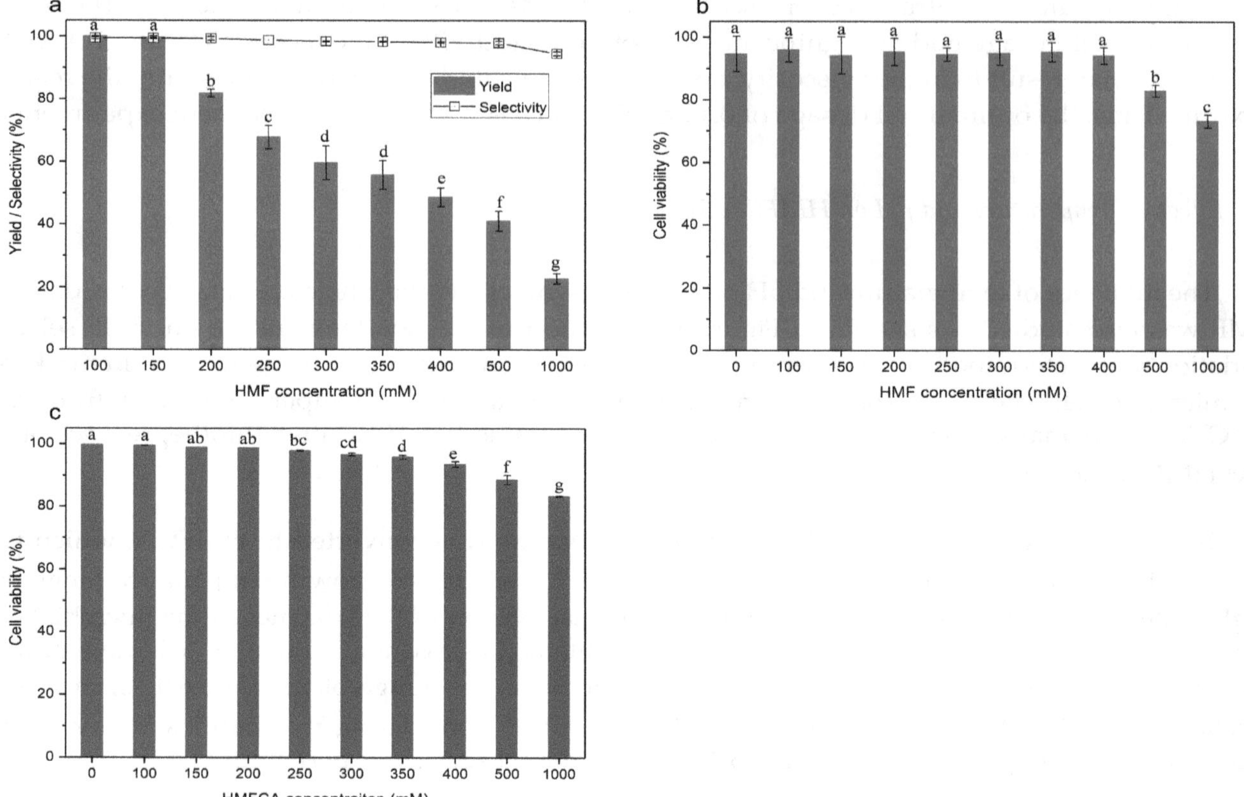

Figure 2. Effects of HMF concentration on (**a**) HMFCA synthesis and (**b**) cell viability. Reaction conditions: 0.12 g/mL microbial cells, in phosphate buffer (100 mM,pH 7.0) under 850 rpm and 35 °C. Fresh harvested cells incubated without HMF/HMFCA under identical conditions were used as a control. Reaction periods: 12 h for 100 mM 150 mM; 24 h for 200 mM, 250 mM; and 36 h for 300 mM, 350 m, 400 mM, 500 mM, 1000 mM. (**c**) Effects of the product concentration on cell viability. HMFCA was incubated for 24 h under the same conditions.

It is well known that the cytotoxicity of HMF to microbial cells is a key parameter in whole-cell biocatalytic conversion of HMF. As shown in Figure 2b, the effect of substrate concentration on cell viability was investigated by using a cell viability assay. The microbial cells were incubated with varying concentrations of HMF under the usual reaction conditions, and the cell viability was subsequently measured using an Annexin V-FITC/PI Apoptosis Detection Kit. Microbial cells incubated in buffer without substrate under the same reaction condition were used as a control. Good cell viability (94%) was unexpectedly obtained in the presence of 400 mM substrate, which was not significantly different than that of the control. This result suggests that the microbial cells can tolerate as much as 400 mM of HMF substrate without losing viability. Further increasing HMF concentration to 500 mM led only to a slight decrease in cell viability to 83%. When 1000 mM of HMF substrate was used, 73% cell viability was still observed, indicating that *D. wulumuqiensis* R12 is extremely tolerant to HMF substrate. However, the conversion of HMF in the oxidation reaction decreased dramatically with increased substrate concentration. Since detailed toxic mechanisms of HMF to microbial cells are not completely understood, nor the reason for the extreme resistance of *D. wulumuqiensis* R12 cells to harsh environmental factors, the present results are not considered surprising.

2.5. Inhibitory and Toxic Effect of the HMFCA Product

In the oxidation of HMF to HMFCA, the product is an acidic compound harboring a carboxylic and hydroxyl group. In our work, the pH of the reaction system decreased over the reaction time due to the accumulation of the HMFCA product. This situation could conceivably become critical when high concentrations of substrate are produced. Thus, it would be of great interest and importance to investigate the possible inhibition and toxicity of the product towards microbial cells. Microbial cells incubated in phosphate buffer without the product was used as the control for such investigations. Based on the high cell viability (>93%), which was comparable to that of the control, there appeared to be no significant toxicity towards the microbial cells at HMFCA product concentrations less than 400 mM, as shown in Figure 2c. However, a product amount of 500 mM resulted in slight toxicity, as the cell viability value decreased to 87%. A further increased product concentration of 1000 mM led to significant toxicity of the microbial cells, with a viability of 84%. To our surprise, the data showed that the product toxicity to the viability of microbial cells was not as high as expected, even at extremely high concentrations. Considering that *D. wulumuqiensis* R12 is a robust strain, able to grow in a broad range ofpH values, one should not be surprised that the microbial cells are highly resistant to HMFCA production, with excellent cell viability at extremely high product concentrations. Further product toxicity tests with still higher concentrations were not performed, as the conversion of HMF was already very low at the substrate concentration of 1000 mM.

2.6. Manufacture of HMFCA Under Optimized Conditions

Obtaining large amounts of HFMCA is highly desired in biocatalytic oxidation of HMF by whole-cell biocatalysis, with great potential applications in industrial production. Therefore, further enhancement of the catalytic performance of *D. wulumuqiensis* R12 was investigated by optimizing the biocatalytic parameters of the conversion process. It was found that the HMFCA product yields were affected significantly by increasing the substrate concentration. For example, due to the known negative effect of HMF, the yield of HMFCA decreased significantly when HMF concentrations were higher than 300 mM (Figure 2a). Based on the catalytic properties of this strain, increasing the dosage of microbial cells in the reaction may further enhance the HMFCA yield. Recently, Zhang et al. reported that improved synthesis of HFMCA from HMF was obtained by tuning the pH of the reaction mixture using NaOH solution during the catalytic process [28]. Thus, we decided to employ the same strategy. In addition, it has been reported that adding furfural and furfural alcohol as inducers during cultivation of microbial cells for biocatalysis can trigger the expression of the enzymes responsible for HMF oxidation, which can facilitate HMFCA production. Therefore, three strategies (increasing microbial cell dosage, using inducing cells and tuningpH of reaction mixture) were applied in subsequent studies.

Increasing the dosage of microbial cells proved to be effective for enhancing the yield of HMFCA (Figure S4a). For example, in the presence of 300 mM HMF, the yield of HMFCA increased considerably from 59% to 71% when the concentration of cells increased from 0.12 g/mL to 0.2 g/mL. A further increase in cells dosage was not performed considering the negative effect of higher viscosity in the reaction mixture. On the other hand, influencing the expression of cells by the use of furfural and furfural alcohol for enhancing the yield of HMFCA proved not to be effective (Figure S4b). The reason for this might be that enzymes in *D. wulumuqiensis* R12 responsible for HMF oxidation are expressed constitutively. Finally,pH tuning was found to be an effective method for improving the yield of HMFCA (Figure S4c). Thus, the pH of the reaction mixture was tuned to approximately 7.0 using a NaOH solution. Compared with the control withoutpH tuning, the HMFCA yield improved from 66% to 83% in the presence of 300 mM HMF substrate, and from 48% to 65% at an HMF concentration of 500 mM.

Therefore, both increasing the dosage of microbial cells and pH tuning was applied together for enhancing the production of HMFCA. As shown in Figure 3, at a high HMFCA concentration, a yield of 90% was achieved after 36 h when the substrate concentration was 300 mM. This demonstrates the considerable effectiveness of the combined strategy. In addition, the oxidative conversion of HMF to HMFCA reached 80% in the presence of 350 mM HMF after 48 h. When the concentration of HMF was set to 500 mM, 66% of the substrate was still converted after 48 h, but further prolonged reaction times did not lead to an increase in HMFCA yield.

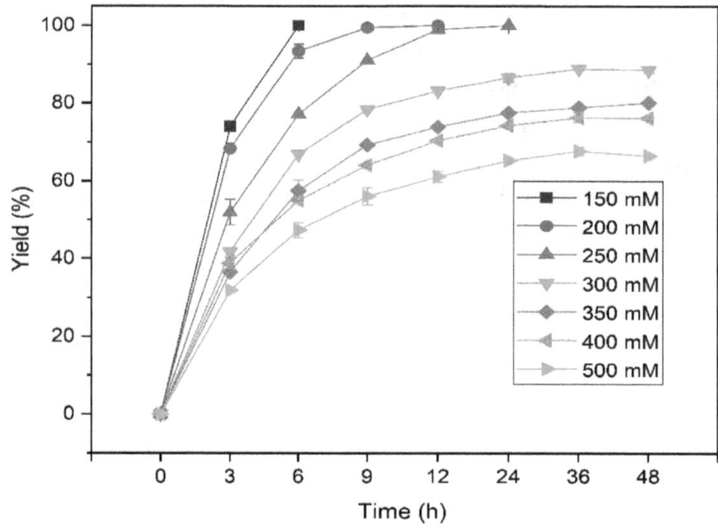

Figure 3. Synthesis of HMFCA under optimized conditions. Reaction conditions: HMF of the designated concentration, 0.2 g/mL microbial cells, 5 mL phosphate buffer (100 mM,pH 7.0), 35 °C, and 850 rpm. Tuning the pH of the reaction system to approximately 7.0 occurred every 3 h in the first 12 h, and then every 12 h in the 36 h that followed.

Compared with previous reported biocatalysis results, the data obtained in this work proved to be more efficient and selective because of a higher substrate concentration and simpler catalytic process. As shown in Table 1, the substrate concentrations used in the reported biocatalytic routes were still very low and co-enzymes were usually required when isolated enzymes were applied. Although *C. testosterone* SC1588 cells also display good selectivity and high HMF tolerance, its catalytic performance is highly sensitive topH [28]. Thus, a considerable amount of histidine co-substrate is required for efficient selective oxidation of HMF. The extreme environment-derived *D. wulumuqiensis* R12 strain used in this work exhibited excellent resistance to highpH and temperatures, and proved to be a robust biocatalyst for HMFCA synthesis by way of selective oxidation of HMF.

Table 1. HMFCA synthesis via HMF oxidation by various biocatalytic systems.

Biocatalysts	Reaction Conditions	t (h)	Yield (%)	Ref.
Chloroperoxidase	50 mM HMF, 1 equiv H_2O_2 per 2 h	2.5	25–40	[31]
Serratia liquefaciens LF14	10 mM HMF, 18.2 mg/mL dry cells, in phosphate buffer	1	97	[30]
Immobilized lipase B	50 mM HMF, 10 mg/mL catalase, addition of aqueous H_2O_2 (30% v/v) hourly, reaction media: acyl butyrate/tBuOH (1:1 v/v), 40 °C	24	76	[32]
Xanthine oxidase (XO)	26 mM HMF, 5.6 U *E. coli* XO, 1.1 mg catalase, phosphate buffer, 37 °C, 150 rpm, air bubbling	7	94	[29]
Comamonas testosterone SC1588	160 mM HMF, 30 mg/mL induced microbial cells, phosphate buffer, 20 mM histidine, 150 rpm, 30 °C, tuningpH of the reaction mixture to approximately 7.0 every 24 h.	36	98	[28]
Aldehyde dehydrogenases	20 mM HMF, 10 μM catalase, 5 μM [NOx], 100 μg/mL [DTT], 0.5 mM [NAD^+] , phosphate buffer, 40 °C, 180 rpm	24	91	[53]
D. wulumuqiensis R12	300 mM HMF, 200 mg/mL microbial cells, phosphate buffer, 850 rpm, 35 °C, tuningpH of the reaction mixture to approximately 7.0 every 3 h.	36	90	This work

2.7. *Efficient Synthesis of HMFCA by a Fed-Batch Strategy*

As mentioned above, excellent yields of HMFCA from selective oxidation of HMF were obtained under optimized conditions. It is highly desirable to manufacture HMFCA on a large scale in an effort to create a practical biocatalytic process. Thus, by applying a fed-batch strategy, in which HMF substrate was added continuously, the accumulation of high concentrations of product was achieved.

Figure 4 shows the results of biocatalytic synthesis of high concentrations of HMFCA. It was found that 511 mM of product was produced within 20 h after three-batch feeding of HMF, affording a total yield of 85% and a productivity of approximately 44 g/L per day. Only 4 mM of BHMF was observed as the sole byproduct (<1%). Chemoselectivity towards the target product reached more than 99%. In addition, a decrease in yield of HMFCA in each batch feeding was observed, indicating possible substrate and/or product inhibition in the whole-cell biocatalyst. An attempt to improve the yield of HMFCA further in this fed-batch process was performed by prolonging the reaction time, but no significant improvement was observed (Data not shown).

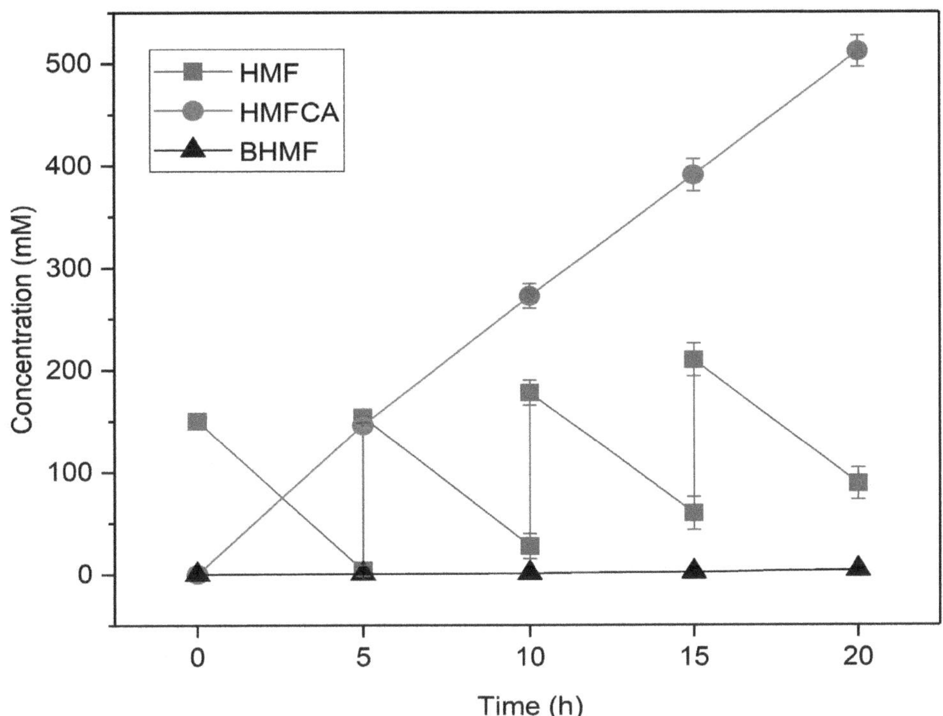

Figure 4. Synthesis of HMFCA by a fed-batch method. Reaction conditions: 150 mM HMF, 0.2 g/mL microbial cells, 5 mL phosphate buffer (100 mM,pH 7.0), 35 °C, and 850 rpm. In each cycle of 5 h, 0.75 mmoL of HMF was added.

2.8. Exploring the Substrate Scope of D. wulumuqiensis R12

In order to examine the substrate spectrum of this novel whole-cell biocatalyst, a set of structurally unique aldehyde compounds was applied in the oxidation reactions catalyzed by *D. wulumuqiensis* R12 cells (Scheme 3). Considering possible solubility and toxicity effects, proper concentrations of these aldehyde compounds were applied. The results showed that the whole cells of the strain readily accept furfural as a substrate, thereby enabling an efficient synthesis of furoic acid (>99% yield). Furoic acid can be used in the pharmaceutical, agrochemical, and flavor industries [9]. In the case of aldehydes containing an additional hydroxyl group, the microbial cells proved to be strictly chemoselective for the aldehyde group, affording the corresponding carboxylic acids with good to excellent yields (Table 2). For example, oxidation of DFF and FFCA to FDCA was achieved with 100% and 63% yields, respectively. Due to solubility issues, higher DFF substrate concentrations were not applied. We also discovered that the aldehyde group of vanillin could be selectively oxidized by the microbial cells to vanillic acid. Vanillic acid has important applications in the pharmaceutical industry, and also as a monomer in polyester synthesis [54]. In addition, terephthalic acid and p-hydroxybenzoic acid were also prepared by selective oxidation of the corresponding aldehydes. In short, the results showed that *D. wulumuqiensis* R12 as a biocatalyst has an amazingly broad substrate spectrum.

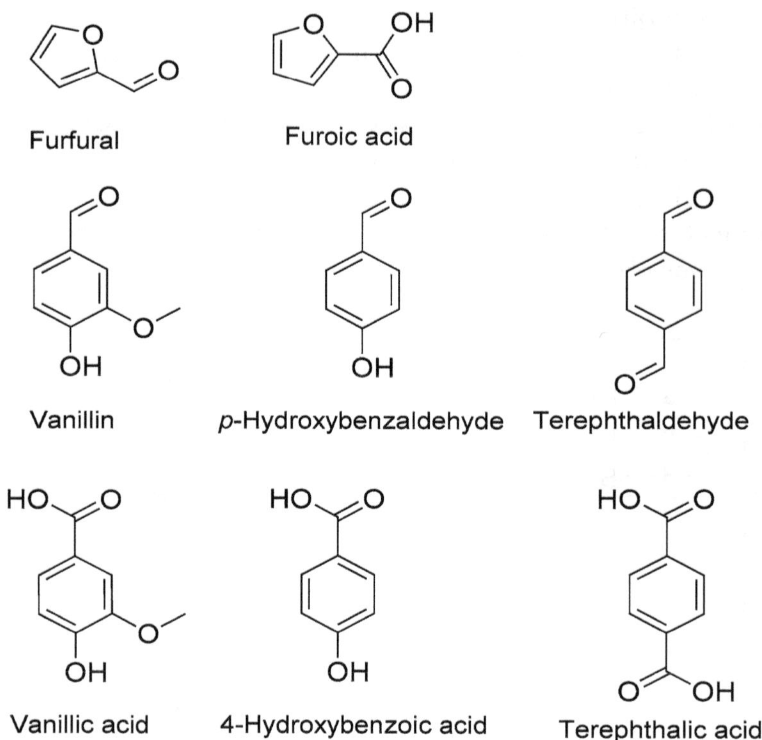

Scheme 3. Aldehyde substrates used and products obtained in biocatalytic oxidation by *D. wulumuqiensis* R12 whole cells.

Table 2. Whole-cell biocatalytic oxidation of aldehydes.

Entry	Substrate	Substrate Concentration (mM)	Product	Yield (%)
1	Furfural	100	Furoic acid	>99
2	DFF	30	FDCA	>99
3	FFCA	100	FDCA	63
4	Vanillin	5	Vanillic acid	>99
5	*p*-Hydroxybenzaldehyde	5	4-Hydroxybenzoic acid	70
6	Terephthaldehyde	2	Terephthalic acid	>99

Reaction conditions: 0.12 g/mL microbial cells, 5 mL of phosphate buffer (100 mM,pH 7.0), 850 rpm, 35 °C for 12 h.

3. Materials and Methods

3.1. Chemicals and Strains

Extremely radiation resistant strains *D. wulumuqiensis* R12 (DSM 28115T), *D. radiodurans* R1 (ATCC NO.13939), and *D. xibeiensis* R13 (NRBC 105666T) were acquired from Zhi-Dong Zhang at the Institute of Microbiology, Xinjiang Academy of Agricultural Sciences in China and stored in our laboratory. The GenBank accession number for the 16S rDNA sequence was KJ784486, while the whole genome sequence was APCS01000000.

HMF (97%) and HMFCA (97%) were purchased from Macklin Biochemical Co., Ltd (Shanghai, China). FDCA (98%), DFF (98%) and furfural (98%) were obtained from Aladdin Biochemical Technology Co., Ltd (Shanghai, China). BHMF (98%) was purchased from Ark Pharm, Inc (Arlington Heights, IL, USA). HMFCA (97%) was obtained from J&K Scientific Ltd (Beijing, China). Both furfuryl alcohol (99.5%) and furoic acid (98%) were obtained from TCI (Shanghai, China). Annexin V-FITC/PI Apoptosis Detection Kit was purchased from Nanjing KeyGen Biotech. Co. Ltd. (Jiangsu, China) for the cell viability assay.

3.2. Cultivation of *D. wulumuqiensis* R12 Cells

The glycerol stock of *D. wulumuqiensis* R12 was pre-cultivated at 30 °C, 200 rpm for 24 h in tryptone glucose yeast extract (TGY) medium containing 0.5% tryptone, 0.1% glucose, and 0.3% yeast extract. Then, 1% of the overnight preculture was transferred to fresh TGY medium. The culture was incubated at 30 °C, 200 rpm for 48 h and was centrifugated under 5000 rpm for 10 min to harvest cells. The cell pellet was washed twice with 100 mM phosphate buffer (pH 7.4) and resuspended in phosphate buffer with a final cell concentration of 0.12 g/mL (cell wet weight).

3.3. General Procedure for the Biocatalytic Oxidation of Aldehyde Substrates

Five milliliters of phosphate buffer (0.1 M,pH 7.0) containing the designated amounts of microbial cells (cell wet weight) and substrates was incubated at 35 °C and 850 rpm for a given reaction time. Aliquots of the reaction mixture were withdrawn at specified reaction times and diluted with the phosphate buffer prior to high-performance liquid chromatography (HPLC) assays. The conversion of HMF and other aldehydes by biocatalytic oxidation was defined as the percentage of the consumed substrate amount in the initial amount. The selectivity of the reaction was defined as the ratio of HMFCA product amount to the sum of all the products. The yield was defined as the percentage of the measured product amount in the theoretical product amount based on the initial amount of HMF.

$$\%\ yield = \frac{\text{Actual yield}}{\text{Theoretical yield}} \times 100\%$$

3.4. Analytical Method

The reaction products were analyzed by HPLC following a previously reported method with slight modifications [28]. Briefly, a reverse-phase HPLC (Thermo Fisher ultimate 3000), equipped with Sepax GP-C18 column (4.6 mm × 250 mm, 5 μm), was used at 25 °C. The mobile phase was the gradient of acetonitrile in 20 mM KH_2PO_4 (pH 6.0) at a flow of 1.0 mL min^{-1}, increasing from 10% to 24% within 7 min and from 24% to 10% within 3 min. The HPLC retention time of the HMFCA product and HMF were 2.90 min and 6.20 min, respectively. All experiments were performed in triplicate and mean values are presented. Data are expressed as the mean ± standard deviation. Duncan's multiple range test (using SPSS software 16.0, Chicago, IL, USA) was used to analyze the statistical significance of differences between the groups. A significance difference was judged to exist at a level of $p < 0.05$. HPLC runs are shown in Figure S1.

3.5. Cell Viability Assay

Cell viability assay experiments were performed using an Annexin V-FITC/PI Apoptosis Detection Kit and flow cytometry following the manufacturers' instructions. Cell viability was determined using ACEA NovoCyte Flow Cytometer (ACEA Biosciences, Inc., San Diego, CA, USA) with the excitation light and emission light wavelengths set at 488 nm and 530 nm, respectively. Data were collected and analyzed using NovoExpress software. The cell viability of *D. wulumuqiensis* R12 when using HMF as the substrate is presented as the percentage of living cells to the total amount of cells.

3.6. Synthesis of HMFCA by the Substrate Fed-Batch Feeding Process

Five milliliters of 100 mM phosphate buffer,pH 7.0, which contained 150 mM HMF substrate and 0.2 g/mL of microbial cells, was incubated at 35 °C and 850 rpm. After 5 h, 0.75 mmol of HMF was repeatedly added to the reaction mixture. During the whole biocatalytic process, the pH of the reaction mixture was adjusted to the range of 7.0–8.0 with NaOH solution and the concentration of substrate and products was analyzed by HPLC.

4. Conclusions

Herein, we successfully explored for the first time the use of a radiation resistant *D. wulumuqiensis* R12 strain as a whole-cell biocatalyst for the efficient synthesis of HMFCA from HMF. The whole cells of this strain proved to be highly tolerant to HMF and the product, HMFCA. The whole-cell system is an excellent biocatalyst for the selective oxidation of HMF. An excellent yield of HMFCA of up to 90% was achieved within 36 h in the presence of 300 mM HMF substrate under optimized conditions. A yield of 80% to 66% was obtained when the substrate concentration increased from 350 mM to 500 mM, while the selectivity towards HMFCA remained at approximately 98%. In addition, up to 511 mM of HMFCA was synthesized in 20 h via a fed-batch method, resulting in a productivity of 44 g/L per day. Thus, *D. wulumuqiensis* R12 cells are a promising catalyst in the biocatalytic process of HMF upgrading. Moreover, the cells were able to transform a set of structurally different aldehyde compounds into their corresponding carboxylic acids with good to excellent selectivity. Since the genome sequence of this strain has been sequenced, exploring the genes that encode the enzymes responsible for HMFCA synthesis from HMF has become feasible in future work. The catalytic properties of these microbial cells can also be further engineered by introduction of other oxidases to form a cell factory for HMF biotransformation. Furthermore, this strain may also have potential applications for the biodetoxification of lignocellulosic hydrolysates in the process of biofuel production. Discovery of *D. wulumuqiensis* R12 as an efficient biocatalyst broadens the toolbox of biocatalysts for the biotransformation of HMF into value-added derivatives and will further facilitate the utilization of biomass for the production of useful chemicals and biofuels.

Author Contributions: Conceptualization, Z.-G.Z.; investigation, R.C., G.Y. and L.-Q.S.; funding acquisition, Z.-G.Z.; resources, Z.-G.Z.; writing—original draft preparation, Z.-G.Z.; writing—review and editing, H.H. and Z.-D.Z.

Acknowledgments: We thank Professor Dr. Manfred T. Reetz (Max-Planck-Institut für Kohlenforschung, Germany) for critical reading of the manuscript and helpful comments.

References

1. Sheldon, R.A. Green and sustainable manufacture of chemicals from biomass: State of the art. *Green Chem.* **2014**, *16*, 950–963. [CrossRef]
2. Wu, L.; Moteki, T.; Gokhale, A.A.; Flaherty, D.W.; Toste, F.D. Production of Fuels and Chemicals from Biomass: Condensation Reactions and Beyond. *Chem* **2016**, *1*, 32–58. [CrossRef]
3. Tuck, C.O.; Pérez, E.; Horváth, I.T.; Sheldon, R.A.; Poliakoff, M. Valorization of Biomass: Deriving More Value from Waste. *Science* **2012**, *337*, 695–699. [CrossRef] [PubMed]
4. Zhang, Z.; Deng, K. Recent Advances in the Catalytic Synthesis of 2,5-Furandicarboxylic Acid and Its Derivatives. *ACS Catal.* **2015**, *5*, 6529–6544. [CrossRef]
5. Christensen, C.H.; Rass-Hansen, J.; Marsden, C.C.; Taarning, E.; Egeblad, K. The Renewable Chemicals Industry. *ChemSusChem* **2008**, *1*, 283–289. [CrossRef] [PubMed]
6. Gallezot, P. Conversion of biomass to selected chemical products. *Chem. Soc. Rev.* **2012**, *41*, 1538–1558. [CrossRef] [PubMed]
7. Isikgor, F.H.; Becer, C.R. Lignocellulosic biomass: A sustainable platform for the production of bio-based chemicals and polymers. *Polym. Chem.* **2015**, *6*, 4497–4559. [CrossRef]
8. Chheda, J.N.; Román-Leshkov, Y.; Dumesic, J.A. Production of 5-hydroxymethylfurfural and furfural by dehydration of biomass-derived mono- and poly-saccharides. *Green Chem.* **2007**, *9*, 342–350. [CrossRef]
9. Van Putten, R.-J.; van der Waal, J.C.; de Jong, E.; Rasrendra, C.B.; Heeres, H.J.; de Vries, J.G. Hydroxymethylfurfural, A Versatile Platform Chemical Made from Renewable Resources. *Chem. Rev.* **2013**, *113*, 1499–1597. [CrossRef] [PubMed]
10. Teong, S.P.; Yi, G.; Zhang, Y. Hydroxymethylfurfural production from bioresources: Past, present and future. *Green Chem.* **2014**, *16*, 2015–2026. [CrossRef]

11. Bozell, J.J.; Petersen, G.R. Technology development for the production of biobased products from biorefinery carbohydrates—the US Department of Energy's "Top 10" revisited. *Green Chem.* **2010**, *12*, 539–554. [CrossRef]
12. Rosatella, A.A.; Simeonov, S.P.; Frade, R.F.M.; Afonso, C.A.M. 5-Hydroxymethylfurfural (HMF) as a building block platform: Biological properties, synthesis and synthetic applications. *Green Chem.* **2011**, *13*, 754–793. [CrossRef]
13. Koopman, F.; Wierckx, N.; de Winde, J.H.; Ruijssenaars, H.J. Efficient whole-cell biotransformation of 5-(hydroxymethyl)furfural into FDCA, 2,5-furandicarboxylic acid. *Bioresour. Technol.* **2010**, *101*, 6291–6296. [CrossRef] [PubMed]
14. Hu, L.; He, A.; Liu, X.; Xia, J.; Xu, J.; Zhou, S.; Xu, J. Biocatalytic Transformation of 5-Hydroxymethylfurfural into High-Value Derivatives: Recent Advances and Future Aspects. *ACS Sustain. Chem. Eng.* **2018**, *6*, 15915–15935. [CrossRef]
15. Gandini, A.; Silvestre, A.J.D.; Neto, C.P.; Sousa, A.F.; Gomes, M. The furan counterpart of poly(ethylene terephthalate): An alternative material based on renewable resources. *J. Polym. Sci. Part A Polym. Chem.* **2009**, *47*, 295–298. [CrossRef]
16. Sousa, A.F.; Vilela, C.; Fonseca, A.C.; Matos, M.; Freire, C.S.R.; Gruter, G.-J.M.; Coelho, J.F.J.; Silvestre, A.J.D. Biobased polyesters and other polymers from 2,5-furandicarboxylic acid: A tribute to furan excellency. *Polym. Chem.* **2015**, *6*, 5961–5983. [CrossRef]
17. Hirai, H. Oligomers from Hydroxymethylfurancarboxylic Acid. *J. Macromol. Sci. Part A Chem.* **1984**, *21*, 1165–1179. [CrossRef]
18. Tamura, G. Antitumor Activity of 5-Hydroxy-methyl-2-furoic Acid AU—Munekata, Masanobu. *Agric. Biol. Chem.* **1981**, *45*, 2149–2150. [CrossRef]
19. Braisted, A.C.; Oslob, J.D.; Delano, W.L.; Hyde, J.; McDowell, R.S.; Waal, N.; Yu, C.; Arkin, M.R.; Raimundo, B.C. Discovery of a Potent Small Molecule IL-2 Inhibitor through Fragment Assembly. *J. Am. Chem. Soc.* **2003**, *125*, 3714–3715. [CrossRef]
20. Casanova, O.; Iborra, S.; Corma, A. Biomass into Chemicals: Aerobic Oxidation of 5-Hydroxymethyl-2-furfural into 2,5-Furandicarboxylic Acid with Gold Nanoparticle Catalysts. *ChemSusChem* **2009**, *2*, 1138–1144. [CrossRef]
21. Gorbanev, Y.Y.; Klitgaard, S.K.; Woodley, J.M.; Christensen, C.H.; Riisager, A. Gold-Catalyzed Aerobic Oxidation of 5-Hydroxymethylfurfural in Water at Ambient Temperature. *ChemSusChem* **2009**, *2*, 672–675. [CrossRef]
22. Davis, S.E.; Houk, L.R.; Tamargo, E.C.; Datye, A.K.; Davis, R.J. Oxidation of 5-hydroxymethylfurfural over supported Pt, Pd and Au catalysts. *Catal. Today* **2011**, *160*, 55–60. [CrossRef]
23. Zhang, Z.; Liu, B.; Lv, K.; Sun, J.; Deng, K. Aerobic oxidation of biomass derived 5-hydroxymethylfurfural into 5-hydroxymethyl-2-furancarboxylic acid catalyzed by a montmorillonite K-10 clay immobilized molybdenum acetylacetonate complex. *Green Chem.* **2014**, *16*, 2762–2770. [CrossRef]
24. Zhou, B.; Song, J.; Zhang, Z.; Jiang, Z.; Zhang, P.; Han, B. Highly selective photocatalytic oxidation of biomass-derived chemicals to carboxyl compounds over Au/TiO_2. *Green Chem.* **2017**, *19*, 1075–1081. [CrossRef]
25. Subbiah, S.; Simeonov, S.P.; Esperança, J.M.S.S.; Rebelo, L.P.N.; Afonso, C.A.M. Direct transformation of 5-hydroxymethylfurfural to the building blocks 2,5-dihydroxymethylfurfural (DHMF) and 5-hydroxymethyl furanoic acid (HMFA) via Cannizzaro reaction. *Green Chem.* **2013**, *15*, 2849–2853. [CrossRef]
26. Kang, E.-S.; Chae, D.W.; Kim, B.; Kim, Y.G. Efficient preparation of DHMF and HMFA from biomass-derived HMF via a Cannizzaro reaction in ionic liquids. *J. Ind. Eng. Chem.* **2012**, *18*, 174–177. [CrossRef]
27. Domínguez de María, P.; Guajardo, N. Biocatalytic Valorization of Furans: Opportunities for Inherently Unstable Substrates. *ChemSusChem* **2017**, *10*, 4123–4134. [CrossRef]
28. Zhang, X.-Y.; Zong, M.-H.; Li, N. Whole-cell biocatalytic selective oxidation of 5-hydroxymethylfurfural to 5-hydroxymethyl-2-furancarboxylic acid. *Green Chem.* **2017**, *19*, 4544–4551. [CrossRef]
29. Qin, Y.-Z.; Li, Y.-M.; Zong, M.-H.; Wu, H.; Li, N. Enzyme-catalyzed selective oxidation of 5-hydroxymethylfurfural (HMF) and separation of HMF and 2,5-diformylfuran using deep eutectic solvents. *Green Chem.* **2015**, *17*, 3718–3722. [CrossRef]
30. Mitsukura, K.; Sato, Y.; Yoshida, T.; Nagasawa, T. Oxidation of heterocyclic and aromatic aldehydes to the corresponding carboxylic acids by *Acetobacter* and *Serratia* strains. *Biotechnol. Lett* **2004**, *26*, 1643–1648. [CrossRef]

31. Van Rantwijk, F.; Sheldon, R.A. Chloroperoxidase-Catalyzed Oxidation of 5-Hydroxymethylfurfural. *J. Carbohydr. Chem.* **1997**, *16*, 299–309. [CrossRef]
32. Krystof, M.; Pérez-Sánchez, M.; Domínguez de María, P. Lipase-Mediated Selective Oxidation of Furfural and 5-Hydroxymethylfurfural. *ChemSusChem* **2013**, *6*, 826–830. [CrossRef] [PubMed]
33. Ni, Y.; Xu, J.-H. Biocatalytic ketone reduction: A green and efficient access to enantiopure alcohols. *Biotechnol. Adv.* **2012**, *30*, 1279–1288. [CrossRef] [PubMed]
34. Ku, S. Finding and Producing Probiotic Glycosylases for the Biocatalysis of Ginsenosides: A Mini Review. *Molecules* **2016**, *21*, 645. [CrossRef] [PubMed]
35. Trudgill, P.W. The metabolism of 2-furoic acid by *Pseudomanas* F2. *Biochem. J.* **1969**, *113*, 577–587. [CrossRef]
36. Koenig, K.; Andreesen, J.R. Xanthine dehydrogenase and 2-furoyl-coenzyme A dehydrogenase from *Pseudomonas putida* Fu1: Two molybdenum-containing dehydrogenases of novel structural composition. *J. Bacteriol.* **1990**, *172*, 5999–6009. [CrossRef]
37. López, M.J.; Nichols, N.N.; Dien, B.S.; Moreno, J.; Bothast, R.J. Isolation of microorganisms for biological detoxification of lignocellulosic hydrolysates. *Appl. Microbiol. Biotechnol.* **2004**, *64*, 125–131. [CrossRef]
38. Koopman, F.; Wierckx, N.; de Winde, J.H.; Ruijssenaars, H.J. Identification and characterization of the furfural and 5-(hydroxymethyl)furfural degradation pathways of *Cupriavidus basilensis* HMF14. *Proc. Natl. Acad. Sci. USA* **2010**, *107*, 4919–4924. [CrossRef]
39. Nikel, P.I.; Chavarría, M.; Danchin, A.; de Lorenzo, V. From dirt to industrial applications: *Pseudomonas putida* as a Synthetic Biology chassis for hosting harsh biochemical reactions. *Curr. Opin. Chem. Biol.* **2016**, *34*, 20–29. [CrossRef]
40. Hoover, R.B.; Tang, J. Microbial Extremophiles at the Limits of Life AU - Pikuta, Elena V. *Crit. Rev. Microbiol.* **2007**, *33*, 183–209. [CrossRef]
41. Madigan, M.T.; Orent, A. Thermophilic and halophilic extremophiles. *Curr. Opin. Microbiol.* **1999**, *2*, 265–269. [CrossRef]
42. Schiraldi, C.; De Rosa, M. The production of biocatalysts and biomolecules from extremophiles. *Trends Biotechnol.* **2002**, *20*, 515–521. [CrossRef]
43. Van den Burg, B. Extremophiles as a source for novel enzymes. *Curr. Opin. Microbiol.* **2003**, *6*, 213–218. [CrossRef]
44. Elleuche, S.; Schröder, C.; Sahm, K.; Antranikian, G. Extremozymes—Biocatalysts with unique properties from extremophilic microorganisms. *Curr. Opin. Biotechnol.* **2014**, *29*, 116–123. [CrossRef] [PubMed]
45. Wang, W.; Mao, J.; Zhang, Z.; Tang, Q.; Xie, Y.; Zhu, J.; Zhang, L.; Liu, Z.; Shi, Y.; Goodfellow, M. *Deinococcus wulumuqiensis sp. nov.*, and *Deinococcus xibeiensis sp. nov.*, isolated from radiation-polluted soil. *Int. J. Syst. Evol. Microbiol.* **2010**, *60*, 2006–2010. [CrossRef] [PubMed]
46. Hong, S.; Farrance, C.E.; Russell, A.; Yi, H. Reclassification of *Deinococcus xibeiensis* Wang et al. 2010 as a heterotypic synonym of *Deinococcus wulumuqiensis* Wang et al. 2010. *Int. J. Syst. Evol. Microbiol.* **2015**, *65*, 1083–1085. [CrossRef]
47. Battista, J.R.; Earl, A.M.; Park, M.-J. Why is *Deinococcus radiodurans* so resistant to ionizing radiation? *Trends Microbiol.* **1999**, *7*, 362–365. [CrossRef]
48. Xu, X.; Tian, L.; Xu, J.; Xie, C.; Jiang, L.; Huang, H. Analysis and expression of the carotenoid biosynthesis genes from *Deinococcus wulumuqiensis* R12 in engineered Escherichia coli. *AMB Express* **2018**, *8*, 94. [CrossRef]
49. Xu, X.; Jiang, L.; Zhang, Z.; Shi, Y.; Huang, H. Genome Sequence of a Gamma- and UV-Ray-Resistant Strain, *Deinococcus wulumuqiensis* R12. *Genome Announc.* **2013**, *1*, e00206–e00213. [CrossRef]
50. Liao, Z.; Zhang, Y.; Luo, S.; Suo, Y.; Zhang, S.; Wang, J. Improving cellular robustness and butanol titers of *Clostridium acetobutylicum* ATCC824 by introducing heat shock proteins from an extremophilic bacterium. *J. Biotechnol.* **2017**, *252*, 1–10. [CrossRef]
51. Wierckx, N.; Koopman, F.; Ruijssenaars, H.J.; de Winde, J.H. Microbial degradation of furanic compounds: Biochemistry, genetics, and impact. *Appl. Microbiol. Biotechnol.* **2011**, *92*, 1095–1105. [CrossRef] [PubMed]
52. Palmqvist, E.; Hahn-Hägerdal, B. Fermentation of lignocellulosic hydrolysates. II: Inhibitors and mechanisms of inhibition. *Bioresour. Technol.* **2000**, *74*, 25–33. [CrossRef]
53. Knaus, T.; Tseliou, V.; Humphreys, L.D.; Scrutton, N.S.; Mutti, F.G. A biocatalytic method for the chemoselective aerobic oxidation of aldehydes to carboxylic acids. *Green Chem.* **2018**, *20*, 3931–3943. [CrossRef]
54. Pang, C.; Zhang, J.; Zhang, Q.; Wu, G.; Wang, Y.; Ma, J. Novel vanillic acid-based poly(ether–ester)s: From synthesis to properties. *Polym. Chem.* **2015**, *6*, 797–804. [CrossRef]

Response Surface Methodology Approach for Optimized Biodiesel Production from Waste Chicken Fat Oil

Fatima Shafiq [1], Muhammad Waseem Mumtaz [1,*], Hamid Mukhtar [2], Tooba Touqeer [1], Syed Ali Raza [3], Umer Rashid [4,*], Imededdine Arbi Nehdi [5,6] and Thomas Shean Yaw Choong [7]

1 Department of Chemistry, University of Gujrat, Gujrat 50700, Pakistan; fatimashafiq369@gmail.com (F.S.); tuba.toqir@gmail.com (T.T.)
2 Institute of Industrial Biotechnology, Government College University, Lahore 54000, Pakistan; hamidwaseer@yahoo.com
3 Department of Chemistry, Government College University, Lahore 54000, Pakistan; chemstone@yahoo.com
4 Institute of Advanced Technology, Universiti Putra Malaysia, UPM Serdang, Selangor 43400, Malaysia
5 Chemistry Department, College of Science, King Saud University, Riyadh 1145, Saudi Arabia; inahdi@ksu.edu.sa
6 Laboratoire de Recherche LR18ES08, Chemistry Department, Science College, Tunis El Manar University, Tunis 2092, Tunisia
7 Department of Chemical and Environmental Engineering, Universiti Putra Malaysia, UPM Serdang, Selangor 43400, Malaysia; csthomas@upm.edu.my
* Correspondence: muhammad.waseem@uog.edu.pk (M.W.M.); umer.rashid@upm.edu.my (U.R.)

Abstract: Biodiesel is gaining acceptance as an alternative fuel in a scenario where fossil fuel reserves are being depleted rapidly. Therefore, it is considered as the fuel of the future due to its sustainability, renewable nature and environment friendly attributes. The optimal yield of biodiesel from cheap feed stock oils is a challenge to add cost effectiveness without compromising the fuel quality. In the current experiment, waste chicken fat oil was taken as the feedstock oil to produce biodiesel through the chemical and enzymatic route of transesterification. The process of chemical transesterification was performed using KOH and sodium methoxide, while enzymatic transesterification was done by using free *Aspergillus terreus* lipase and *Aspergillus terreus* lipase immobilized on functionalized Fe_3O_4 nanoparticles (Fe_3O_4_PDA_Lipase) as biocatalysts. The physico-chemical properties of the understudy feedstock oil were analyzed to check the feasibility as a feedstock for the biodiesel synthesis. The feedstock oil was found suitable for biodiesel production based upon quality assessment. Optimization of various reaction parameters (the temperature and time of reaction, catalyst concentration and methanol-to-oil mole ratio) was performed based on the response surface methodology (RSM). The maximum yield of biodiesel (90.6%) was obtained from waste chicken fat oil by using Fe_3O_4_PDA_Lipase as an immobilized nano-biocatalyst. Moreover, the above said optimum yield was obtained when transesterification was done using 6% Fe_3O_4_PDA_Lipase with a methanol-to-oil ratio of 6:1 at 42 °C for 36 h. Biodiesel production was monitored by FTIR spectroscopic analysis, whereas compositional profiling was done by GC–MS. The measured fuel properties—cloud point, pour point, flash point, fire point and kinematic viscosity—met the biodiesel specifications by American Society for Testing and Materials (ASTM).

Keywords: biodiesel; transesterification; immobilized lipase; RSM; fuel properties

1. Introduction

The rapid industrial growth and population explosion have built an immense pressure on natural resources, including fossil fuels. The whole world is determined to find suitable solutions in context with the forthcoming energy crisis. The world is in search of alternate sources of fuel to reduce its dependency on conventional fuels. Biodiesel has emerged as a promising alternative fuel in recent years due to its renewable nature and environment friendly attributes. Biodiesel may be characterized as alkyl esters of fatty acids and may be utilized easily in diesel engines without major alterations [1]. The emissions of CO and NOx from diesel burning are issues of keen interest as both are greenhouse gases and responsible for tropospheric ozone formation. It is an established fact based on the work of many researchers that, comparative to conventional diesel, combustion of biodiesel produces less CO and unburnt hydrocarbons but higher NOx emissions, probably due to a higher oxygen content in biodiesel [2,3].

Initially, synthesis of biodiesel was extensively carried out using vegetable oils and seed oils of non-edible origin. Usually, the production of biodiesel from edible oils is not cost-effective and these vegetable oils are used in food, hence are valuable. To avoid the problems associated with cost, edibility and food shortage, biodiesel production from non-edible fractions of food and related wastes is gaining sound gravity. Similarly, the biodiesel preparation from non-edible seeds like *Jatropha* is not completely feasible as the cultivation of non-edible seed oil plants may create competition with edible crops on shrinking fertile agriculture land [4]. Recently many studies were carried out for biodiesel synthesis from low cost vegetal and animal-based feed stocks like waste cooking oils and animal fats by reducing their viscosity through the transesterification process [5]. The chemical and enzymatic transesterification processes are adopted to convert fatty acids of feedstock into their alkyl esters. Both transesterification modes have their own modalities and advantages but may be optimized for high quality biodiesel [6].

Chicken fat is a poultry waste that can be used to produce biodiesel. The fat content in chicken is about 10% by weight, which is very high, and its cost is low. Commercial broiler chicken meat was reported to have relatively high contents of polyunsaturated lipids as compared with organic chicken [7]. Researchers have reported that chicken fat constitute about 25% to 35% saturated and 40% to 75% unsaturated fatty acids. Palmitic acid, along with stearic acid, linoleic acid and oleic acid, are major fatty acids in chicken fat [8,9]. The fats can be converted into alkyl esters by the process of transesterification. In an alkali-catalyzed transesterification reaction, both the glyceride and alcohol should be extensively free of water contents as the water compels the reaction to partly change into a saponification reaction, resulting in soap formation [6]. Sodium hydroxide and potassium hydroxide are commonly used as alkali catalysts, but they result in water formation during transesterification, that is why sodium and potassium methoxides are preferred for biodiesel production. Alkali catalysts are good especially for those feed stocks that contain minimal acid value. However, if the acid value of the feed stock is high, then it is recommended to perform pre-treatment acid esterification to reduce the free fatty acid contents before performing base-catalyzed transesterification of the feedstock [10].

On the other hand, enzyme-catalyzed transesterification is gaining acceptance and is considered technically comparable to alkali transesterification. This method normally employs lipase as a catalyst. Lipase-catalyzed transesterification of feedstock oils with a relatively higher free fatty acid content can be carried out without performing any pre-treatment acid esterification step that is normally required in case of alkaline transesterification [11]. However enzymatic transesterification is a high-cost process, because enzymes can be denatured easily in the presence of short-chain alcohols and it is difficult to recover [12]. To cope with these problems, enzymes are immobilized on various supports to enhance their durability. Immobilized enzymes are adoptable to harsh conditions as compared to the free enzymes and are easy to recover. Immobilization of enzymes on the matrix and beads may reduce the enzyme activity by blocking its active site and lowering the mass transfer. However, due to very small size and Brownian movement of nanoparticles, these are a potent choice for enzyme immobilization [13]. There are few reports on enzymatic transesterification of chicken fat oil [14,15].

In the present work the transesterification process was optimized to synthesize biodiesel from a cheaper source in the form of waste chicken fat. The relative effects of various catalysts and their concentration were studied and optimized for improved yields of biodiesel by involving the methanol-to-oil ratio along with reaction time and temperature. The synthesized biodiesel was also analyzed for fuel properties to check its feasibility for use in a compression ignition (CI) engine.

2. Results and Discussion

2.1. Physico-Chemical Characterization of Waste Chicken Fat Oil (WCFO)

The pre-analysis tests of WCFO revealed that the acid value of the oil was 6.56 ± 0.05 mg KOH/g, saponification value 200 ± 7.50 mg KOH/g, refractive index 1.46 ± 0.01, density 0.85 ± 0.07 g/cm^3, iodine number 75 ± 10.70 g iodine/100 g and the peroxide value was computed as (5.5 ± 0.50 meqO_2/kg). These values were depicted comparable with that reported by previous studies [16]. Chicken fat oil has a high acid value, which is why the acid esterification of the feedstock was done prior to the alkaline transesterification to reduce the free fatty acid content and avoid saponification.

2.2. Optimization of Biodiesel Production Process

The experimental results obtained after performing reactions as per CCRD were statistically analyzed to select the most appropriate model from the linear, 2F1, cubical and quadratic models. The model that was best suited was chosen by considering the *p*-values, R^2 values, lack-of-fit tests and adjusted R^2 values. It was observed that the quadratic model was most suited for both the chemical and enzymatic routes of biodiesel production (Table 1).

Table 1. Summary of selected quadratic models.

Feedstock	Catalysts/Biocatalysts		Selected Models	Sequential *p*-Value	Lack-of-Fit *p*-Value	Adjusted R-Squared
WCFO	Enzymes	Fe_3O_4_PDA_Lipase	Quadratic	<0.0001	0.0701	0.9713
WCFO		*Aspergillus terreus* lipase	Quadratic	<0.0001	0.1276	0.9679
WCFO	Chemicals	CH_3ONa	Quadratic	<0.0001	0.4021	0.9519
WCFO		KOH	Quadratic	<0.0001	0.0916	0.9518

The summary statistics clearly determined the fitness of quadratic models for chemical as well as enzymatic biodiesel production process for WCFO.

2.3. Graphs of Predicted vs. Actual Values

The predicted vs. actual value graphs for biodiesel yield depicts the fitness of the selected quadratic model. The graphs of predicted vs. actual values are shown in Figure 1, where Figure 1a–d describes the predicted vs. actual graphs based on experimental data about yield of biodiesel obtained through the transesterification of waste chicken fat oil by Fe_3O_4_PDA_Lipase (Figure 1a), *Aspergillus terreus* lipase (Figure 1b), sodium methoxide (Figure 1c) and KOH (Figure 1d). The distribution of the data along the straight line and the small difference between the predicted and actual value reveals the fitness of the quadratic model for all four experimental designs.

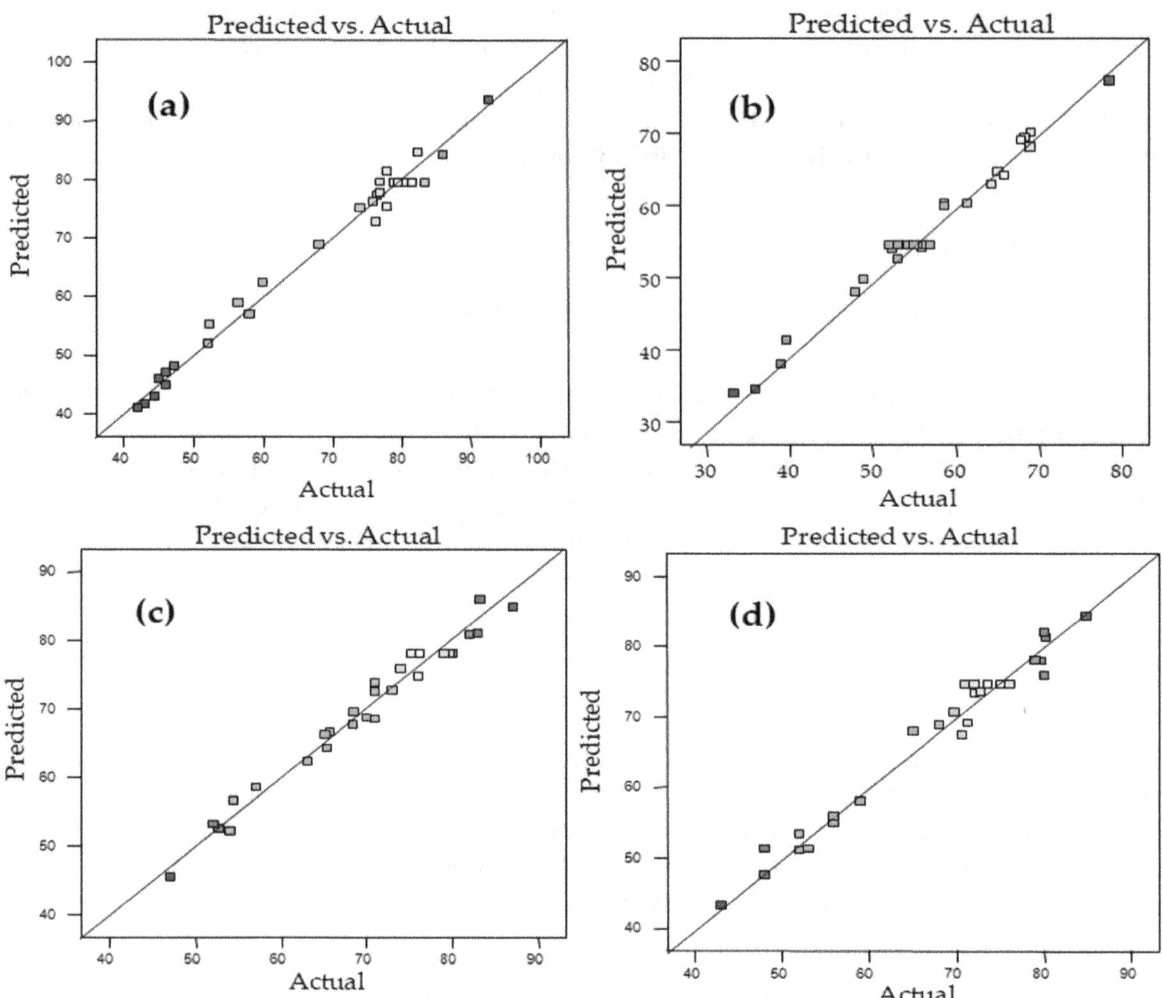

Figure 1. Graphs of predicted vs. actual yield for waste chicken fat oil-based biodiesel by Fe_3O_4_PDA_Lipase (**a**), *Aspergillus terreus* lipase (**b**), Sodium methoxide (**c**) and KOH (**d**).

2.4. Optimization of Reaction Parameters for Manufacturing of Biodiesel Using Chicken Fat Oils

The enzymatic transesterification of waste chicken oil using Fe_3O_4_PDA_Lipase as a bio-catalyst resulted in optimal biodiesel yield when transesterification reactions were performed by employing 6% Fe_3O_4_PDA_Lipaseconcentration with a 6:1 molar ratio of methanol to oil, at 42 °C for 36 h. While in case of enzymatic transesterification by *Aspergillus terreus* lipase, a 1% enzyme concentration, methanol-to-oil ratio of 6:1 and reaction temperature of 35 °C for 36 h were the optimal process conditions. However, when the sodium methoxide-catalyzed transesterification of WCFO was conducted, the optimum conditions for the reaction were a 1% catalyst level and a 6:1 methanol:oil mole ratio at 60 °C for a 1.25 h reaction time (Table 2). The optimum biodiesel yield in case of a potassium hydroxide (KOH)-catalyzed reaction was obtained at a 1% catalyst concentration, 1 h of reaction time, a methanol-to-oil ratio of 6:1 and 60 °C. Highest biodiesel yield was obtained for the nano-biocatalyst (Fe_3O_4_PDA_Lipase), which might be due to the high stability and activity of the immobilized enzyme at an elevated temperature and adoptability towards harsh conditions [17]. Moreover, the lipase can also convert the free fatty acids present in the feedstock to FAMEs. Lower yield obtained by the free lipase can be explained by the reduction of enzyme activity due to denaturation of the free enzymes at a higher temperature, which is required for biodiesel production from chicken fat oil, and the presence of short-chain alcohol [18]. For chemical transesterification, sodium methoxide was proven to be better than KOH, because sodium methoxide did not produce water, which might be responsible for saponification, and the separation of glycerol from biodiesel could be difficult, thus reducing process efficiency. Comparable results for enzymatic and chemical transesterification of waste chicken fat oil have been reported in the published literature. Coppini et al. has reported a 90.61% biodiesel yield from chicken fat by using a 0.3 wt % NS-40116 enzyme, 1.5 of methanol:oil and 1.5 wt % water at 45 °C for 24 h [11]. Da Silva et al. has reported a 77% esterification yield by using 0.3 wt % lipase, 1:4.5 methanol:oil and 2 wt % water at 30 °C i n 2 4 h [15].

Alptekin et al. has reported an 87.4% biodiesel yield from waste chicken fat using a 1% concentration of a KOH catalyst and a 6:1 methanol-to-oil ratio at 60 °C [19]. Mata et al. has reported a 76.8% biodiesel yield by transesterification of chicken fat using a 0.8% KOH catalyst, 6:1 methanol:oil at 60 °C f o r 2 h [20]. The few variations in the results are probably due to the different fatty acid profiles of chicken fats and different enzyme sources.

Table 2. Optimized factors for biodiesel synthesis via enzymatic and chemical modes of transesterification of chicken fat oil.

Feedstock Oil	Catalysts/Biocatalysts	Reaction Time (Hours)	Reaction Temperature °C	CH_3OH:Oil Molar Ratio	Catalyst's Concentration (%)	Biodiesel Yield (%)
WCFO	Fe_3O_4_PDA_Lipase	36	42	6:1	6	90.6
WCFO	*Aspergillus terreus* Lipase	36	35	6:1	1	78.4
WCFO	CH_3ONa	1.25	60	6:1	1	87.1
WCFO	KOH	1	60	6:1	1	84.8

2.5. *ANOVA for Transesterification Data of WCFO*

The influence of various reaction parameters such as linear factors, 1st order interactions and quadratic expressions on percentage biodiesel yield are described in the ANOVA table (Table 3). The terms (a)–(d) represents the quadratic models based on findings of Fe_3O_4_PDA_Lipase, *Aspergillus terreus* lipase, sodium methoxide and KOH-catalyzed transesterification of WCFO, respectively. The statistical analysis depicted that the linear term, A—reaction time, had a significant impact for models a, b and c on biodiesel yield (p < 0.0001, 0.0003 and 0.0003, respectively), which were <0.05, while for model d it was not significant. The linear term B–reaction temperature, showed p values of 0.1743, <0.0001, 0.0004 and <0.0001 for models a, b, c and d, respectively.

The Fe_3O_4_PDA_Lipase catalyzed transesterification was not affected significantly by temperature change in the selected range. Reaction temperature significantly affected the biodiesel yield for *Aspergillus terreus* lipase, which was temperature sensitive. The p values for the linear term C—CH_3OH:Oil, was <0.05 for model (a) and (c) but it was 0.1524 for model (b) and 0.7970 for model (d), which is >0.05. D—catalysts/biocatalysts concentration, was proven to have a significant effect on biodiesel yield for all the four models.

A previous report on *Jatropha curcas* seed oil transesterification showed the significant impact of catalyst concentration, methanol-to-oil molar ratio, reaction temperature and reaction time on biodiesel yields [21] In case of Model (a), the 1st order interaction terms AC, AD and CD were found to be significant having p-values of 0.0007, 0.0001 and 0.0018, respectively, which were less than 0.05; however, for Model (b) only BD and CD were found significant. In case of Model (c), the 1st order interaction variables, i.e., AD and CD, were significant with p-values of 0.0065 and 0.0017 being less than 0.05; for Model (d), only AC 1st order interactions were imparting a significant impact on biodiesel yield with p-values lower than 0.05. Where the quadratic terms C^2 and D^2 were significant for Models (a) having a $p < 0.05$, for Model (b) the statistical significance was noted among the quadratic terms B^2, C^2 and D^2. In Model (c), B^2 and C^2 were significant, while in the case of Model (d), A^2 and D^2 were significantly affecting the biodiesel yield with $p < 0.05$.

Table 3. RSM-based ANOVA for transesterification of waste chicken fat oil (WCFO).

Source	Df	SS (MS) [a]	F-Value (p Value) [a]	SS (MS) [b]	F Value (p Value) [b]	SS (MS) [c]	F Value (p Value) [c]	SS (MS) [d]	F Value (p Value) [d]
Model	14	7213.87 (515.28)	70.99 (< 0.0001)	3018.69 (215.62)	63.48 (< 0.0001)	3159.24 (225.66)	41.95 (< 0.0001)	3806.96 (271.93)	41.87 (< 0.0001)
A—Reaction Time	1	310.24 (310.24)	42.74 (< 0.0001)	76.08 (76.08)	22.40 (0.0003)	114.01 (114.01)	21.19 (0.0003)	4.84 (4.84)	0.75 (0.4014)
B—Reaction Temperature	1	14.76 (14.76)	2.03 (0.1743)	1656.84 (1656.84)	487.79 (< 0.0001)	107.62 (107.62)	20.01 (0.0004)	268.40 (268.40)	41.33 (< 0.0001)
C—CH_3OH:Oil	1	188.55 (188.55)	25.97 (0.0001)	7.72 (7.72)	2.27 (0.1524)	376.29 (376.29)	69.95 (< 0.0001)	0.45 (0.45)	0.069 (0.7970)
D—Catalyst/Biocatalyst Concentration	1	3164.57 (3164.57)	435.97 (< 0.0001)	277.26 (277.26)	81.63 (< 0.0001)	460.40 (460.40)	85.59 (< 0.0001)	2557.47 (2557.47)	393.81 (< 0.0001)
AB	1	9.79 (9.79)	1.35 (0.2637)	15.05 (15.05)	4.43 (0.0525)	0.46 (0.46)	0.085 (0.7750)	0.45 (0.45)	0.070 (0.7952)
AC	1	129.39 (129.39)	17.83 (0.0007)	8.70 (8.70)	2.56 (0.1303)	16.61 (16.61)	3.09 (0.0993)	105.50 (105.50)	16.25 (0.0011)
AD	1	104.67 (104.67)	14.42 (0.0018)	0.83 (0.83)	0.25 (0.6277)	53.66 (53.66)	9.97 (0.0065)	0.13 (0.13)	0.021 (0.8880)
BC	1	3.05 (3.05)	0.42 (0.5264)	2.16 (2.16)	0.64 (0.4376)	1.27 (1.27)	0.24 (0.6346)	1.56 (1.56)	0.24 (0.6313)
BD	1	0.81 (0.81)	0.11 (0.7424)	222.82 (222.82)	65.60 (< 0.0001)	17.21 (17.21)	3.20 (0.0939)	1.17 (1.17)	0.18 (0.6774)
CD	1	193.91 (193.91)	26.71 (0.0001)	47.01 (47.01)	13.84 (0.0021)	77.88 (77.88)	14.48 (0.0017)	12.23 (12.23)	1.88 (0.1901)
A^2	1	2.49 (2.49)	0.34 (0.5670)	11.18 (11.18)	3.29 (0.0897)	5.19 (5.19)	0.096 (0.3416)	115.85 (115.85)	17.84 (0.0007)
B^2	1	1.19 (1.19)	0.16 (0.6913)	22.09 (22.09)	6.50 (0.0222)	79.02 (79.02)	14.69 (0.0016)	0.15 (0.15)	0.023 (0.8811)
C^2	1	524.04 (524.04)	72.19 (< 0.0001)	112.00 (112.00)	32.97 (< 0.0001)	1011.94 (1011.94)	188.12 (< 0.0001)	1.20 (1.20)	0.19 (0.6731)
D^2	1	96.56 (96.56)	13.30 (0.0024)	112.82 (112.82)	33.21 (< 0.0001)	2.48 (2.48)	0.46 (0.5078)	117.67 (117.67)	18.12 (0.0007)
Residual	15	108.88 (7.26)	-	50.95 (3.40)	-	80.69 (5.38)	-	97.41 (6.49)	-
Lack of Fit	10	96.75 (9.67)	3.99 (0.0701)	30.77 (5.13)	2.29 (0.1276)	58.46 (5.85)	1.31 (0.4021)	85.11 (8.51)	3.46 (0.0916)
Pure Error	5	12.13 (2.43)	-	20.18 (2.24)	-	22.24 (4.45)	-	12.30 (2.46)	-
Cor Total	29	7322.75	-	3069.64	-	3239.93	-	3904.37	-

Note: Fe_3O_4_PDA_Lipase (a), *Aspergillus terreus* lipase (b), sodium methoxide (c) and KOH (d). SS stands for sum of squares and MS is mean square.

The 3D surface plots of the significant 1st order interaction terms are presented in Figure 2. Figure 2a–c presents the significant 1st order interaction terms of Model (a). Figure 2a shows the 3D surface plot between the methanol-to-oil ratio and reaction time; it reveals that the yield increases with an increase in reaction time and methanol:oil, but further increases in the methanol-to-oil ratio resulted in a decreased biodiesel yield. The joint impact of time and concentration to increase the biodiesel yield is given in Figure 2b. Figure 2c presents the 3D plot between bio-catalyst/enzyme concentration and methanol:oil for Model (a). The plot shows that enzyme concentration directly increases the biodiesel yield but the increase of metahonl:oil after a specific limit decreases the biodiesel yield.

Figure 2d,e presents the 3D response surface plots for Model (b). Figure 2d reveals the relation between catalyst concentration and reaction temperature; an increase in temperature decreases the biodiesel yield probably due to the denaturation of free enzyme. Figure 2e presents possible impact of methanol:oil and bio-catalyst/enzyme concentration on biodiesel yield.

Figure 2f,g are the 3D plots of the significant 1st order interaction terms of Model (c). Figure 2f presents the relation between catalyst concentration and reaction time. It is observed that increase in both parameters increases the biodiesel yield. Figure 2g shows the relation between the methanol-to-oil ratio and catalyst concentration. The catalyst concentration increased the biodiesel yield but by further increasing the methanol-to-oil ratio up to certain level, however beyond optimal level, a decrease in biodiesel yield was noted.

Figure 2h presents the response of surface plot on the only significant interaction term of Model (d), which is between the reaction time and methanol-to-oil ratio. It showed that biodiesel yield increased with time but after a specific period any further increase in time was not effective.

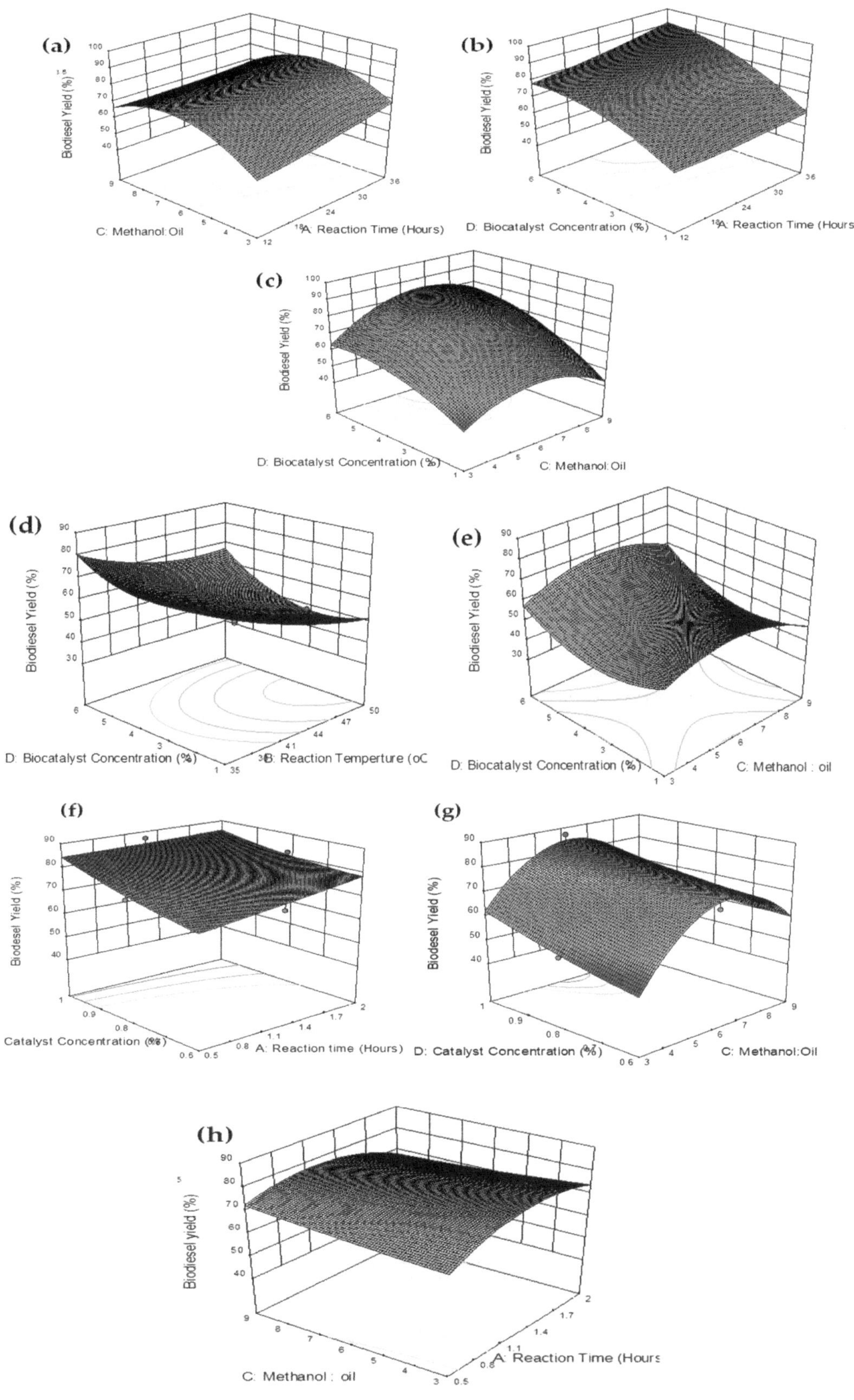

Figure 2. Response surface graphs for the significant 1st order interaction terms of Model (**a**) (A × C), (**b**) (A × D), (**c**) (C × D); Model (b), (**d**) (B × D), (**e**) (C × D); Model (c), (**f**) (A × D), (**g**) (C × D); Model (d), (**h**) (A × C).

2.6. FTIR Spectroscopic Analysis of Feedstock Oil, Biodiesel and Composition of Fatty Acid Methyl Esters

Asymmetric bending of the CH_3 group was observed in the region between 1425 and 1447 cm^{-1} and in the region ranging from 1188 to 1200 cm^{-1} which were basic characteristic peaks of biodiesel. While the C=O stretch vibrations observed in the region between 1700 and 1800 cm^{-1} and CH_2 asymmetric and symmetric stretching vibrations appeared at 2900–3100 cm^{-1} were present in FTIR spectra of both feedstock oil and synthesized biodiesel samples. However, signals in the 1390–1400 cm^{-1} region confirmed the O–CH_2 group and in the 1095–1101 cm^{-1} region defined the asymmetric axial stretching of O–CH_2–C for WCFO in FTIR spectra; however, these bands were absent in their respective biodiesel spectra. The above spectroscopic observations were according with the findings of a previous study [22]. The palmitic acid methyl ester (C16:0) 17.96%, stearic acid methyl ester (C18:0) 20.85%, oleic acid methyl ester (C18:1) 42.92% and linoleic acid methyl ester (C18:2) 16.54%, respectively, were the major FAMEs (Table 4). The current findings were found comparable with those reported by a previous study [16].

Table 4. The fatty acid methyl ester (ME) composition of the synthesized biodiesel.

Biodiesel Type	Palmitic Acid (C16:0) ME %	Stearic Acid (C18:0) ME %	Oleic Acid (C18:1, cis) ME %	Linoleic Acid (C18:2, cis)
WCFAO	17.96	20.85	42.92	16.54

2.7. Fuel Characteristics of WCFO Biodiesel

Fuel analysis plays a vital role in the evaluation of the manufactured biodiesel for its technical compatibility with conventional diesel. Fuel analysis of the understudy biodiesel samples was carried out in accordance with ASTM standard methods and the findings are mentioned below.

Kinematic viscosity is considered as one of the most significant fuel properties, as it is related to the resistance of flow that mainly occurs due to internal friction. If a biofuel contains greater values of kinematic viscosity, it will result in poor fuel atomization or incomplete combustion. Flash point and fuel volatility are inversely related to each other. Similarly, high values of cloud point generally result in problems such as fuel-line clogging. The kinematic viscosity (mm^2/s), flash point (°C), fire point (°C), pour point (°C) and cloud point (°C) values for WCFO biodiesel are given in Table 5. The mentioned fuel-quality parameters were found comparable with those from previous studies [19,20].

Table 5. Fuel characteristics of WCFO biodiesel.

Properties	WCFOB
Kinematic viscosity (mm^2/s) at 40 °C	4.9 ± 0.55
Flash point °C	171 ± 2.51
Fire point °C	187 ± 3.51
Pour point °C	3.0 ± 2.0
Cloud point °C	6.3 ± 2.37

3. Materials and Methods

Chemicals and reagents of analytical grade were utilized during study and were procured from Sigma-Aldrich (Saint Louis, MO, USA) and Merck (Darmstadt, Germany). Lipase from *Aspergillus terreus* was produced through fermentation at the Institute of Industrial Biotechnology, GC University, Lahore, Pakistan. The nano-biocatalyst (Fe_3O_4_PDA_Lipase) prepared and characterized in our previous work has been used as the immobilized lipase for biodiesel production from chicken fat oil [17]. The chicken fat was collected from the local market of Gujrat City, Pakistan.

3.1. *Pre-Treatment of Feedstock*

The collected waste chicken fat was heated at 100 °C to convert it into liquid. The liquid was then filtered to remove the solid waste. Since the chicken fat oil contains a high free fatty acid (FFA) content, alkaline catalysts are not suitable for un-treated chicken fat oil. In this case, acid esterification was used to reduce the acid value before alkaline transesterification. For this purpose, the chicken fat oil was taken in three neck flasks equipped with a thermometer and a glass condenser. The third neck was used to withdraw the sample. Oil was homogenized by heating and stirring at 600 rpm. Briefly, 50 mg concentrated sulfuric acid and 2.25 g methanol for each gram of FFA present in the oil was mixed in a beaker. The acid value of the sample was checked after specific time intervals by taking small aliquots. The process was carried out till the acid number reduced to the desired value. After completion of the process, the mixture was put in a separating funnel. Three layers were formed after some time. The top layer consisted of unreacted methanol and the lower layer was water while the middle layer was fatty acid methyl esters (FAMEs) and oil. The middle layer was collected for chemical transesterification [23]. The enzymatic transesterification, however, was done without pre-acid esterification. The collected oil was subjected to analysis for some basic parameters, including saponification value, acid value (AV) and peroxide value. The degree of unsaturation of feedstock oil was determined by iodine number. Similarly, the specific gravity was also determined along with density and the refractive index.

3.2. *Experimental Design*

The Central Composite Response Surface Methodology (CCRD–RSM) was used to evaluate the impact of the different conditions, namely A) reaction time, B) reaction temperature, C) CH_3OH: oil and D) the catalyst/biocatalyst's concentration on percentage biodiesel yield, for different transesterification routs using catalysts (KOH, CH_3ONa, free *Aspergillus terreus* lipase and Fe_3O_4_PDA_Lipase). The ranges of the selected parameters for the four models are presented in Table 6.

Table 6. The ranges of reaction parameters: reaction time, temperature, CH_3OH-to-oil ratio and enzyme concentration used for optimization studies of enzymatic and chemical transesterification.

Design	Catalysts/Biocatalysts	A—Reaction Time (h)	B—Reaction Temp (°C)	C—CH_3OH:oil	D—Catalyst/Biocatalyst Concentration (%) (with Respect to Substrate)
(a)	Fe_3O_4_PDA_Lipase	12–36	35–50	3:1–9:1	1–6
(b)	*Aspergillus terreus* lipase	12–36	35–50	3:1–9:1	1–6
(c)	CH_3ONa	0.5–2	40–60	3:1–9:1	0.6–1
(d)	KOH	0.5–2	40–60	3:1–9:1	0.6–1

In each case, thirty experiments were carried out as per CCRD factorial design. Chemically, this process was performed in three neck flasks equipped with a temperature regulator. A stirrer and reflux condenser were also attached with the flask. The reactions were accomplished at 500 RPM.

3.2.1. Chemical Transesterification

Briefly, 50 g of pre-treated waste chicken fat oil (WCFO) in a flat-bottom three-neck flask was subjected to pre-heating for moisture removal. The molecular weight of the understudy chicken fat oil (873.72 g/mol) was calculated as per a previously reported method [24]. The next step involved the mixing of known amounts of methanol and catalysts, respectively. The resultant mixture was transferred gently to the flask. The whole mixture present in the flask was kept for heating with stirring. According to the CCRD design, the reaction conditions were maintained. The reaction was allowed to proceed, and on completion the mixture was separated into two layers. The upper layer was taken and further processed to obtain the refined biodiesel.

3.2.2. Enzymatic Transesterification

For enzymatic biodiesel production, firstly waste chicken oil was mixed with methanol. A specific amount of immobilized lipase (Fe_3O_4_PDA_Lipase) for design (a) and free *Aspergillus terreus* lipase for design (b) was introduced to the oil/methanol solution and the reaction mixture was subjected to orbital shaking at 200 rpm with a 0.5% water content (with respect to oil), for a specific time period [25]. The CCRD was followed to set the alcohol to oil molar ratio, enzyme concentration, reaction temperature and time. After completion of the reaction, the glycerol was removed to obtain crude biodiesel, which was purified to get refined biodiesel. Magnetic nano-biocatalyst was separated from biodiesel and glycerol by magnetic decantation.

For the optimization studies, suitable statistical models based on experimental data were employed. Linear, 2FI, cubical and quadratic models were tested. The lack-of-fit test values, model significance (p-value), the R^2 and adjusted R^2 values provided the base to select most appropriate statistical model. Finally, the response surface graphs were utilized to check the influence of the studied reaction conditions on the yield of biodiesel.

3.3. *Quantification and Characterization of Synthesized Biodiesel*

For the FTIR spectroscopic study, a Carry660 FTIR spectrophotometer (Agilent Technologies, Santa Clara, CA, United States) was used and FTIR spectra were drawn over 400–4000 cm^{-1} scanning range.

The biodiesel from waste chicken oil was subjected to GC–MS analysis in order to evaluate the fatty acid methyl esters content (FAMEs). For this purpose, the GC–MS (QP 2010) instrument with a dB 5 column (Shimadzu, Japan) having diameter of 0.15 mm was used. The sample size (1.0 μL) was taken with a split ratio of 1:100, while a source of carrier gas, helium, was used having a 1.20 ml/min flow rate. The oven temperature was kept at 160.0 to 260.0 °C with a ramp rate of 4 °C per minute. The scanning of mass was done from 40.0 to 560.0 *m/z*. The detection of FAMEs was ascertained by comparing the relative retention time of each discrete FAMEs with reliable standards of FAMEs and by comparison with the NIST mass spectral library.

Fuel characteristics of the biodiesel were evaluated by some test experiments utilizing the ASTM standard procedure, i.e., cloud point (ASTM D 2500), viscosity (ASTM D 455), pour point (ASTM D 97) and flash point (ASTM D 93) [19]. The measurements were made in triplicate and the results were analyzed with the help of statistical tools.

4. Conclusions

The waste chicken fat oil was transformed into biodiesel by alkaline and enzymatic transesterifications. The reaction time, temperature, methanol:oil ratio and catalyst concentration were selected for the process optimization. Among all the catalysts and enzymes used, Fe_3O_4_PDA_Lipase-catalyzed transesterification of the studied feedstock oil was proved to be the most efficient to give maximum biodiesel yield. On the other hand, in case of chemical catalysis, CH_3ONa was proved to be better than KOH when chicken fat oil was used as the feedstock. FTIR spectroscopy and GC/MS characterization further confirmed biodiesel formation. The compositional profiles and fuel characteristics of the synthesized biodiesel showed a promising compatibility of WCFO as a potential candidate for biodiesel production for future fuel regimes.

Author Contributions: The idea and concept for the current research was floated by, F.S., T.T., M.W.M., H.M. and U.R.; the methodology was also designed by F.S., M.W.M. and U.R.; the enzyme production was carried out by H.M.; the writing of the original draft was prepared by F.S. with the help and supervision of M.W.M., H.M., S.A.R., T.S.Y.C.; and I.A.N. reviewed the manuscript for improvement of the final version. All authors have read and agreed to the published version of the manuscript.

Acknowledgments: One of the authors acknowledges his gratitude to King Saud University (Riyadh, Saudi Arabia) for the support/technical assistance of this research through Researchers Supporting Project number (RSP-2019/80).

References

1. Berchmans, H.J.; Hirata, S. Biodiesel production from crude *Jatropha curcas* L. seed oil with a high content of free fatty acids. *Bioresour. Technol.* **2008**, *99*, 1716–1721. [CrossRef] [PubMed]
2. Niculescu, R.; Clenci, A.; Iorga-Siman, V. Review on the use of diesel–biodiesel–alcohol blends in compression ignition engines. *Energies* **2019**, *12*, 1194. [CrossRef]
3. Rehan, M.; Gardy, J.; Demirbas, A.; Rashid, U.; Budzianowski, W.M.; Pant, D.; Nizami, A.S. Waste to biodiesel: A preliminary assessment for Saudi Arabia. *Bioresour. Technol.* **2018**, *250*, 17–25. [CrossRef]
4. Talebian-Kiakalaieh, A.; Amin, N.A.S.; Mazaheri, H. A review on novel processes of biodiesel production from waste cooking oil. *Appl. Energy* **2013**, *104*, 683–710. [CrossRef]
5. Jamil, F.; Al-Muhtaseb, A.H.; Al-Haj, L.; Al-Hinai, M.A.; Hellier, P.; Rashid, U. Optimization of oil extraction from waste "Date pits" for biodiesel production. *Energ. Conver. Manag.* **2016**, *117*, 264–272. [CrossRef]
6. Lotero, E.; Goodwin, J.G., Jr.; Bruce, D.A.; Suwannakarn, K.; Liu, Y.; Lopez, D.E. The catalysis of biodiesel synthesis. *Catalysis* **2006**, *19*, 41–83.
7. Gürü, M.; Koca, A.; Can, Ö.; Çınar, C.; Şahin, F. Biodiesel production from waste chicken fat based sources and evaluation with Mg based additive in a diesel engine. *Renew. Energy* **2010**, *35*, 637–643. [CrossRef]
8. Chiu, M.C.; Gioielli, L.A. Solid fat content of abdominal chicken fat, its stearins and its binary mixtures with bacon. *Food Sci. Technol.* **2002**, *22*, 151–157.
9. Gugule, S.; Fatimah, F.; Rampoh, Y. The utilization of chicken fat as alternative raw material for biodiesel synthesis. *Anim. Prod.* **2011**, *13*, 115–121.
10. Liu, K.S. Preparation of fatty acid methyl esters for gas-chromatographic analysis of lipids in biological materials. *J. Am. Oil Chem. Soc.* **1994**, *71*, 1179–1187. [CrossRef]
11. Coppini, M.; Magro, J.D.; Martello, R.; Valério, A.; Zenevicz, M.C.; Oliveira, D.D.; Oliveira, J.V. Production of methyl esters by enzymatic hydroesterification of chicken fat industrial residue. *Braz. J. Chem. Eng.* **2019**, *36*, 923–928. [CrossRef]
12. Christopher, L.P.; Kumar, H.; Zambare, V.P. Enzymatic biodiesel: Challenges and opportunities. *Appl. Energy* **2014**, *119*, 497–520. [CrossRef]
13. Dumri, K.; Hung Anh, D. Immobilization of lipase on silver nanoparticles via adhesive polydopamine for biodiesel production. *Enzym. Res.* **2014**, *2014*, 389739. [CrossRef]
14. Aryee, A.N.; Simpson, B.K.; Cue, R.I.; Phillip, L.E. Enzymatic transesterification of fats and oils from animal discards to fatty acid ethyl esters for potential fuel use. *Biomass Bioenergy* **2011**, *35*, 4149–4157. [CrossRef]
15. da Silva, J.R.P.; Nürnberg, A.J.; da Costa, F.P.; Zenevicz, M.C.; Lerin, L.A.; Zanetti, M.; Valério, A.; de Oliveira, J.V.; Ninow, J.L.; de Oliveira, D. Lipase NS40116 as catalyst for enzymatic transesterification of abdominal chicken fat as substrate. *Bioresour. Technol. Rep.* **2018**, *4*, 214–217. [CrossRef]
16. Feddern, V.; Kupski, L.; Cipolatti, E.P.; Giacobbo, G.; Mendes, G.L.; Badiale, F.E.; de Souza-Soares, L.A. Physico-chemical composition, fractionated glycerides and fatty acid profile of chicken skin fat. *Eur. J. Lipid Sci. Technol.* **2010**, *112*, 1277–1284. [CrossRef]
17. Touqeer, T.; Mumtaz, M.W.; Mukhtar, H.; Irfan, A.; Akram, S.; Shabbir, A.; Rashid, U.; Nehdi, I.A.; Choong, T.S.Y. Fe_3O_4-PDA-Lipase as surface functionalized nano biocatalyst for the production of biodiesel using waste cooking oil as feedstock: Characterization and process optimization. *Energies* **2019**, *13*, 177. [CrossRef]
18. Andrade, M.F.; Parussulo, A.L.; Netto, C.G.C.M.; Andrade, L.H.; Toma, H.E. Lipase immobilized on polydopamine-coated magnetite nanoparticles for biodiesel production from soybean oil. *Biofuel Res. J.* **2016**, *3*, 403–409. [CrossRef]
19. Alptekin, E.; Canakci, M. Optimization of pretreatment reaction for methyl ester production from chicken fat. *Fuel* **2010**, *89*, 4035–4039. [CrossRef]
20. Mata, T.M.; Cardoso, N.; Ornelas, M.; Neves, S.; Caetano, N.S. Evaluation of two purification methods of biodiesel from beef tallow, pork lard, and chicken fat. *Energy Fuels* **2011**, *25*, 4756–4762. [CrossRef]
21. Kamel, D.A.; Farag, H.A.; Amin, N.K.; Zatout, A.A.; Ali, R.M. Smart utilization of jatropha (*Jatropha curcas* Linnaeus) seeds for biodiesel production: Optimization and mechanism. *Ind. Crops. Prod.* **2018**, *111*, 407–413. [CrossRef]

22. Mumtaz, M.W.; Mukhtar, H.; Anwar, F.; Saari, N. RSM based optimization of chemical and enzymatic transesterification of palm oil: Biodiesel production and assessment of exhaust emission levels. *Sci. World J.* **2014**, *2014*. [CrossRef] [PubMed]
23. Chai, M.; Tu, Q.; Lu, M.; Yang, Y.J. Esterification pretreatment of free fatty acid in biodiesel production, from laboratory to industry. *Fuel Process. Technol.* **2014**, *125*, 106–113. [CrossRef]
24. Marulanda, V.F.; Anitescu, G.; Tavlarides, L.L. Investigations on supercritical transesterification of chicken fat for biodiesel production from low-cost lipid feedstocks. *J. Supercrit. Fluids* **2010**, *54*, 53–60. [CrossRef]
25. Katiyar, M.; Ali, A. Onepot lipase entrapment within silica particles to prepare a stable and reusable biocatalyst for transesterification. *J. Am.OilChem. Soc.* **2015**, *92*, 623–632. [CrossRef]

PERMISSIONS

All chapters in this book were first published in MDPI; hereby published with permission under the Creative Commons Attribution License or equivalent. Every chapter published in this book has been scrutinized by our experts. Their significance has been extensively debated. The topics covered herein carry significant findings which will fuel the growth of the discipline. They may even be implemented as practical applications or may be referred to as a beginning point for another development.

The contributors of this book come from diverse backgrounds, making this book a truly international effort. This book will bring forth new frontiers with its revolutionizing research information and detailed analysis of the nascent developments around the world.

We would like to thank all the contributing authors for lending their expertise to make the book truly unique. They have played a crucial role in the development of this book. Without their invaluable contributions this book wouldn't have been possible. They have made vital efforts to compile up to date information on the varied aspects of this subject to make this book a valuable addition to the collection of many professionals and students.

This book was conceptualized with the vision of imparting up-to-date information and advanced data in this field. To ensure the same, a matchless editorial board was set up. Every individual on the board went through rigorous rounds of assessment to prove their worth. After which they invested a large part of their time researching and compiling the most relevant data for our readers.

The editorial board has been involved in producing this book since its inception. They have spent rigorous hours researching and exploring the diverse topics which have resulted in the successful publishing of this book. They have passed on their knowledge of decades through this book. To expedite this challenging task, the publisher supported the team at every step. A small team of assistant editors was also appointed to further simplify the editing procedure and attain best results for the readers.

Apart from the editorial board, the designing team has also invested a significant amount of their time in understanding the subject and creating the most relevant covers. They scrutinized every image to scout for the most suitable representation of the subject and create an appropriate cover for the book.

The publishing team has been an ardent support to the editorial, designing and production team. Their endless efforts to recruit the best for this project, has resulted in the accomplishment of this book. They are a veteran in the field of academics and their pool of knowledge is as vast as their experience in printing. Their expertise and guidance has proved useful at every step. Their uncompromising quality standards have made this book an exceptional effort. Their encouragement from time to time has been an inspiration for everyone.

The publisher and the editorial board hope that this book will prove to be a valuable piece of knowledge for researchers, students, practitioners and scholars across the globe.

LIST OF CONTRIBUTORS

Hendrik Puetz and Frank Hollmann
Department of Biotechnology, Delft University of Technology, van der Maasweg 9, 2629 HZ Delft, The Netherlands

Eva Puchl'ová and Kvetoslava Vranková
Axxence Slovakia s.r.o, Mickiewiczova 9, 81107 Bratislava, Slovakia

Mitul P. Patel, Nathaneal T. Green, Jacob K. Burch and Robert M. Hughes
Department of Chemistry, East Carolina University, Greenville, NC 27858, USA

Kimberly A. Kew
Department of Biochemistry and Molecular Biology, Brody School of Medicine, East Carolina University, Greenville, NC 27834, USA

Christin Burkhardt, Christian Schäfers and Garabed Antranikian
Institute of Technical Microbiology, Hamburg University of Technology (TUHH), Kasernenstr. 12, 21073 Hamburg, Germany

Jörg Claren and Georg Schirrmacher
Group Biotechnology, Clariant Produkte (Deutschland) GmbH, Semmelweisstr. 1, 82152 Planegg, Germany

Mai-Lan Pham, Suwapat Kittibunchakul and Thu-Ha Nguyen
Food Biotechnology Laboratory, Department of Food Science and Technology, BOKU-University of Natural Resources and Life Sciences, A-1190 Vienna, Austria

Anh-Minh Tran
Food Biotechnology Laboratory, Department of Food Science and Technology, BOKU-University of Natural Resources and Life Sciences, A-1190 Vienna, Austria
Department of Biology, Faculty of Fundamental Sciences, Ho Chi Minh City University of Medicine and Pharmacy, 217 Hong Bang, Ho Chi Minh City, Vietnam

Tien-Thanh Nguyen
School of Biotechnology and Food Technology, Hanoi University of Science and Technology, 1 Dai Co Viet, Hanoi, Vietnam

Geir Mathiesen
Faculty of Chemistry, Biotechnology and Food Science, Norwegian University of Life Sciences (NMBU), N-1432 Ås, Norway

Zhihai Liu
Agricultural Bio-pharmaceutical Laboratory, College of Chemistry and Pharmaceutical Sciences, Qingdao Agricultural University, Qingdao 266109, China
Beijing Advanced Innovation Center for Food Nutrition and Human Health, College of Veterinary Medicine, China Agricultural University, Beijing 100193, China

Dejun Liu, Wan Li, Yang Wang and Jianzhong Shen
Beijing Advanced Innovation Center for Food Nutrition and Human Health, College of Veterinary Medicine, China Agricultural University, Beijing 100193, China

Alessandra Piccirilli
Dipartimento di Scienze Cliniche Applicate e Biotecnologiche, Università degli Studi dell'Aquila, 67100 L'Aquila, Italy

Thais S. Milessi-Esteves and Willian Kopp
Department of Chemical Engineering, Federal University of São Carlos, Rodovia Washington Luiz, km 235, 13565-905, São Carlos, SP, Brazil

Felipe A.S. Corradini
Graduate Program of Chemical Engineering, Federal University of São Carlos (PPGEQ-UFSCar), Rodovia Washington Luiz, km 235, 13565-905, São Carlos, SP, Brazil

Teresa C. Zangirolami, Paulo W. Tardioli, Roberto C. Giordano and Raquel L.C. Giordano
Department of Chemical Engineering, Federal University of São Carlos, Rodovia Washington Luiz, km 235, 13565-905, São Carlos, SP, Brazil
Graduate Program of Chemical Engineering, Federal University of São Carlos (PPGEQ-UFSCar), Rodovia Washington Luiz, km 235, 13565-905, São Carlos, SP, Brazil

Huixia Yang and Weiwei Zhang
State Key Laboratory of High-efficiency Utilization of Coal and Green Chemical Engineering, School of Chemistry and Chemical Engineering, Ningxia University, Yinchuan 750021, China

Paula Bracco, Nelleke van Midden, Epifanía Arango, Guzman Torrelo and Ulf Hanefeld
Biokatalyse, Afdeling Biotechnologie, Technische Universiteit Delft, Van der Maasweg 9, 2629 HZ Delft, The Netherlands

Valerio Ferrario and Lucia Gardossi
Dipartimento di Scienze Chimiche e Farmaceutiche, Università degli Studi di Trieste, Via Licio Giorgieri 1, 34127 Trieste, Italy

Lynn Sophie Schwardmann, Sarah Schmitz, Volker Nölle and Skander Elleuche
Miltenyi Biotec B.V. & Co. KG, Friedrich-Ebert-Straße 68, 51429 Bergisch Gladbach, Germany

Najme Gord Noshahri and Jamshid Fooladi
Department of Biotechnology, Faculty of Biology Science, Alzahra University, 1993893973 Tehran, Iran

Christoph Syldatk, Ulrike Engel and Jens Rudat
BLT_II: Technical Biology, Karlsruhe Institute of Technology (KIT), Fritz-Haber-Weg 4, 76131 Karlsruhe, Germany

Majid M. Heravi
Department of Chemistry, School of Sciences, Alzahra University, 1993891176 Tehran, Iran

Mohammad Zare Mehrjerdi
Higher Education Complex of Shirvan, 9468194477 Shirvan, Iran

Juan Pinheiro De Oliveira Martinez, Guiqin Cai, Matthias Nachtschatt, Laura Navone, Zhanying Zhang and Robert Speight
Science and Engineering Faculty, Queensland University of Technology, Brisbane, QLD 4000, Australia

Karen Robins
Science and Engineering Faculty, Queensland University of Technology, Brisbane, QLD 4000, Australia
Sustain Biotech, Sydney, NSW 2224, Australia

Laure Guillotin, Zeinab Assaf, Pierre Lafite and Richard Daniellou
Institut de Chimie Organique et Analytique, ICOA, Université d'Orléans, CNRS, UMR 7311, BP6759 Rue de Chartres, F-45067 Orléans CEDEX 2, France

Salvatore G. Pistorio and Alexei V. Demchenko
Department of Chemistry and Biochemistry, University of Missouri-St. Louis, One University Boulevard, St. Louis, MO 63121, USA

David Aregger, Christin Peters and Rebecca M. Buller
Competence Center for Biocatalysis, Institute of Chemistry and Biotechnology, Department of Life Sciences and Facility Management, Zurich University of Applied Sciences, Einsiedlerstrasse 31, 8820 Waedenswil, Switzerland

Ran Cang, Li-Qun Shen, Guang Yang and Zhi-Gang Zhang
School of Pharmaceutical Sciences, Nanjing Tech University, 30 Puzhu Road(S), Nanjing 211816, China

He Huang
School of Pharmaceutical Sciences, Nanjing Tech University, 30 Puzhu Road(S), Nanjing 211816, China
State Key Laboratory of Materials-Oriented Chemical Engineering, Nanjing Tech University, 30 Puzhu Road(S), Nanjing 211816, China

Zhi-Dong Zhang
Institute of Microbiology, Xinjiang Academy of Agricultural Sciences, 403 Nanchang Rd, Wulumuqi 830091, China

Fatima Shafiq, Muhammad Waseem Mumtaz and Tooba Touqeer
Department of Chemistry, University of Gujrat, Gujrat 50700, Pakistan

Hamid Mukhtar
Institute of Industrial Biotechnology, Government College University, Lahore 54000, Pakistan

Syed Ali Raza
Department of Chemistry, Government College University, Lahore 54000, Pakistan

Umer Rashid
Institute of Advanced Technology, Universiti Putra Malaysia, UPM Serdang, Selangor 43400, Malaysia

Imededdine Arbi Nehdi
Chemistry Department, College of Science, King Saud University, Riyadh 1145, Saudi Arabia
Laboratoire de Recherche LR18ES08, Chemistry Department, Science College, Tunis El Manar University, Tunis 2092, Tunisia

Thomas Shean Yaw Choong
Department of Chemical and Environmental Engineering, Universiti Putra Malaysia, UPM Serdang, Selangor 43400, Malaysia

Index

Printed in the USA
CPSIA information can be obtained
at www.ICGtesting.com
JSHW051402091023
49903JS00006B/243